**Cell Migration
in Development
and Disease**

*Edited by
Doris Wedlich*

Related Titles

A. Ridley, M. Peckham, P. Clark

Cell Motility: From Molecules to Organisms

2004

ISBN 0-470-84872-3

R. A. Meyers

Encyclopedia of Molecular Cell Biology and Molecular Medicine, Second Edition

2004

ISBN 3-527-30542-4

E. A. Nigg

Centrosomes in Development and Disease

2004

ISBN 3-527-30980-2

M. Schliwa

Molecular Motors

2003

ISBN 3-527-30594-7

Cell Migration
in Development and Disease

Edited by
Doris Wedlich

WILEY-VCH

WILEY-VCH Verlag GmbH & Co. KGaA

Editor:

Prof. Dr. Doris Wedlich
University of Karlsruhe
Institute of Zoology II
Kaiserstr. 12
76131 Karlsruhe
Germany

Library of Congress Card No.: Applied for

British Library Cataloging-in-Publication Data:
A catalogue record for this book is available
from the British Library

**Bibliographic information published by
Die Deutsche Bibliothek**
Die Deutsche Bibliothek lists this publication
in the Deutsche Nationalbibliografie; detailed
bibliographic data is available in the
Internet at <http://dnb.ddb.de>.

© 2005 WILEY-VCH Verlag GmbH & Co.
KGaA, Weinheim

Printed in the Federal Republic of Germany.
Printed on acid-free paper.

Cover design SCHULZ Grafik-Design,
Fußgönheim

Typesetting K+V Fotosatz GmbH,
Beerfelden
Printing betz-druck gmbh, Darmstadt
Bookbinding Litges & Dopf Buchbinderei
GmbH, Heppenheim

ISBN 3-527-30587-4

Contents

Cell Migration. Edited by Doris Wedlich
Copyright © 2005 WILEY-VCH Verlag GmbH & Co. KGaA, Weinheim
ISBN: 3-527-30587-4

Preface

In 1675, when Antonie van Leeuwenhoeks, the pioneer of microscopy, described the movements of bacteria in a letter sent to the Royal Society of London, the fascinating world of cell motility was born. However, at that time nobody could imagine how important – more than 325 years later – the molecular understanding of this process would be for medicine.

Cell migration is a fundamental process for unicellular but also multicellular organisms. It plays a key role in chemotaxis, development, infection, immunity, tissue regeneration and cancer. These processes were studied independently over a long period by zoologists, cell biologists and medical scientists – each discipline interpreting its subject as unique. Apart from advances in methods, such as the improvement of microscopic techniques and the introduction of GFP labeling, the largest impact in the field in recent years resulted from communication between the different disciplines and the ability of people to free themselves in their thinking.

This book will encourage students, young scientists and newcomers in cell migration research to keep an eye on motility factors, signaling pathways and molecules regulating cytoskeleton rearrangements, adhesion complexes and gene transcription in different organisms, and to look out for homologies or analogies in these systems.

Who would have expected that primordial germ cells use the same guiding molecules as lymphocytes? Jumping from Chapter 10 by Raz directly to Chapter 14 by Lipp you will be surprised by the evolutionary conservation of chemokine receptors. You might feel a similar "aha experience" when learning about the repulsive mechanism of roundabout/ slit interaction during glial cell migration in the fly (Chapter 8 by Edenfeld and Klämbt) which is also used in vertebrates as a hindrance for trunc neural crest to enter the gut (Chapter 9 by Bronner-Fraser). Further examples are the signaling pathways repeatedly used in different systems. The JAK/Stat pathway involved in zebrafish gastrulation movement (Chapter 5 by Castanon and Heisenberg) also regulates border cell migration in the Drosophila egg chamber (Chapter 7 by MacDonald and

Cell Migration. Edited by Doris Wedlich
Copyright © 2005 WILEY-VCH Verlag GmbH & Co. KGaA, Weinheim
ISBN: 3-527-30587-4

Montell). HGF/c-Met signaling driving pre-myoblast migration in limb development is often up-regulated in tumors with high invasive potential (Chapters 11 and 16). The most important motility factors and their signaling cascades, which were identified by extensive genetic screens and characterized by functional studies in different model organisms, are nowadays targets of tumor therapy, tissue regeneration and stem cell therapy. Thus, investigating cell migration in embryos has inspired the understanding of tumor invasion, regeneration and wound healing. However, there is also an impact from clinical studies on regulatory mechanisms of cell migration in the embryo. For many years now, the stimulatory effect of bioactive lipids on tumor cell dissemination has been known (Chapter 16), with their role in embryogenesis now coming to light.

Cell migration represents a complex program of cellular behavior including sensing, polarization, cytoskeletal reorganization, changes in adhesion and shape. As long as a cell migrates, these single behaviors underlie a continuous regulation and feedback control to direct and coordinate orientated cell movements. A perfect motility system has been evolved in unicellular organisms: sensor clustering and signaling used for orientation combined with a set of protein polymers, the cytoskeleton, allowing rapid changes of the cell shape in response to the environment. These basic principles of cell motility are best studied in single cells like fibroblasts or Dictyostelium and, as an introduction, these general aspects of cell migration are touched on in the first part of the book. The perfect motility system has been taken over, however, by multicellular organisms, a complex regulatory network was added to coordinate cell movements over short or long distances or to guide single cells or cohorts to their different destinations during embryogenesis. In the second part of the book, important examples are given on how cell migration drives morphogenetic movements and contributes to tissue differentiation in development. We learn that signaling cascades predominantly turn the adjusting knobs of the cellular motility system, the small GTPase, GEFs and GAPs. In the last part of the book cell migration is considered from the medical viewpoint. Previously mentioned molecular principles are re-addressed in a different context and we are surprised by the "tricks" of bacteria and invasive tumor cells in converting the cellular motility system for efficient dissemination of their own.

The journey through the molecular world of cell migration presented in this book was made possible because the authors invested much time to summarize and comment on important results from their research fields. I would like to thank them for their contributions. My gratitude is also extended to Andreas Sendtko from Wiley-VCH, who always kept me in a good mood when I saw the project sinking. I wish to thank Alexandra Schambony, Swetlana Kalinina and Ira Röder from my group for their support in preparing figures and manuscripts.

The roots of this book go back to the initiation of a priority program on Cell Migration funded over six years (1998–2004, SP1049) by the German Research Foundation (DFG). Annual group meetings and international conferences kept the discussions on cell migration in motion. Without the enthusiasm and commitment shown by Walter Birchmeier, Dietmar Vestweber, and Rudolf Tauber in bringing together the worldwide experts on cell migration at our meetings, such a cross-talk between cell and developmental biologist and medical scientist would never have evolved.

I would also like to mention and to thank Ingrid Ehses and Dorette Breitkreuz from the DFG, who constantly helped me in managing the priority program.

At the end of the priority program and with the "outcome" of this book, I hope to stimulate young scientists to draw their attention and research activities to the topic of cell migration, because we have just lifted the curtain in understanding this process. New tools developed in molecular biology and biophysics and fascinating new techniques of live microscopy are waiting in the wings to be used for more in-depth investigations.

Karlsruhe, October 2004   Doris Wedlich

SMALL, JOHN VICTOR
Institute of Molecular
Biotechnology of the Austrian
Academy of Sciences
Dr Bohr-Gasse 7
1030 Vienna
Austria

STRADAL, THERESIA E. B.
German Research Centre
for Biotechnology (GBF)
Department of Cell Biology
Mascheroder Weg 1
38124 Braunschweig
Germany

WEDLICH, DORIS
University of Karlsruhe (TH)
Institute of Zoology
Kaiserstr. 12
76131 Karlsruhe
Germany

WEHLAND, JÜRGEN
German Research Centre
for Biotechnology (GBF)
Department of Cell Biology
Mascheroder Weg 1
38124 Braunschweig
Germany

YU LONG
Cell Biology and Anatomy
Johns Hopkins University School
of Medicine
725 N. Wolfe St.
520 WBSB
Baltimore, MD 21205
USA

Color Plates

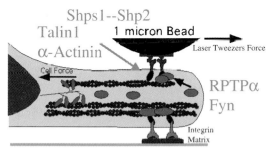
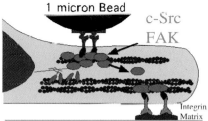

Force-Dependent Assembly   **Contact Disassembly**

Fig. 1.2 Model for integrin–cytoskeleton linkage strengthening. The force-dependent strengthening of the linkages between fibronectin–integrin complexes and the actin-rich cytoskeleton are summarized. This is only a partial list of the components involved but these appear to have critical roles in the process. (This figure also appears on page 10.)

Cell Migration. Edited by Doris Wedlich
Copyright © 2005 WILEY-VCH Verlag GmbH & Co. KGaA, Weinheim
ISBN: 3-527-30587-4

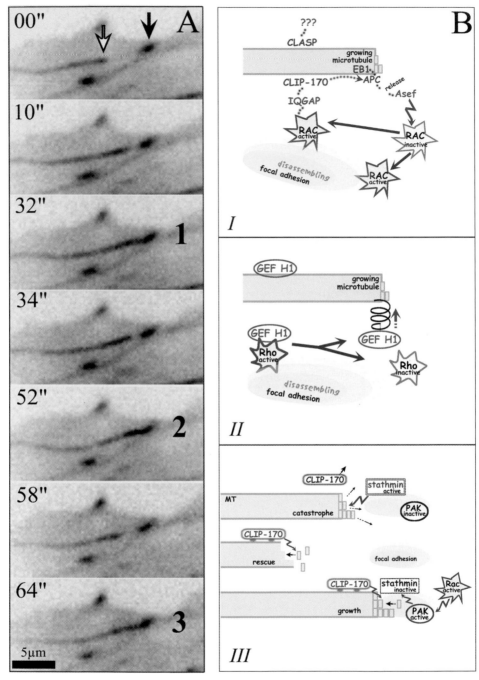

Fig. 2.2 A) The targeting of focal adhesions by microtubules. Panels show video frames of the periphery of a goldfish fibroblast that was transfected with EGFP–tubulin and EGFP–zyxin. From [28], with permission. B) Some possible signaling scenarios associated with the targeting of adhesion complexes by microtubules. (The full legend to this figure appears on page 19.)

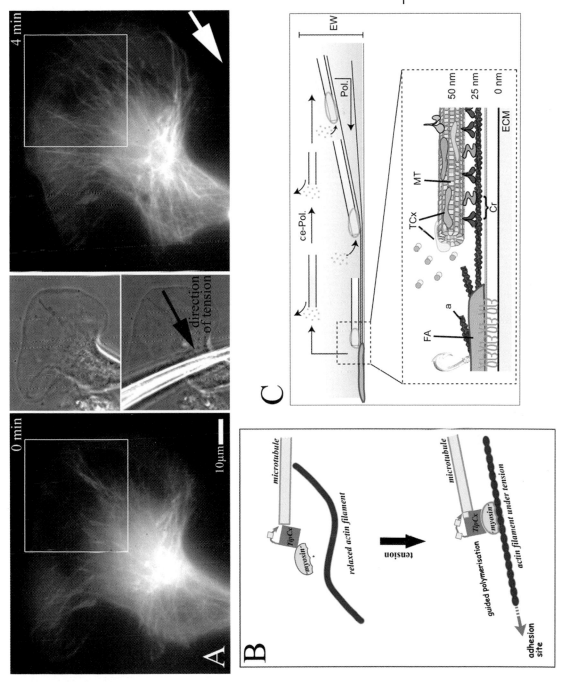

Fig. 2.4 A) Mechanical tension in the actin cytoskeleton stimulates the outgrowth of microtubules towards the cell edge. From [44], with permission. B) Mechano-sensor model of the guidance of microtubules to adhesion sites. From [28], with permission. C) The guidance scenario, as suggested by data from total internal reflection fluorescence microscopy (TIRFM). From [28], with permission. (The full legend to this figure appears on page 27.)

Unpolarized cell exposed to uniform stimulus:

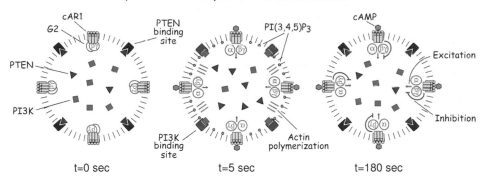

t=0 sec   t=5 sec   t=180 sec

Unpolarized cell   **Polarized cell**
exposed to a gradient: **exposed to a gradient:**

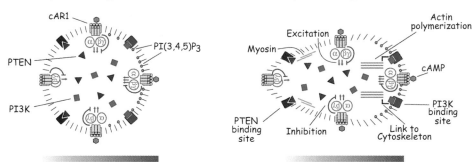

Gradient   Gradient

Fig. 3.1 Cartoon depicting the distribution and activity of signaling molecules in un-polarized and polarized cells during temporal and spatial stimulation. Chemo-attractant receptors (cAR1), receptor occupancy, associated and dissociated (activated) G protein $\alpha\beta\gamma$-subunits, excitation, inhibition, PI3Ks, PTEN, PI3K and PTEN binding sites, PI(3,4,5)P$_3$, F-actin, and myosin are indicated. Upper diagram illustrates an unpolarized cell at times before and during stimulation. In resting cells, PTEN is bound to the membrane and PI3Ks are in the cytosol (t=0 s). An increase in receptor occupancy by chemo-attractant (orange hexagons) triggers, through the heterotrimeric G proteins, a rapid increase in excitation (green arrow) which leads to binding of PI3Ks (light blue squares) to their binding sites (dark blue squares) at the membrane and causes PTEN (maroon triangles) to dissociate from binding sites (purple squares) at the membrane (t=5 s). The combined effect causes a large increase in PI(3,4,5)P$_3$ (green lollipops). At longer times, inhibition (red line) increases and eventually balances excitation. PI3Ks return to the cytosol, PTEN returns to the membrane, and PI(3,4,5)P$_3$ returns to pres-timulus levels (t=180 s). Lower diagram shows the response of a cell treated with latrunculin A (left) and of a polarized cell (right) in a spatial gradient. The appearance of the polarized cell would be similar in a uniform concentration of attractant (see text and Fig. 3.5). The "global" inhibition (red line) is equal at the two ends of the cell. "Local" excitation is slightly higher at the front causing the binding of PI3K to and the loss of PTEN from the membrane at the front. This leads to a large steady-state accumulation of PI(3,4,5)P$_3$ selectively at the front, and in untreated cells, actin polymerization polari-zation and directed motility. (This figure also appears on page 36.) (Figure reprinted from [14] with permission from the American Society of Biochemistry and Molecular Biology.)

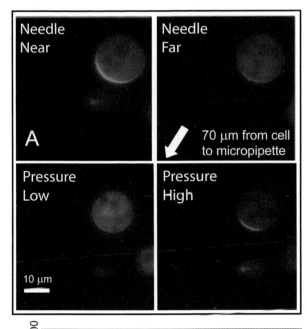

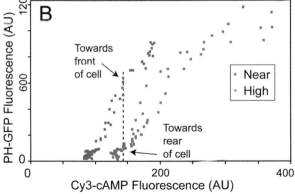

Fig. 3.2 Response to chemo-attractant gradients of vary-ing steepness and absolute concentrations. A) Cells were exposed to steep gradients (needle near) and shallow gradients (needle far). The mid-point concentration of chemo-attractant was varied by changing the pressure in the micropipette (pressure high and low). B) Shown are the Cy3-cAMP and PH-GFP fluorescence levels for the same cell in two different gradients. Following the dotted line it is clear that the same Cy3-cAMP concentration elicited vastly different PH-GFP responses.

(This figure also appears on page 38.) (Figure reprinted from [53] with permission from the National Academy of Sciences of the United States of America.)

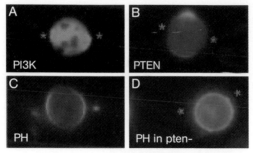

Fig. 3.4 Response to multiple gradients. Two micropipettes were brought in close proximity to latrunculin-treated cells creating cAMP profiles with two sharp gradients on either side. A) PI3K-GFP localized to both ends of the cells. B) PTEN-GFP re-localized to the cell membrane at the point of lowest cAMP concentration. C) Cells expressing PH-GFP responded at both ends similar to that of PI3K-GFP. D) pten⁻ cells expressing PH-GFP are incapable of responding with two sharp crescents as in panel C. (This figure also appears on page 39.) (Figure reprinted from [53] with permission from the National Academy of Sciences of the United States of America.)

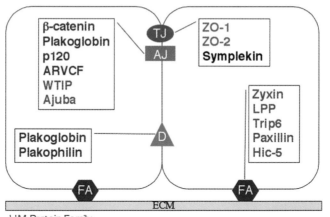

Fig. 4.1 Dual location proteins found in various cell adhesion plaques. A schematic of two epithelial cells in contact with each other and the substratum is shown to illustrate the four major cell adhesions and their respective dual location proteins. FA, focal adhesions; D, desmosome; AJ, adherens junction; TJ, tight junction. Families of dual location proteins are indicated by their respective colored fonts. Not shown is the basolateral complex of SAP97 and CASK. (This figure also appears on page 48.)

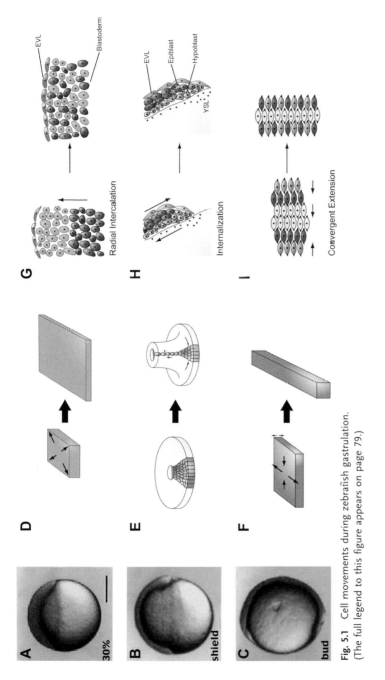

Fig. 5.1 Cell movements during zebrafish gastrulation. (The full legend to this figure appears on page 79.)

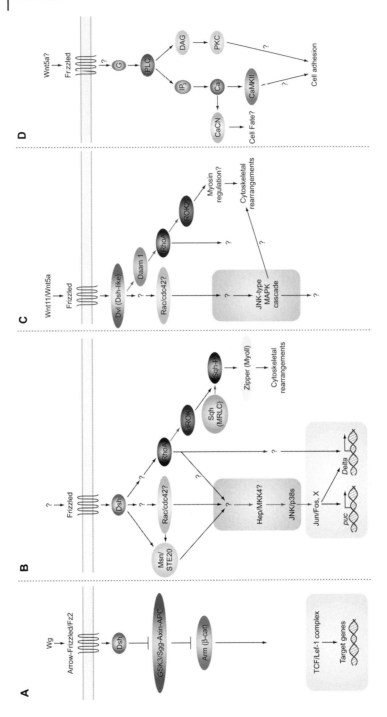

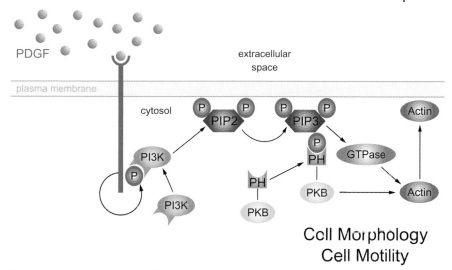

PDGF

extracellular
space

plasma membrane

cytosol

PI3K

PIP2

PIP3

Actin

P

PI3K

PH

PKB

PH
PKB

GTPase

PKB

Actin

Cell Morphology
Cell Motility

Fig. 5.3 Schematic diagram of the PI3Kinase signaling pathway during zebrafish gastrulation. PDGF (Platelet Derived Growth Factor) binds and activates its receptor. The receptor becomes phosphorylated and binds phosphoinositol-3-kinase (PI3K), which then converts Phosphoinositide-biphosphate (PI2P) to Phosphoinositide-triphosphate (PIP3) at the plasma membrane. Protein Kinase B (PKB) binds to PIP3 via its Pleckstrin Homology (PH) domain, triggering the localized accumulation of Actin and subsequent polarization, process formation and cell motility. (This figure also appears on page 86.)

Fig. 5.2 The Wnt/PCP pathway. A) A simplified version of the canonical Wnt/β-catenin pathway for comparison. B), C) The Wnt/PCP cascades regulating planar cell polarity (PCP) in *Drosophila* (B) and convergence extension movements in zebrafish (C). C) Like the Wnt/β-catenin signaling, the Wnt/PCP pathway involves the cytoplasmic signal transduction protein Dishevelled. Dishevelled regulates the activity of members of the Rho family of small GTPases. RhoA, which is linked to Dishevelled by Daam 1, activates the effector kinase Rok, which in turn regulates the actin cytoskeleton. Additionally, Dishevelled can also activate the small GTPase Cdc42 and further downstream the JNK signaling pathway, which regulates transcription of target genes. D) The Frizzled receptors can also stimulate calcium influx and the activation of the calcium-sensitive kinases PKC and CamKII, which determine cellular adhesion. (This figure also appears on page 83.)

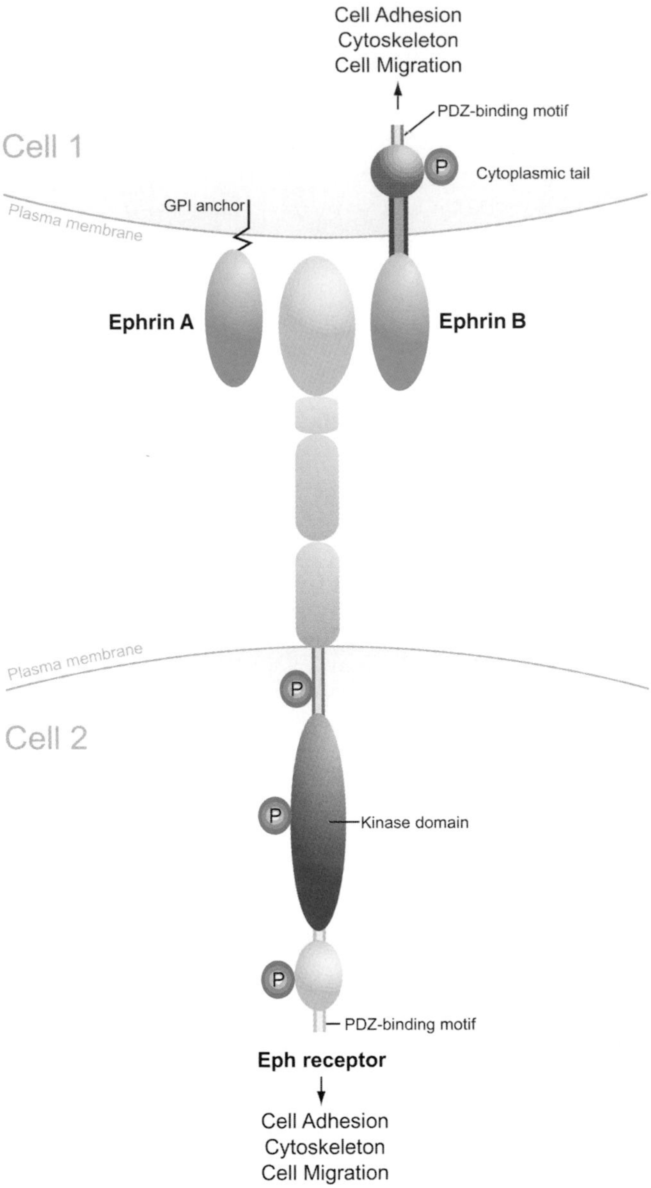

Cell Adhesion
Cytoskeleton
Cell Migration

PDZ-binding motif

Cell 1

Cytoplasmic tail

P

GPI anchor

Plasma membrane

Ephrin A   Ephrin B

Plasma membrane

Cell 2

P

P   Kinase domain

P

PDZ-binding motif

Eph receptor

Cell Adhesion
Cytoskeleton
Cell Migration

Fig. 5.4 Schematic diagram summarizing general features of Ephrin/Eph signaling. Cell 1 expresses class A and class B Ephrins. Class A Ephrins are bound to the plasma membrane through a GPI anchor, while Class B Ephrins are transmembrane proteins with a cytoplasmic tail and a PDZ-binding motif. Cell 2 expresses the Eph receptor characterized by the presence of a kinase domain and a PDZ-binding motif. The PDZ motif in both Ephrins and Eph receptor serves as a binding site for other proteins such as PI3K. The Ephrin-expressing cell interacts with the Eph-expressing cell, and this interaction is required for cell sorting and compartment boundary formation through its effects on cell adhesion, cytoskeleton organization, and cell motility. (This figure also appears on page 88.)

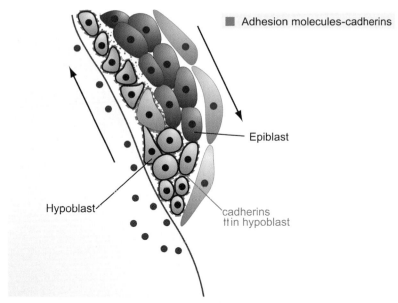

Adhesion molecules-cadherins

Epiblast

Hypoblast

cadherins
↑↑in hypoblast

Fig. 5.5 Schematic diagram illustrating the (potential) role of differential cell adhesion in germ layer formation and separation at the onset of gastrulation. (The full legend to this figure appears on page 91.)

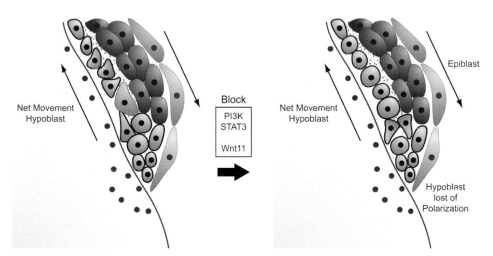

Net Movement Hypoblast

Block

PI3K
STAT3

Wnt11

Net Movement Hypoblast

Epiblast

Hypoblast lost of Polarization

Fig. 5.6 Schematic diagram illustrating the (potential) role of cell polarization and directed cell migration for germ layer formation at the onset of gastrulation. (The full legend to this figure appears on page 95.)

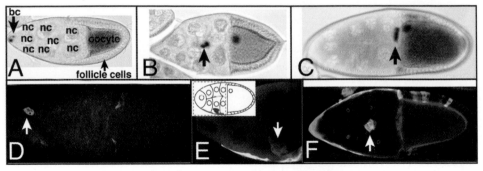

Fig. 7.1 Border cell migration in wild type and mutant egg chambers. (A–C) Nomarski optics of egg chambers stained for β-galactosidase activity from the *PZ6356* enhancer trap, which stains the border cells (arrows) and the oocyte nucleus (B–C). A) Stage 8 egg chamber in which the border cells have not migrated yet and are located at the anterior tip of the egg chamber. The border cells (bc), nurse cells (nc), oocyte, and follicle cells are labeled. B) Stage 9 egg chamber in which the border cells are in the process of migrating. C) Stage 10 egg chamber in which the border cells have completed their migration. D) Stage 10 egg chamber expressing DN-EcR in the border cells. The border cells (arrow) fail to migrate when any member of the ecdysone pathway is disrupted. Genotype shown is *slbo-GAL4*, UAS-mCD8-GFP/ UAS-DN-EcR (F645A) [18]. E) Example of a stage 10 egg chamber with misguided border cells in which follicle cells misexpress PVF1 ligand (green). The border cells (red, arrow) are located on the side of the egg chamber, adjacent to the cells misexpressing PVF1. Inset shows a schematic of the egg chamber. Genotype shown is *hs-flp, Pvf1^{EP1624}; AyGAL4, UAS-lacZ/UAS-Pvf1*. Reprinted from [19]. F) Stage 10 egg chamber in which the border cells (green) are mutant for *Pvr*. Loss of *Pvr*, using the *Pvr^{c2195}* allele, results in mild border cell (arrow) migration defects [4, 19]. Anterior is located to the left in all panels. (This figure also appears on page 124.)

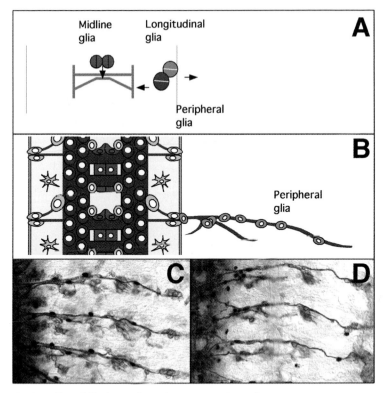

Fig. 8.1 A) and B) show schematic representations of ventral nerve cord development. A) In early development first axons form the segmental commissures (grey). The midline glia (red) start to migrate posteriorly whereas the progeny of the lateral glioblast migrate towards the forming commissures. The peripheral glial cells (green) migrate along the segmental nerves as soon as axons have left the CNS (B). B) In a mature embryonic nervous system the midline glia (red) has covered the commissures, the longitudinal glia covers the longitudinal connectives and the peripheral glial cells (green) are found in the PNS. Some additional glial cells are indicated in grey. C, D) *Drosophila* embryonic PNS axons labeled anti-Futsch and subsequent HRP immunohistochemistry; subset of peripheral glia nuclei visualized via an enhancer trap insertion into *gliotactin* (blue). C) Note the stereotyped positions of the glial nuclei along the peripheral nerves. D) In *pilage* mutant embryos, glial cells stall at the CNS/PNS transition zone and do not migrate into the periphery. (This figure also appears on page 143.)

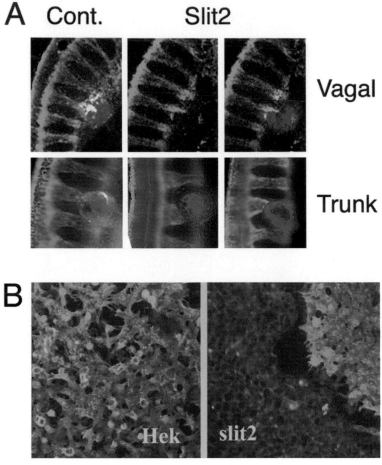

Fig. 9.6 Slit inhibits trunk neural crest migration.
A) Injection of cells producing Slit2 (red) at trunk levels
results in formation of a border between neural crest
cells, recognized with the HNK-1 antibody (green), and
Slit2 cells (red); in contrast, similar injections at vagal
levels lead to intermixing of Slit2 (red) and HNK-1 positive cells (green). B) HNK-positive neural crest cells
(green) intermix with control HEK cells, identified by
DAPI staining (blue). In contrast, neural crest cells
avoid and form a border with Slit-2 transfected cells
(blue = DAPI staining) cells (adapted from deBellard
et al., 2003). (This figure also appears on page 166.)

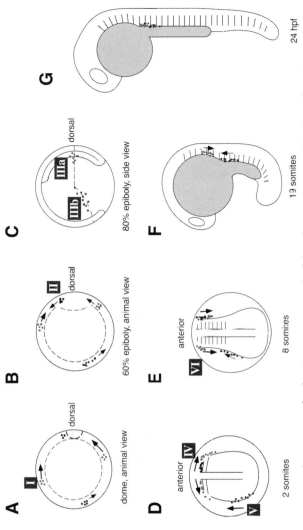

Fig. 10.1 The six steps of early PGC migration in zebrafish. Schematic drawings of embryos from dome stage (4.5 h post fertilization (hpf)) to 24 hpf showing the positions and movements of the four PGC clusters. At dome stage, four clusters of PGCs are found close to the blastoderm margin in a symmetrical 'square' shape. All possible orientations of the square relative to the dorsal side of the embryo can be observed. At 24 hpf, the PGC clusters are located at the anterior end of the yolk extension, which corresponds to the 8th to 10th somite level. In most embryos, all PGCs have reached this region by the end of the first day of development. (This figure also appears on page 174.)

wild-type	spadetail

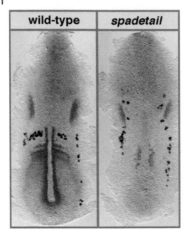

Fig. 10.2 At the 3-somite stage (11 h post fertilization), alignment of PGCs at their intermediate target at level of the 1st somite as seen in wild-type embryos (the two groups of cell perpendicular to the body axis) but is lost in *spadetail* mutants in which the differentiation of somatic cells in the target region is defective. Germ cells in blue (*vasa*) and somatic expression of *myoD*, *papc* and *pax8* in red. (This figure also appears on page 175.)

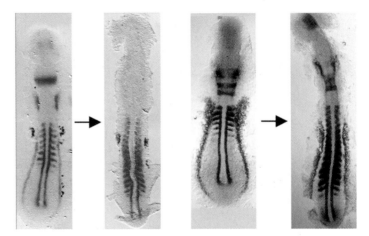

Fig. 10.5 During somitogenesis stages the PGC migrate as cell clusters from the level of the first 3 somites to the level of the 8th somite (left two panels, purple labeled cells). At the same time, *sdf*-1a expression pattern changes in a similar pattern (right two panels, blue labeled are on either side of the embryo). Reproduced with permission from [13] and [3]. (This figure also appears on page 179.)

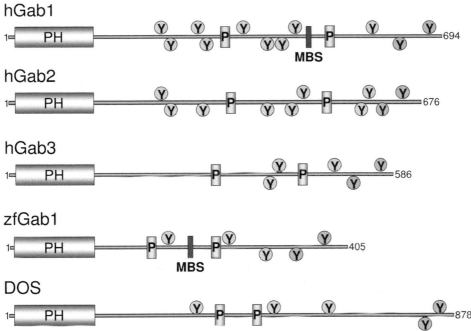

Fig. 11.4 Domain structure of Gab1-related docking proteins. Grb2 associated binder (Gab1) family members are substrates of tyrosine kinases. They are harbor an N-terminal pleckstrin homology domain (PH), and they contain multiple tyrosine phosphorylation sites that can function as docking sites (Y) for signaling proteins, for instance PI(3)K, Crk/CrkL and Shp2. Consensus binding sites for Shp2 are marked in green, PI(3)K binding sites in orange, and Crk/CrkL consensus binding sites in blue. Proline-rich regions (P) of Gab1 mediate interaction with Grb2. A c-Met binding site (MBS) is present in human Gab1 and in zebrafish Gab1 (zfGab1), but not found in Gab2 and Gab3 or *Drosophila* Dos. (This figure also appears on page 196.)

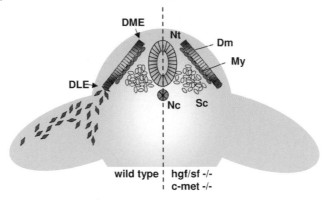

Fig. 11.5 Origin of limb muscle progenitor cells. This figure was modified from Ref. [15, 56]. (Reproduced with permission from Nature Reviews Molecular Cell Biology, Macmillan Magazines Ltd. and Curr Opin Cell Biol., Elsevier.) (The full legend to this figure appears on page 198.)

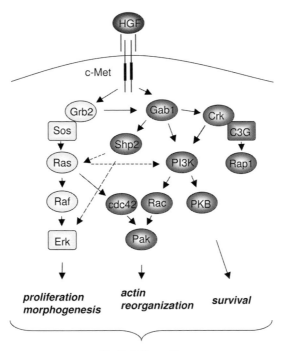

Cell Migration

Fig. 11.6 Signaling pathways downstream of Gab1 and Grb2 are implicated in cell migration. (The full legend to this figure appears on page 200.)

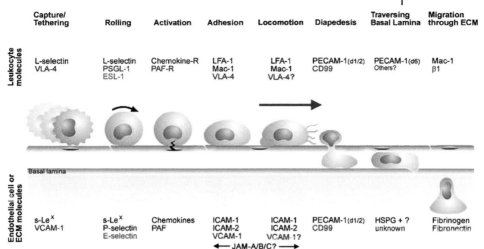

Fig. 13.1 Sequential steps in leukocyte emigration are controlled by specific adhesion molecules on leukocytes and endothelial cells. The various steps of leukocyte emigration described in the text are depicted schematically here. For each step the interacting pairs of adhesion molecules, ligands, or counter-receptors expressed by the leukocytes and endothelial cells or extracellular matrix are shown in the same color. This diagram is not inclusive, and other molecules may mediate each of these events for distinct leukocyte types under different inflammatory conditions. For the Capture/Tethering step, the protrusions on the leukocyte surface are meant to represent the microvilli that bear L-selectin or VLA-4. The lightning bolt at the Activation step represents the triggering of inside-out activation of leukocyte integrins by signals from the endothelium and endothelial surface via G protein coupled receptors. The arrow under Locomotion indicates translational movement across the endothelial surface. Question marks next to some molecules indicate that their role in this step is postulated, but not documented. The basal lamina is depicted as separate from the remainder of the extracellular matrix (ECM) since migration across the subendothelial basal lamina appears to be a separate step controlled by distinct molecules [34, 50]. However, these steps may not be molecularly distinct (see text.) Since the exact β_1 integrin(s) that are involved in migration through ECM via fibronectin interactions have not been defined, the leukocyte molecules are designated simply as β_1. Abbreviations: s-Le^x = sialyl-Lewis^x carbohydrate antigen; PSGL-1 = P-selectin glycoprotein ligand 1; ESL-1 = E-selectin ligand 1; PAF = platelet activating factor; PAF-R = PAF receptor; PECAM-1 (d1/2) = interaction involves immunoglobulin domains 1 and/or 2 of PECAM-1; PECAM-1 (d6) = interaction involves immunoglobulin domain 6 of PECAM-1; HSPG = heparan sulfate proteoglycan.
(This figure also appears on page 238.)

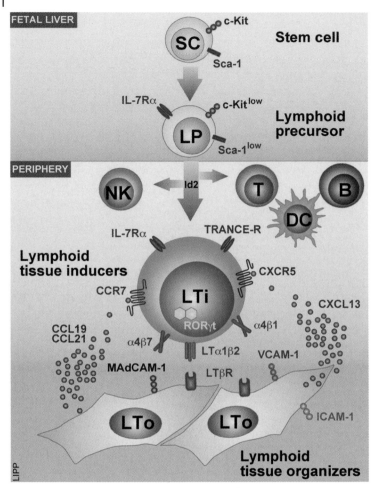

Fig. 14.1 Model of lymph node and Peyer's patch development by lymphoid tissue inducer (LTi) cells expressing the chemokine receptors CXCR5 and CCR7. (The full legend to this figure appears on page 253.)

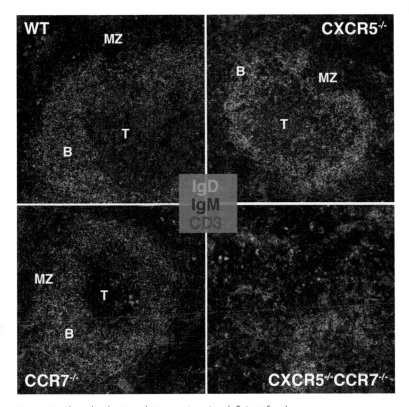

Fig. 14.2 Altered splenic architecture in mice deficient for the chemokine receptors CXCR5 and CCR7. Cryostat sections of the spleen of C57BL/6 mice were stained with PE-labeled anti-CD3 (red), FITC-labeled anti-IgD (green) and biotinylated anti-IgM/ Streptavidin-Cy5 (blue) to visualize the architecture of primary B cell follicles and of T cell-rich areas. In wild-type mice, the T cell zone is typically situated around a central arteriole with B cell rich areas eccentrically positioned at the periphery of the T cell zone. Marginal zone (MZ) B cells (IgM^{hi}IgD^{lo}) appear in wild-type mice as a rim around the polarized clusters of follicular B cells, which are mainly IgM^{lo}IgD^{hi}. In CXCR5$^{-/-}$ mice, B cells fail to organize in polarized clusters. The T cell zone is instead surrounded by a small ring of IgM^{-}IgD^{hi} B cells, which is in turn encompassed by a thickened ring of IgM^{hi}IgD^{lo} marginal zone B cells. CCR7$^{-/-}$ mice generally lack prominent T cell zones in the white pulp. The boundaries between the rudimentary T cell areas and the B cell zones are poorly demarcated. In addition, clusters of IgM^{hi}IgD^{lo} B cells are frequently visible close by the central arteriole. In contrast, T cell zones and B cell zones are largely disorganized in CXCR5$^{-/-}$CCR7$^{-/-}$ mice. Similar to the spleen of mice lacking the expression of CCR7, T cells are randomly distributed throughout the red pulp. (WT, wild type; T, T cell zone; B, B cell zone; MZ, marginal cell zone; Original magnification: ×200). (This figure also appears on page 257.)

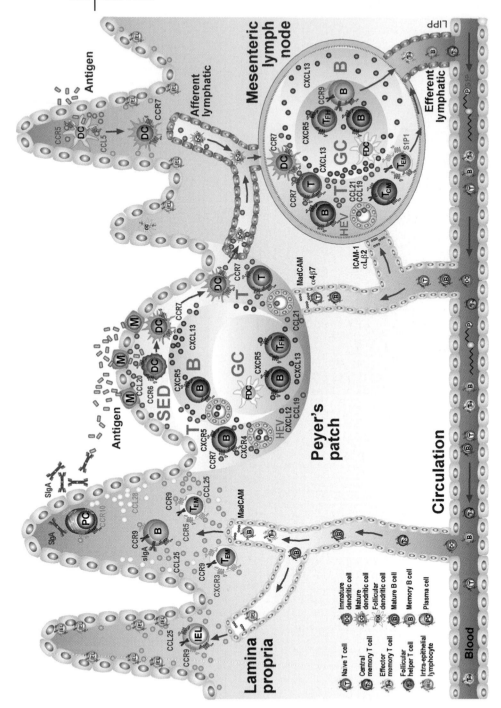

Fig. 14.3 Chemokine-driven lymphoid cell migration in mucosal immunity. In terms of immune functions the tissue of the small intestine can be divided into the initiation compartment defined as gut-associated lymphoid tissue (GALT), which includes the organized structures of Peyer's patches (PP) and mesenteric lymph nodes (MLN), and into the effector compartment consisting of the lamina propria (LP) with scattered effector T and B cells, and of the intra-epithelial lymphocytes (IEL) embedded in the epithelial cell layer of the intestinal wall.

The cell type and tissue-specific expression of chemokines and their corresponding receptors plays a decisive role in lymphocyte and dendritic (DC) cell homing to mucosal tissues. The entry of naïve, recirculating lymphocytes into Peyer's patches (PP) is mediated by the homeostatic/lymphoid chemokines CXCL12, CXCL13, CCL19 and CCL21. Naïve, recirculating T cells leave the bloodstream in high endothelial venules, (HEVs) crossing the T cell rich area of PPs. T cell entry into the T cell rich area depends on expression of the chemokine receptors CCR7 (which binds CCL19 and CCL21) and CXCR4 (which binds CXCL12) on T cells. In contrast, naïve B cells predominantly employ CXCR5 (which binds CXCL13) to leave the bloodstream in follicular high endothelial venules [38]. B cells may also utilize CXCR4 or CCR7 to enter PPs; however, both receptors cannot compensate for a lack of CXCR5. Moreover, CCR7 and CXCR5 are responsible for the organization of secondary lymphoid tissue into B cell-rich and T cell-rich zones as the ligands for these receptors are expressed in the T- and B cell zone, respectively [33].

DCs depend on CCR6 to enter the sub-epithelial dome (SED) region of PPs [9]. The ligand for CCR6, CCL20, is constitutively expressed by epithelial cells. Thereby, CCL20 directs CD11c$^+$CD11b$^+$ DCs in close proximity to M cells, that actively acquire antigen from the intestinal lumen. Following antigen uptake, maturing DCs upregulate CCR7, which enables these cells to migrate via the afferent lymphatic vessels into the inter-follicular T cell rich regions of PPs or the T cell zones of MLN.

T cell activation by DCs confers to T cells the ability to home to non-lymphoid organs. However, Peyer's Patch DCs establish gut tropism in activated memory/effector T cells which preferentially home to mucosal effector sites [7]. Homing of central memory/effector memory T cells to the lamina propria of the small intestine involves CCR9, which is expressed by virtually all T cells present in the LP. The ligand for CCR9, CCL25, is constitutively expressed by epithelial cells of the lower villi and crypts within the small intestine and deposited on the luminal surface of epithelial cells in intestinal venules. In addition, expression of receptors for inflammatory chemokines, such as CXCR3 and CCR5, contribute to the homing of effector T cells (T$_{EM}$) to the LP.

IgA-secreting plasma cells (PC), which are generated in germinal centers (GC) of PP and MLN, utilize CCR9 to home to the LP of the small intestine [39]. Once PC have entered the mucosal effector tissue, CCR9 becomes downregulated by these cells; therefore CCL25 is not necessary for the retention of plasmablasts in the LP of the small intestine. In addition, PC may use CCR10 to enter the LP. Most IgA-secreting PC express CCR10 and show a chemotactic response towards CCL28, a ligand for CCR10 that is constitutively expressed by epithelial cells in the small and large intestine. A subset of circulating memory surface IgA$^+$ B cells developed in GC of PP and MLN expresses CCR9; therefore CCR9 might be involved in memory B cell homing to the small intestine. Together with receptors for inflammatory chemokines, CCR9 may also play a role for the recruitment of intra-epithelial lymphocytes (IEL) to the epithelium, which are more frequently CD8$^+$ T cells.

After activation in lymphoid organs, lymphocytes must again return to circulation to reach effector sites. Upregulation of the sphingosine-1-phosphate receptor S1P1 in memory/effector T and B cells in course of ongoing immune responses enables these cells to egress from secondary lymphatic organs and enter the circulation [28]. (This figure also appears on page 262.)

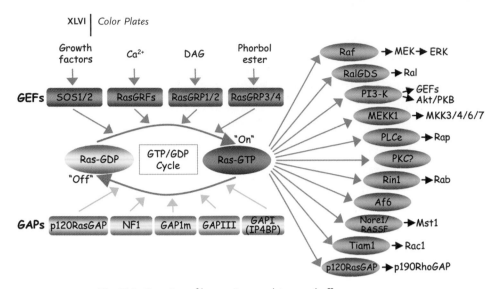

Fig. 16.1 Overview of known Ras regulators and effectors. Ras proteins are activated by multiple extracellular stimuli, such as growth factors activating receptor tyrosine kinases and G protein-coupled receptors, or by second messengers (DAG: diacylglycerol). They induce the activation of various guanine-nucleotide exchange factors (GEFs; orange rectangle) which mediates the exchange of GDP to GTP. Several GTPase-activating proteins (GAPs; yellow rectangle) are involved in downregulating Ras by enhancing the hydrolysis of GTP to GDP. GTP-bound Ras interacts with a variety of downstream effector proteins (green ovals) thereby inducing transcription, cell-cycle progression, differentiation, apoptosis or survival, cell motility and phagocytosis. For details concerning individual GEFs, GAPs or effectors see: [5] and [7].
(This figure also appears on page 301.)

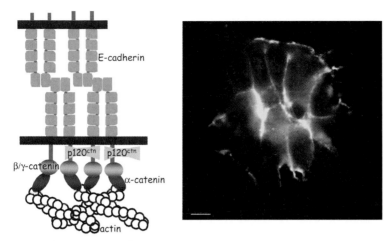

Fig. 16.2 Formation of cell–cell adhesion complexes in
epithelial cells by E-cadherin and catenins (left). Immuno-
fluorescence localization of E-cadherin (green), β-catenin
(red) and DNA (blue) in cultured epithelial cells
(PaTu8988s). Co-localization of E-cadherin and β-catenin
is indicated by the yellow color. The bar represents
10 μm. (This figure also appears on page 307.)

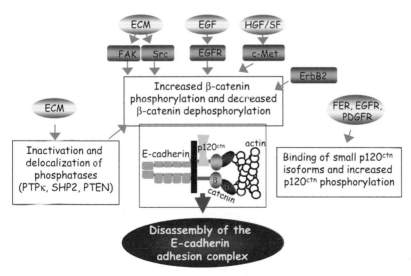

Fig. 16.3 Regulatory processes can contribute to a dissociation of E-cadherin adhesion complexes, reduction of
cellular adhesion of epithelial cells, and induction of the
dissemination of tumor cells. (This figure also appears
on page 312.)

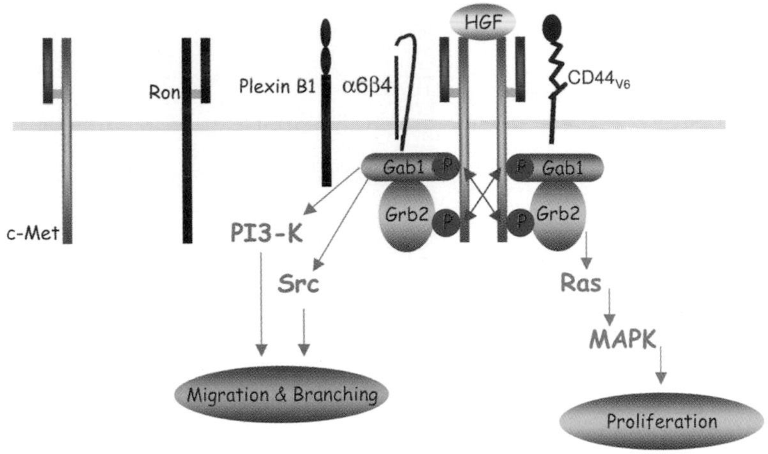

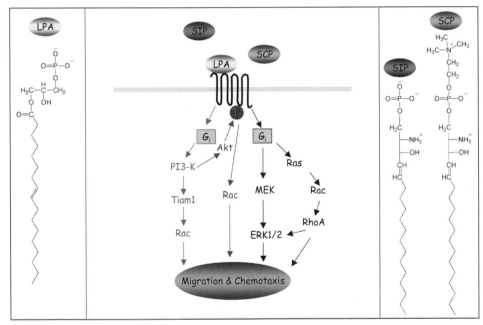

Fig. 16.4 Signaling pathway of HGF/SF via c-Met responsible for migration and branching. Putative co-receptors are labeled in magenta. (This figure also appears on page 314.)

Fig. 16.5 Structure of bioactive lipids and the signaling pathways leading to stimulation of migration and chemotaxis. Pathways described in [235] (green), [236] (red) and [52] (magenta). (This figure also appears on page 319.)

I

Cell Shape Modulations and Cell Surface-Nucleus Connections: Prerequisites for Cell Migration

Cell Migration. Edited by Doris Wedlich
Copyright © 2005 WILEY-VCH Verlag GmbH & Co. KGaA, Weinheim
ISBN: 3-527-30587-4

1
Functional Phases in Cell Attachment and Spreading

Michael P. Sheetz, Benjamin J. Dubin-Thaler, Gregory Giannone, Guoying Jiang, and Hans-Günther Döbereiner

Abstract

There are many types of cellular motility and inherent in each motile process is a series of steps that must be performed for the cell to accomplish the desired function. Underlying each motility step is a set of proteins that must be activated in the right place for the correct period. In the case of cell attachment and spreading, cells perform several tasks that appear to occur in series rather than in parallel, which results in the cell transitioning between a number of distinct phases. For each phase, there are significant differences in protein activities, which belie differences in function for each of the phases. In the isotropic mode of attachment and spreading, we observed four distinct phases: suspension, early spreading, contractile spreading and fully spread. Suspension cells often exhibit a basal level of motility, in which they extend and retract large finger-like projections presumably to explore the environment. In early isotropic spreading, cells have committed to spreading on the surface and there is stimulated actin assembly with relatively little contraction of the assembled filaments. Over a very short period, cells transition to contractile spreading that is characterized by periodic contractions that test the rigidity of the surface. When the cell is fully spread, extension activity is significantly decreased and focal complexes start to assemble near the cell periphery. Transitions between the phases occur quite rapidly with dramatic changes in the activity of many cellular components. The highly reproducible characteristics of behavior in each of the phases indicates that the cells have only a few modes of spreading behavior and comparisons between cellular activities should be made between cells in the same phase.

1.1
Introduction

The process of cell attachment and spreading onto a matrix-coated surface has been studied at many different levels. At a basic level, the binding of integrins to the matrix molecules underlies the process because anti-integrin antibodies or RGD-peptides block cell spreading or cause cells to be released from a surface. Blocking actin filament assembly blocks spreading but allows weak adhesion. A number of treatments such as calpain inhibition also inhibit spreading indicating that many proteins are involved directly or indirectly in the process [1]. Sorting out the functional roles of the various proteins is difficult because of the co-ordination between different activities and the redundancy in different functions. Recent quantitative analyses of spreading have revealed several distinct steps or phases in the process that represent new aspects of the process that were not evident previously.

With the development of rapid digital methods to analyze images, it is now possible to re-examine the behavior of cells at a submicron level on a relatively continuous basis. This review considers several recent studies of cellular motile behavior primarily using total internal reflection fluorescence (TIRF) and differential interference contrast (DIC) microscopy. The findings are consistent with observations using other microscopic methods but provide new insights into the motile process. Although it is useful justification for studying a protein to say that without it an animal will die, a tissue will be malformed or a cell will not undergo a process, the understanding of the cellular process involves a much more detailed description. We need to borrow logic from the engineering of complex systems to come up with a complete description. Part of the process of design engineering is to describe the functional requirements for a given process and the dependent parameters [2]. Several aspects of this paradigm are not particularly suited for the dynamics of biological systems. However, it is a useful exercise to define all of the requirements for a given process to be accomplished, e.g. energy source, protein synthesis, ion movement, filament assembly or motor activity. Those requirements can only be achieved by satisfying certain dependent parameters. In the case of energy, there typically must be an ATP source that is linked in some way to ATP hydrolysis. This sort of accounting often seems mundane or trivial; however, it provides the basis for putting the understanding of cell function in context. Otherwise, there is no obvious way to organize the numerous cell functions and the various interactions that they have into a cohesive description.

1.2
Fibroblast Spreading on Matrices

The process of fibroblast spreading on an extracellular matrix is necessary for fibroblast survival and is a process that can be easily quantified using total internal reflection fluorescence (TIRF) microscopy. TIRF microscopy can be used to follow the regions of close contact (< 200 nm) between the glass surface and the cells. Teleological arguments favor the idea that fibroblasts have a major role in generating force on collagen and other matrices during wound repair and the maintenance of tendons or other connective tissues. If a given cell cannot generate force, then it is in the wrong place in the body and should die, which explains the rapid apoptosis on soft or non-adherent substrata [3]. Upon detailed analysis, spreading for fibroblasts involves the generation of physical force and it is force on matrix ligands in addition to not matrix binding that is required for survival and growth.

1.3
Summary of Spreading Process

The initial steps of cell spreading involve the binding of matrix ligands and activation of edge extension by progression through a series of distinct phases of cell behavior (see list below [22]). However, even in the absence of a substrate or hormone signals, cells exhibit motility that is characterized by the extension and retraction of actin-rich processes. An explanation for this basal phase of motility is that the cell is continually probing its environment (Fig. 1.1). If the environment has a matrix mole-

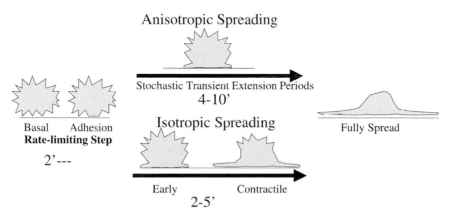

Fig. 1.1 Phases of fibroblast spreading. The distinct phases of cell spreading are shown for cells adhering to a fibronectin coated surface (adapted from [5, 6]).

cule and that matrix attaches to a rigid substrate, then the cell can generate force on the matrix, which activates further spreading. There are two distinct modes of spreading, isotropic and anisotropic. In the case of anisotropic spreading, there are multiple extension and retraction events that are limited to small regions of the edge. In the case of isotropic spreading, most of the edge extends rapidly for 1–2 min. Since the isotropic spreading is favored in serum-starved cells, it may be similar to the type of motility stimulated by epidermal growth factor [4]. After an initial period of extension with little retraction, then the activation of contraction will stimulate multiple cycles of extension and contraction that test the rigidity of the substratum. Gradually the extension activity decreases until the cell is fully spread and focal complexes start to form. In the majority of cases, the cell waits until it is fully spread before it polarizes and starts to move in a given direction.

1.3.1
Steps in Cell Spreading

Phases and Phase Transitions in Spreading (Functional Roles)

1) Basal motility of suspended cells
2) Initial attachment (cell recognizes and binds to surface without spreading)
3) Transition to spreading (integrate signal from surface to activate spreading)
4) Spreading (two different early spreading phases are seen, isotropic and anisotropic)
   (a) Anisotropic spreading (series of spreading and contractile events that test the surface rigidity often involving filopodia. Continues until cell is spread)
   (b) Isotropic spreading (nearly continuous spreading of whole edge initially. Favored in serum starvation)
      (i) Early isotropic spreading (rapid spreading to get a large contact area)
      (ii) Transition to contractile spreading (significant contact area, about 700 square microns, start of testing of environment stiffness)
      (iii) Contractile spreading (periodic contractions to test stiffness)
5) Transition to fully spread form (surface area to volume ratio, signal from surface, and low level of motility).

1.3.1.1 **Basal Motility Phase (Cells in Suspension)**
When cells are in suspension, they can be quiescent but often they exhibit motility by extending large finger-like processes for 5–10 µ. This ba-

sal motility appeared normal in that the rate of extension was approximately the same as during edge extension on a matrix-coated surface and the time of extension was typically less than a minute. If the extended processes bound to extracellular matrix (ECM) ligands, the cell would try to spread on the ECM-coated surface as described below. Alternatively, cell motility was stimulated by hormone activation, which increased activity for 1–2 min. In both states (basal and enhanced), motility involved cycles of extension and retraction that lasted 20–60 s.

1.3.1.2 Adhesion to the Surface

When cells settle onto a matrix-coated surface, they probe the surface before they extend a small lamella on the surface. During the time lag, the cell touches the surface multiple times and possibly requires an internal activation event to adhere. Even centrifugation of some cells onto the surface does not result in immediate adhesion. When cells are in a suspended state, the process of forming a stable adhesion is still slow and does not occur with each touch of the cell to the surface even at high fibronectin concentrations on the surface.

1.3.1.3 Initiation of Actin Assembly and Spreading (Rate-limiting Step)

As the cell starts to spread, the initial lamella originates from the cortical (actin-rich) region at the periphery of the cell. Because there is no organized myosin component in such peripheral regions, the centripetal, inward, movement of the actin is limited. Rather, we see the "aging" of the filaments in the central region. The actin assembly near the glass surface is dependent upon signals from the matrix binding to stimulate further assembly (although this is not totally clear, because it is difficult to catch cells at this very early stage). In normal gravity, the initiation of spreading is the rate-limiting step and we find dramatic differences in the average time for initiation of spreading dependent upon the concentration of matrix ligand on the glass. Our findings indicate that the concentration of matrix molecules on the surface is inversely related to the time lag before the start of spreading [5]. Two logical explanations for this phenomenon are either, (1) the cell requires the accumulation of X number of binding events before it will spread or (2) the probability of a critical activation event is dependent upon the concentration of matrix (i.e. with a higher concentration, the average time until an event occurs is less). Specific modeling of this under defined integrin and matrix conditions may yield interesting aspects of the mechanism.

1.3.1.4 **Continued Spreading**

Differences between isotropic and anisotropic spreading (Stages 4 a & b)
At this stage, the extension of lamellipodia is essentially continuous in isotropic and discontinuous in anisotropic cells. Although there are oscillations in the rate of extension of the leading edge in later stages of isotropic spreading, the assembly of actin filaments is essentially continuous. In anisotropic spreading, large areas of the cell edge are quiescent, showing neither extension or rearward movement of actin. Further, when the edge does extend, there is rearward transport of actin at early stages. From the earliest stages, there may be a myosin network associated with the lamellipodial actin in anisotropic spreading, which results in greater contraction in the anisotropic case [5]. Inhibition of integrin binding will rapidly block further motility in all cases, indicating that there needs to be continuous input from the integrin binding.

As the actin filaments in the central region of the initial contact zone are disassembled, the central region of the cytoplasm or endoplasm will start to spread on the surface. Initially, the endoplasm is approximately spherical. When the edge of that sphere comes into contact with peripheral regions that are spread on the matrix surface, there must be a breakdown of the endoplasm–ectoplasm boundary and a spreading of the endoplasm on the substrate. We can see from the behavior of the GFP-a-actinin in central regions that there is a phase of little or no contraction when this region expands to fill in behind the spreading wave of lamellipodial actin.

Isotropic activation of and local periodic cycles of contraction (Stage 4 b.ii)
When the endoplasm–ectoplasm boundary reforms in the region adjacent to the glass surface, the contractile machinery reforms at the periphery of the endoplasm. This increases the rate of centripetal movement and logically the force on the substrate is increased. In the area under the center of the cell, there are foci of contraction, which appear as local contractions. The pattern of alpha actinin fluorescence in TIRF undergoes a concerted contraction that appears to signal the formation of a continuous cytoskeletal network to transmit tension from one side of the cell to the other. Once the network is formed the cell is poised to develop force at its periphery and test the rigidity of the substrate. We don't know the composition of the structural elements of the network but fundamentally it is needed for the generation of force on the substrate.

For the cell to extend further, the cell must normally find and test a rigid surface. Contractions pull on the matrix molecules and if they are attached to a rigid substrate, force is generated rapidly and locally. Either rapid or local force generation is interpreted as the signal that the surface is rigid and the cell should extend further on the surface. If a rigidity signal is generated by a contraction, the cell also appears to generate a signal to test the surface again for its rigidity. Thus, we found a periodic cycle of contractions on rigid surfaces. In this cycle, the first contraction not only causes further extension but also generates a factor that activates the next contraction. The period between the contractions is related to the time needed for the actin to move from the cell edge to the back of the lamellipodia (about 25 s to travel 1.5 μm) [6]. When the actin reaches the end of the lamellipodium, it disassembles and releases the agent that then activates the next round of contractions. In this manner a cyclic contraction pattern can be generated, which will repeatedly test the rigidity of the substrate before the cell extends further.

1.3.1.5 Transition to Fully Spread State

For normal spreading, we believe that the cell volume and surface area are constant. As cells transition from the nearly spherical form in suspension to a highly spread form, the excess membrane area is depleted and cell ultimately must stop spreading. In the quantitative analyses of spreading, there is a general decrease in spreading activity over time as the spread area asymptotically approaches the maximal spread area. Often the cells will not polarize until they are maximally spread and polarization then is accompanied by a slight decrease in area. We don't understand what triggers cessation of spreading or sets the final spread area.

1.3.2
Binding to Rigid Matrices Causes Strengthening of Cytoskeleton–Integrin Linkages

An important part of motility is the dynamics of the linkages between the liganded integrins and the cytoskeleton. When force applied to those contacts to enable the cell to move the contact or to move itself, there are characteristic changes that involve activation of growth and signaling to the cell nucleus [7]. As lamellipodia extend to new regions of a surface, new matrix molecules bind to integrins that are at the leading edge. When fibronectin or other ECM ligands bind to integrins at an extending leading edge, there is rapid attachment of the integrins to the cytoskeleton [8]. Preferential attachment to the cytoskeleton at the leading edge is an indication that a specialized set of proteins is concentrated there [9–11]. As the actin in the leading edge moves rearward, there are a number of events that are triggered by the generation of force on the integrin–cytoskeleton bonds. We have a partial list of the components involved and have developed a working model (Fig. 1.2); however, many critical questions remain, including the identification of the roles that many other proteins play in the process.

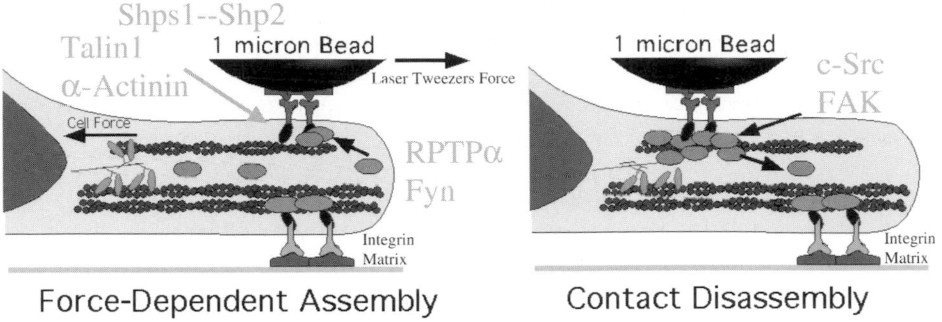

Force-Dependent Assembly   **Contact Disassembly**

Fig. 1.2 Model for integrin–cytoskeleton linkage strengthening. The force-dependent strengthening of the linkages between fibronectin–integrin complexes and the actin-rich cytoskeleton are summarized. This is only a partial list of the components involved but these appear to have critical roles in the process. (This figure also appears with the color plates.)

1.3.2.1 Initial Binding of Fibronectin Multimers at the Leading Edge and Over Actin Cables

Using fibronectin and anti-integrin antibody coated beads, the initial events in $a_v\beta_3$ and possibly $a_5\beta_1$ integrin binding to the cytoskeleton have been defined. Cross-linking of the integrins by multiple fibronectin ligands is a critical first step and the minimum aggregate is a trimer rather than a dimer [10]. Ligand binding and cross-linking initially results in edge binding [9]; and then after 3–5 min, the aggregates accumulate over focal contact points [10]. The steps in between are poorly understood but earlier studies of fibronectin bead binding suggest two models: (1) diffusion-trapping model or (2) directed transport model. In the diffusion-trapping model, the finding that fibronectin beads can diffuse when bound behind the leading edge makes it possible for those liganded integrins to diffuse to the focal contact sites where linkage proteins would be concentrated and would trap those complexes. In a directed transport model, the focal contacts would be considered as contractile foci and the actin network to which the liganded integrins were attached would be drawn to such foci. Once aggregated at such points, additional proteins of the focal contacts would stabilize the liganded integrins in those regions. Smaller foci would be more likely to dissipate and would end up in larger foci.

1.3.2.2 Force-dependent Activation of the ECM–Integrin Complexes

There is considerable evidence that the rearward movement of actin from the leading edge results in force-dependent activation of signaling processes locally that are clearly involved in whole cell responses to adhesion. At leading lamellipodia the responses are different than at the retracting tails of cells. Based upon several recent papers, there is a relatively complex working model for the events that follow from $a_v\beta_3$ integrin binding to fibronectin or vitronectin. Initially, binding of a fibronectin trimer with proper spacing will engage talin1 binding and become linked to the cytoskeleton. The initial linkages can be easily broken with 2 pN of force but will eventually become strengthened. In parallel, there appears to be activation of Src family kinases by RPTPa, which is at the leading edge in complex with $a_v\beta_3$ integrin during the first 30 min of spreading [12]. As beads are pulled rearward against a laser tweezers force, the linkages to the cytoskeleton are strengthened [13]. Deletion of either talin1 [14], RPTPa [12] or Shp2 [15] prevents the strengthening of the cytoskeleton linkages, indicating that the process involves many cellular components.

Our working model for strengthening is that fibronectin binding to $a_v\beta_3$ causes attachment to the cytoskeleton either through talin1 or other components. Force generation on the complex activates RPTPa, which then activates several Src family kinases, including Fyn. Fyn is localized to early focal complexes under liganded integrins and catalyzes the strengthening of the linkages. We speculate that Fyn indirectly causes many focal complex proteins to assemble at the site including paxillin, vinculin and alpha actinin. However, strengthening correlates most strongly with the binding of alpha actinin and not with paxillin [15]. Alpha actinin binding depends upon local FAK inactivation by Shp2 which is in turn downstream of Shps1 [15]. A scaffolding role is postulated for talin1, since its deletion blocks strengthening but does not block activation of Src family kinases or spreading [14]. A major gap in this model involves the steps leading from Fyn activation to the assembly of focal complex components. Finally, we know that the strengthening often only lasts for a period of 1–2 min in the absence of continued force application [16] and the steps leading to stable focal contact formation have not been analyzed in this way, although many of these proteins are thought to have roles in stable focal contacts.

1.3.2.3 **Additional Steps in the Spreading Process**
For the membrane to assemble adjacent to the glass surface, the local integrin binding of ligand must cause activation of actin assembly and the subsequent machinery. One model that fits with many observations is that PLC is activated locally which in turn recruits PKC that causes MARCKS to leave the membrane, revealing more PIP2 than was cleaved. The activation of SFKs by RPTPs is early in the process and may be critical for subsequent steps. Another early step is the activation of GEFs (Dbl family members) that produce active Cdc42, Rac, and Rho for sub-

sequent steps [17]. For example, the activation of WASP and Arp2/3 must follow for actin polymerization. In addition, the activation of PAKs leading to activation of LIM kinase and cofilin for filament disassembly should occur simultaneously. An additional factor that needs activation is a myosin that will draw actin filaments in the lamellipodium rearward. It is obvious that many of these different protein complexes need to be controlled in parallel and the important issue for the future is how can there be such a seamless communication between them.

1.3.3
MTs and Motility

MT activation of Rac1 in a positive feedback through APC [18] and Asef and Rac1 acts through PAK1 to activate motility [19]. Many labs have focused on the role of mDia in the feedback between microtubule polymerization and motility [20, 21]. When we have depolymerized microtubules prior to spreading, spreading has proceeded normally. Thus, we believe that the role for microtubule-dependent motility is primarily in polarization.

1.3.4
Conclusion

Recent quantitative analyses of cell spreading and motility have revealed that the cells have a complicated but characteristic pattern of motility. It is useful to categorize the process in terms of motility phases that occur in a serial fashion for specific motile processes. Both physical feedback from rigidity sensing and the strengthening of integrin-cytoskeleton linkages are important for cell viability and other critical aspects of cell function such as morphology. Future detailed analyses at the quantitative level holds the prospect of defining the protein complexes that accomplish specific functions and the ways that they communicate with other functional complexes.

1.4
References

1 DENKER, S. P., and BARBER, D. L. (2002) *J Cell Biol 159*, 1087–1096.
2 THOMAS, J. D., LEE, T., and SUH, N. (2004) *Ann Rev Biophys Biomol Struct 33*, In press.
3 PELHAM, R. J., JR., and WANG, Y. (1997) *Proc Natl Acad Sci USA 94*, 13661–13665.
4 BAILLY, M., MACALUSO, F., CAMMER, M., CHAN, A., SEGALL, J. E., and CONDEELIS, J. S. (1999) *J Cell Biol 145*, 331–345.

5 DUBIN-THALER, B.J., GIANNONE, G., DOBEREINER, H.G., and SHEETZ, M.P. (**2004**) *Biophys J 86*, 1794–1806.
6 GIANNONE, G., DUBIN-THALER, B.J., DOBEREINER, H.G., KIEFFER, N., BRESNICK, A.R., and SHEETZ, M.P. (**2004**) *Cell 116*, 431–443.
7 CHEN, C.S., MRKSICH, M., HUANG, S., WHITESIDES, G.M., and INGBER, D.E. (**1998**) *Biotechnol Prog 14*, 356–363.
8 FELSENFELD, D.P., CHOQUET, D., and SHEETZ, M.P. (**1996**) *Nature 383*, 438–440.
9 JIANG, G., GIANNONE, G., CRITCHLEY, D.R., FUKUMOTO, E., and SHEETZ, M.P. (**2003**) *Nature 424*, 334–337.
10 COUSSEN, F., CHOQUET, D., SHEETZ, M.P., and Erickson, H.P. (**2002**) *J Cell Sci 115*, 2581–2590.
11 NISHIZAKA, T., SHI, Q., and SHEETZ, M.P. (**2000**) *Proc Natl Acad Sci USA 97*, 692–697.
12 VON WICHERT, G., JIANG, G., KOSTIC, A., DE VOS, K., SAP, J., and SHEETZ, M.P. (**2003**) *J Cell Biol 161*, 143–153.
13 CHOQUET, D., FELSENFELD, D.P., and SHEETZ, M.P. (**1997**) *Cell 88*, 39–48.
14 GIANNONE, G., JIANG, G., SUTTON, D.II., CRITCHLEY, D.R., and Sheetz, M.P. (**2003**) *J Cell Biol 163*, 409–419.
15 VON WICHERT, G., HAIMOVICH, B., FENG, G.S., and SHEETZ, M.P. (**2003**) *EMBO J 22*, 5023–5035.
16 GALBRAITH, C.G., YAMADA, K.M., and SHEETZ, M.P. (**2002**) *J Cell Biol 159*, 695–705.
17 YAN, C., MARTINEZ-QUILES, N., EDEN, S., SHIBATA, T., TAKESHIMA, F., SHINKURA, R., FUJIWARA, Y., BRONSON, R., SNAPPER, S.B., KIRSCHNER, M.W., GEHA, R., ROSEN, F.S., and ALT, F.W. (**2003**) *EMBO J 22*, 3602–3612.
18 MIMORI-KIYOSUE, Y., SHIINA, N., and TSUKITA, S. (**2000**) *J Cell Biol 148*, 505–518.
19 KAWASAKI, Y., SENDA, T., ISHIDATE, T., KOYAMA, R., MORISHITA, T., IWAYAMA, Y., HIGUCHI, O., and AKIYAMA, T. (**2000**) *Science 289*, 1194–1197.
20 WITTMANN, T., and WATERMAN-STORER, C.M. (**2001**) *J Cell Sci 114*, 3795–3803.
21 BURRIDGE, K., and WENNERBERG, K. (**2004**) *Cell 116*, 167–179.
22 DOBEREINER, H.G., DUBIN-THALER, B.J., GIANNONE, G., XENIAS, H.S., and SHEETZ, M.P. (**2004**) *Phys. Rev. Lett. 93*, 108105.

2
Polarized Cell Motility: Microtubules Show the Way

J. Victor Small and Irina Kaverina

2.1
Introduction

2.1.1
The Vasiliev Conundrum

The series of studies we will describe begins in the early 1970s, when Vasiliev and co-workers showed that the destruction of microtubules in fibroblasts rendered them apolar and incapable of directed locomotion
(Fig. 2.1 A). In the context of the knowledge at that time, this was indeed
a surprising result. Contemporary ideas about cell movement were then
influenced by the wealth of biochemical and structural information about
muscle contraction, with actin and myosin filaments as central players.
Actomyosin had also been extracted from non-muscle cells and the actin
antagonist, cytochalasin, was found to inhibit cell motility. So where did
microtubules fit in? Over the following years, as ideas of the cytoskeleton
evolved, Vasiliev and his colleagues further substantiated their findings
and developed a central theme; namely that microtubules exert an overriding influence on the actin cytoskeleton to "stabilize the differentiation
of active (motile) and non-active parts of the cell edge" [1, 2]. The means
by which they exerted this influence remained, however, indefinable.

2.1.2
Cell Polarity and Adhesion

Working on living cells, the diehards of the 1970s showed that locomotion involved the protrusion of a front followed by retraction of the cell
rear. Protrusion was seen to occur in the form of thin sheets and fingers
of cytoplasm, which were termed lamellipodia and filopodia, while retraction was likened to a contractile process [1, 3]. Before moving on, we
should not forget to reflect on the technology available at that time for recording the movements of living cells under the microscope. The luxury
of CCD cameras and digital image acquisition were then unknown. In-

Cell Migration. Edited by Doris Wedlich
Copyright © 2005 WILEY-VCH Verlag GmbH & Co. KGaA, Weinheim
ISBN: 3-527-30587-4

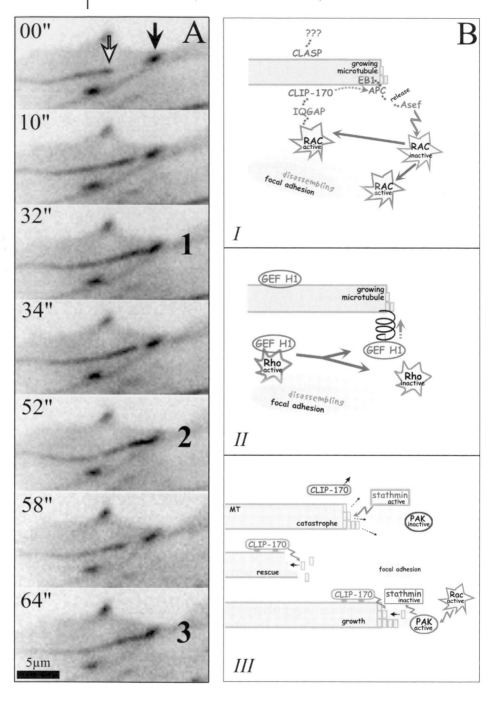

sis of the dynamics of microtubules and substrate adhesions in living cells [9]. With the temporal component now added, it became clear that interactions between microtubules and adhesions were not rare, but frequent, events. Indeed, all focal adhesions formed behind the cell front and those formed through retraction at the cell edge were found to be "targeted" by microtubules [9, 10].

More recent studies, using the technique of evanescent wave microscopy, have demonstrated the intimacy of this cross-talk between microtubules and focal adhesions [11] (Fig. 2.2 A). With this technique, substrate bound cells may be exposed to exciting radiation in a zone that ex-

Fig. 2.2 A) The targeting of focal adhesions by microtubules. Panels show video frames of the periphery of a goldfish fibroblast that was transfected with EGFP–tubulin and EGFP–zyxin. The images were taken using the technique of total internal reflection fluorescence microscopy and show that microtubules interact with adhesion sites within a vertical range of 50 nm. Open arrowhead indicates the tip of a microtubule and the closed arrowhead, an adhesion site. Three targeting events (frames marked 1, 2 and 3) occur in succession during the sequence. Time is in seconds. From [28], with permission. B) Some possible signaling scenarios associated with the targeting of adhesion complexes by microtubules. In *I* and *II*, possible events are depicted in which microtubules could relay factors that would change the local balance in the activity of Rho proteins in the vicinity of focal adhesions and promote their turnover. In *I*, the concentration of active Rac at adhesion foci by microtubules could antagonize Rho and promote adhesion disassembly. This could be effected through a mutual collaboration between the microtubule tip proteins APC and CLIP-170. APC, which is linked to microtubule tips via EB1, binds the Rac-GEF Asef and could concentrate this exchange factor to locally switch Rac from an inactive (brown) to an active state (red). CLIP-170 has been shown to bind IQGAP, which in turn binds the active, GTP-bound form of Rac. Therefore, CLIP-170, through IQGAP, could synergize with APC and Asef to concentrate active Rac at the microtubule tip. The CLIP-associated proteins, CLASPs, could participate in the targeting process. The scenario in *II* is based on the finding that the exchange factor for Rho, GEF-H1, can be inactivated by binding to the surface of microtubules. Through targeting interactions, microtubules could therefore sequester (spring) and locally silence GEF-H1 in the vicinity of focal adhesions, leading to the suppression of Rho (red to brown state) and to adhesion turnover. In *III*, a possible mechanism for the feedback regulation of microtubule polymerization is depicted that could play a role in the repetitive targeting of focal adhesions by microtubules. When dephosphorylated, stathmin promotes the depolymerization of microtubules (dashed arrows pointing away from microtubule), leading to catastrophe and the loss of microtubule tip proteins, including CLIP-170. PAK, which concentrates in focal adhesions, can phosphorylate and inactivate stathmin and could therefore promote microtubule assembly in the region of focal adhesions. The switch from catastrophe to polymerization could be facilitated by the rescue activity of CLIP-170. PAK, in turn, is activated via Rac, so that a local increase in Rac activity at focal adhesions (*I*) would promote the polymerization of microtubules into them. The local balance of Rac and Rho activities would presumably determine the net influence of this route on microtubule dynamics, as reflected by the duration and frequency of targeting of focal adhesions. (This figure also appears with the color plates.)

tends only 100–200 nm above the substrate. For fish fibroblasts expressing GFP tubulin, it was found that microtubules dip down towards the
dorsal cell surface to within 50 nm from the substrate. This dipping activity occurred mainly towards the cell edge and dual labeling with GFP-
tubulin and DsRed zyxin (to mark adhesions) showed that dipping correlated with the targeting of focal adhesions. In other words, microtubule
tips approached focal adhesions within a range of 50 nm in the z-axis,
close enough for the precise exchange of molecular signals.

2.3
Microtubule Targeting Promotes Focal Adhesion Turnover

The results of the simultaneous analysis of microtubule and adhesion dynamics in living cells clearly indicated that the targeting of focal adhesions by microtubules exerts, in fact, a *negative* influence on adhesion assembly [10]. This conclusion was reached from experiments on living
cells in different situations. A first indication that microtubules act as
negative modulators of adhesion came from observations of adhesion formation during the spreading of cells when they first settle on a substrate. In control cells, early adhesions disassemble as subsequent adhesions are created at a greater radius, under the advancing lamellipodia/fi-
lopodia boundary. In contrast, for cells spreading in the presence of nocodazole, early adhesions do not readily disassemble, but tend to elongate as the cell radius increases [10]. This phenomenon explains the re-
tardation of cell spreading by microtubule inhibitors that was noted by
Vasiliev and Gelfand in their early work [1]. Microtubules are therefore
apparently required to promote the turnover of adhesions to facilitate
spreading.

Other experiments showed that the release of focal adhesions at retracting cell edges was preceded and accompanied by multiple targeting
events by microtubules (Fig. 2.3). As already indicated, focal adhesion targeting also occurred in protruding zones and, here, targeting is apparently necessary to limit adhesion growth and to promote turnover to allow further protrusion. This was indicated by experiments in which targeting of focal adhesions in an advancing front was abrogated by low
concentrations of taxol, to freeze microtubule growth: as a consequence,
the size of anterior adhesions grew several-fold larger than in control situations [10], analogous to the situation of cells spreading in nocodazole.
These collected findings support the idea that microtubule–adhesion
cross-talk acts to retard adhesion formation or to promote adhesion turnover.

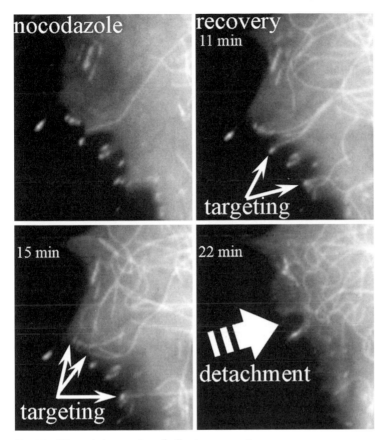

Fig. 2.3 Microtubule targeting of adhesions promotes adhesion disassembly. This example shows adhesion site targeting and retraction after recovery from local application of nocodazole. Cell shown was transfected with zyxin-EGFP and tubulin-EGFP and treated on one edge with 50 µg ml^{-1} nocodazole applied through a micro-needle to locally depolymerize microtubules. Video frames show points of recovery after removal of the needle at time 0'00''. In the video sequence all peripheral adhesions were targeted; arrows in central frames highlight some of these targeting events, prior to cell edge retraction and detachment. From [10], with permission.

2.4
Contractility, the Functional Link

It is noteworthy that Vasiliev and Gelfand [1] had already predicted an interaction of the microtubule cytoskeleton with "tension-producing structures (bundles of microfilaments?)". Moreover, Vasiliev elaborated on the idea that tension plays a role in the maturation and maintenance of focal adhesions [2], which was later substantiated by the work of Burridge and Chrzanowska-Wodnicka [12] and Bershadsky [13, 14]. A link between microtubules and tension was dramatically illustrated by the finding that contractile stress in fibroblasts increases when microtubules are disassembled [15]. This effect correlated with the growth of focal adhesions [13, 16], a reaction that was inhibited by antagonists of myosin II [13]. All these findings indicated that microtubules mediate an antagonism of myosin-linked tension production in the actin cytoskeleton.

To align these findings with the phenomenon of microtubule-adhesion site targeting, we have proposed that the antagonism of contractility by microtubules is highly localized, namely at focal adhesion sites. We suggest that microtubules provide the means to deliver highly localized "relaxing" signals to focal adhesions, via targeting events, that prevent adhesion growth or promote adhesion turnover. Circumstantial evidence for this scenario came from experiments in which local relaxation at a cell edge was induced by the application of a myosin antagonist via a micropipette [10, 17]. With this experimental protocol, the retraction of a cell edge could be induced that mimicked that seen at the trailing edges of migrating cells. Using cells injected with rhodamine-tagged myosin it could be shown (from the spacing between myosin assemblies) that retraction was due to the shortening of the actin bundles in body of the cytoplasm away from the cell edge, while the peripheral actomyosin (in the region of focal adhesions) remained relaxed [17]. Taking this finding one step further, we could show that fibroblasts rendered apolar by nocodazole could be induced to migrate by the application of the myosin inhibitor to one side of the cell [17]. In general terms, this suggests that microtubules induce polarization by establishing a gradient of contractility within the actin cytoskeleton. We suppose that this gradient arises through a differential relaxation of adhesion sites in spatially segregated regions of the cell.

2.5
Kinesin and Signal Transmission

Microtubules serve as cellular highways for molecular motors that transport cargo between the centrosome and the cell periphery [18, 19]. It was therefore pertinent to establish whether or not microtubule motors of the kinesin and dynein families were involved in the transmission of the putative modulatory signal(s) to focal adhesions.

The involvement of a kinesin motor in the control of cell polarization [20] as well as of adhesion [21] was first addressed by Gelfand's and Vasiliev's groups, who used micro-injection of anti-kinesin antibodies and analyzed fixed and stained cells. Their findings revealed that the changes in morphology and adhesion in cells in which kinesin activity was blocked were strikingly similar to those induced by microtubule inhibitors. We subsequently re-investigated the involvement of the major and ubiquitous kinesin, kinesin-1, in adhesion dynamics in living fibroblasts [22]. For these studies, we used another function-blocking antibody as well as a kinesin-1 heavy chain construct carrying a mutation in the motor domain that binds irreversibly to microtubules.

When kinesin activity was blocked in cells labeled with fluorescently tagged adhesion components (zyxin or vinculin), a marked increase in size of focal adhesions was observed (Fig. 2.1 B), in agreement with earlier results on fixed cells [21]. Parallel experiments with cells that additionally express GFP-tubulin, showed that kinesin inhibition had no effect on either microtubule dynamics or on microtubule-adhesion site targeting. In contrast to the results with kinesin, blocking of dynein activity either by an antibody injection or over-expression of dynamitin had no detectable effect on either cell polarity or adhesion dynamics.

The changes in adhesion dynamics induced by kinesin inhibition mimicked those seen in nocodazole and indicated that kinesin was indeed involved in signal transmission. In other studies, the kinesin-1 light chains have been implicated in cargo binding [23]; however expression of the light chain TPR motifs which displace the light chains from kinesin had no detectable effect on adhesion dynamics (unpublished data). We therefore conclude that the kinesin heavy chain participates in the transport of components involved in signaling focal adhesion disassembly.

2.6
Tip Complexes Meet Adhesion Complexes

As indicated above (see also Fig. 2.2 A), evanescent wave microscopy shows that the interaction between microtubules and focal adhesions is very intimate, occurring in a range of tens of nanometres [11]. We can therefore speculate that microtubules serve to catalyze signaling reactions at adhesion foci by bringing molecular complexes to their vicinity. Microtubule-associated regulators can be divided into three groups: (1) those that bind along a microtubule; (2) those transported by molecular motors; and (3) members of the microtubule tip complex [24, 25].

The presence of proteins specifically at microtubule tips was first indicated by the work of Kreis and colleagues [26] who showed that CLIP-170 was recruited to the growing, plus ends of microtubules in living cells. The other microtubule tip proteins so far identified ([25] with no doubt more to come) show essentially the same behavior: they are recruited to growing microtubule ends and dissociate when polymerization ceases. Evanescent wave microscopy also shows that the tip complex targets directly into focal adhesions [11] making members of this complex hot candidates for signaling from microtubules to components of focal adhesions. Based on current information we have suggested possible signaling scenarios which could lead to focal adhesion destabilization and disassembly [27, 28] (Fig. 2.2 B). In one scenario (Fig. 2.2 B, I), the concentration of a Rac exchange factor (Asef) bound to the tip protein APC [29] or a ligand of active Rac/Cdc42 (IQGAP) bound to CLIP-170 [30] could increase Rac activity in the vicinity of focal adhesions. Since Rac antagonizes the activity of Rho [31] this could effect the local inhibition of contractility, leading to adhesion destabilization.

Another scenario (Fig. 2.2 B, II) is suggested by the observation that two exchange factors for Rho bind to microtubules: p190RhoGEF and GEF-H1/Lfc [32–35]. Significantly, GEF-H1 is inactivated by microtubule binding and could be the "stress fiber inducing factor" released when microtubules are depolymerized [16]. Microtubules could then serve to "mop up" GEF-H1 in the vicinity of adhesion sites and thereby reduce Rho activity and promote adhesion turnover.

The question of where kinesin-1 fits in remains open. It has been shown, however, that the Rac/Cdc42 activator, RhoG is inhibited by blocking kinesin activity [36]. Also, APC appears to be accumulated at microtubule tips via a plus end directed motor [37] which has been identified as Kif3 A, B [38]. The dependence of other tip proteins on molecular motors for their tip accumulation remains to be investigated.

2.7
Focal Adhesions Influence Microtubule Dynamics

The mechanism of targeting as such includes growth of a microtubule to the adhesion site and a short association of the microtubule tip with the contact. Three subsequent scenarios are possible: first, the microtubule continues to grow; second, it pauses in the contact; and third, it undergoes catastrophe and shrinks. The delivery of molecules belonging to three groups of microtubule-associated factors (see above) thus depends on microtubule dynamics.

As already noted, an important characteristic of tip binding proteins is that they are released when microtubules depolymerize during phases of catastrophe. For delivery of these factors, repetitive targeting, entailing phases of rescue following catastrophe, would be most efficient. Rescue may depend on the availability of tip proteins, like CLIP-170, which shows rescue-enhancing activity [39]. Also, a switch to shrinkage from adhesion sites could be regulated by a catastrophe factor, such as stathmin [40] (Fig. 2.2 B, III). In this connection, factors associated with Rho-family signaling influence microtubule polymerization dynamics and could therefore act in a feedback mode to modulate microtubule-adhesion foci interactions. In particular, the multifunctional effector of Rac and Cdc42, PAK, is often found concentrated in focal adhesions [41] and via its inhibition of stathmin [42] could promote microtubule polymerization in the region of adhesion foci (Fig. 2.2 B, III).

To add to the complexity of the problem, kinesin-dependent delivery of regulators would be most efficient if a microtubule stayed associated with an adhesion. In this context, we have shown that adhesions possess microtubule-capturing and stabilizing ability [9], that could serve to prolong the pause of a microtubule tip in an adhesion site for signal trafficking.

2.8
Actin Talks Back: Tension and Microtubule Guidance

How is microtuble polymerization directed into adhesion foci? Clues about the mechanism of guidance come from the finding that stress in the actin cytoskeleton has a dramatic influence on microtubule dynamics. This was first illustrated by experiments involving the local application of inhibitors of contractility, mentioned above. In the short term, local relaxation at a cell edge (induced by either BDM or ML-7 in a micropipette) caused rapid, local depolymerization of microtubules [10], that preceded any effect of the inhibitor on adhesion integrity. When the inhibitor was removed, microtubules re-polymerized towards the cell periphery.

The effect of tension on microtubule dynamics was further illustrated using the experimental protocol developed by Bershadsky and co-workers [14]. Using a micro-needle, these authors applied shearing force to the lamella regions of living fibroblasts and in this way induced the formation of focal adhesions at the cell edge, confirming the expectations of Vasiliev [2]. They went on to show (among other data) that the external application of force on the actin cytoskeleton could replace that induced by myosin. When the same type of experiment was performed on melanoma cells expressing either GFP-tubulin alone (Fig. 2.4 A) or both GFP-zyxin and GFP-tubulin, the growth of peripheral adhesions induced by externally applied stress was seen to be accompanied by a 2–3-fold increase in the number of microtubules polymerizing towards the cell periphery [43]. Notably, these microtubules targeted the focal adhesions induced at the cell edge. A third series of experiments, using flexible substrates as growth support, further substantiated the effect of stress on microtubule dynamics. In this case, stress was applied to cells by stretching the underlying substrate. Again, there was an amplification of microtubule polymerization into peripheral zones that came under stress [43].

Results relevant to the question of how microtubules are guided to the cell periphery were also obtained from evanescent wave microscopy of cells expressing GFP-tubulin. Accordingly, it was observed that microtubules dipping into the evanescent wave often followed common routes

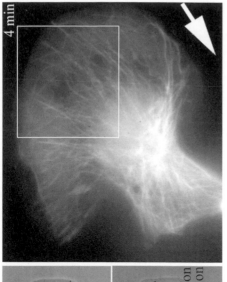

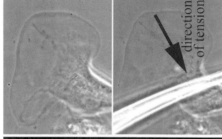

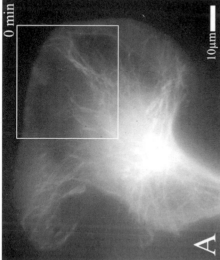

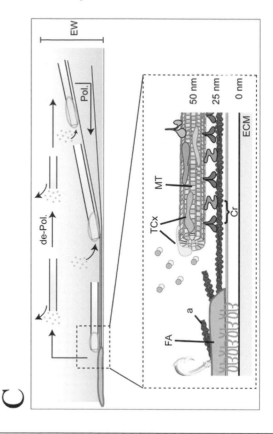

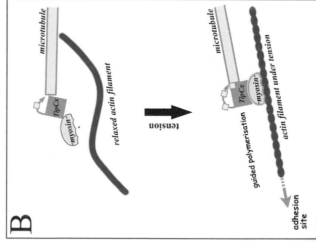

into adhesion foci, suggestive of the existence of a guidance track [11]. Taking these different sets of data together, we have suggested that the microtubule tip complex harbors a mechano-sensor, in the form of an unconventional myosin, that recognizes and docks onto actin filaments that are under tension (Fig. 2.4 B). Further provisos are that such actin filaments link into focal adhesions with their barbed ends (to direct myosin to the periphery) and that they are not (yet) integrated into stress fiber bundles. The latter condition is suggested by the fact that the tracking of microtubule tips along stress fibers has not been observed. This model of guidance (Fig. 2.4 C) is similar to a scenario in budding yeast that already has an experimental basis [44]. In *S. cerevisiae*, guidance of microtubule polymerization into the bud is directed by a myosin V homolog resident in the microtubule tip complex that uses the actin bundles along the cell axis as guidance tracks [44, 45]. In this context, a noteworthy finding from evanescent wave microscopy of fibroblasts expressing GFP-tubulin is that microtubule tips lift away from the substrate when they depolymerize from the cell edge (schematically illustrated in Fig. 2.4 C), consistent with the idea that the loss of the tip complex results in dissociation of microtubules from the putative track [11].

2.9
Conclusions and Perspectives

It is now generally acknowledged that microtubules interact with "cortical" actin in different cells and situations and exert some kind of orientation or guidance role [46]. Our findings on crawling cells indicate that

---

Fig. 2.4 A) Mechanical tension in the actin cytoskeleton stimulates the outgrowth of microtubules towards the cell edge. The tension across a mouse B16 melanoma was increased by drawing the cell body backwards with a micro-needle (centre frames). The outside frames show the microtubule pattern in the same cell (transfected with EGFP tubulin), before and 4 min after tension application; note the increase in numbers of microtubules growing towards the cell periphery. From [44], with permission. B) Mechano-sensor model of the guidance of microtubules to adhesion sites. It is proposed that the microtubule tip complex (TipCx) includes an unconventional myosin that can recognize and dock onto actin filaments that are under tension. At the cell periphery, tensioned filaments are presumed to be linked to adhesion sites. From [28], with permission. C) The guidance scenario, as suggested by data from total internal reflection fluorescence microscopy (TIRFM). Microtubules polymerize (Pol) along actin tracks down to the ventral cell surface, within the shallow evanescent wave (EW) of exciting light created by TIRFM. When depolymerizing (de-Pol) from the adhesion, microtubules lift away from the ventral surface. Abbreviations: MT, microtubule; TCx, microtubule tip complex; FA, focal adhesion; a, actin filament; Cr, crosslinkers (myosin?); ECM, extracellular matrix. From [28], with permission. (This figure also appears with the color plates.)

the interaction between microtubules and the actin cytoskeleton, predicted by Vasiliev and Gelfand [1] occur at specific cortical sites, namely at focal adhesions. This allows us now to focus on the nature of the interactions, their perpetrators and consequences.

The new findings we describe have been largely facilitated by advances in imaging technology and the development of means to tag proteins, almost at will, with fluorescent probes. Only by recording the activities of microtubules and substrate adhesions in time and space was it possible to reveal their mutual interaction. Likewise, the presence of a complex of proteins at the growing tips of microtubules (shown by others) would not have been predicted from images of fixed cells. By exploiting live cell analysis in combination with molecular tinkering, to modify putative adaptors and regulators, the avenues are now open to test ideas and predictions about the cross-talk between microtubule tips and adhesion complexes.

2.10
Acknowledgments

This work was supported by generous grants from the Austrian Science Research Council, P14007-Bio (to JVS) and P6066 (to IK).

2.11
References

1   Vasiliev, J. M. a. G., I. M., *Effects of colcemid on morphogenetic processes and locomotion of fibroblasts*, in *Cell Motility*, T. P. a. J. R. R. Goldmam, Editor. 1976, Cold Spring Harbor Laboratory: Cold Spring Harbor, p. 279–304.

2   Vasiliev, J. M., Spreading of non-transformed and transformed cells. *Biochim Biophys Acta*, **1985**, *780*(1), 21–65.

3   Ambercrombie, M., The crawling movement of metazoan cells. *Proc. Roy. Soc. Lond. B.*, **1980**, *207*, 129–147.

4   Izzard, C. S. and L. R. Lochner, Formation of cell-to-substrate contacts during fibroblast motility: an interference-reflexion study. *J Cell Sci*, **1980**, *42*, 81–116.

5   Small, J. V. a. R., G., Cytostructural dynamics of contact formation during fibroblast locomotion in vitro. *Expl Biol. Med.*, **1985**, *10*, 54–68.

6   Burridge, K., et al., Focal adhesions: transmembrane junctions between the extracellular matrix and the cytoskeleton. *Annu Rev Cell Biol*, **1988**, *4*, 487–525.

7   Hall, A., Rho GTPases and the actin cytoskeleton. *Science*, **1998**, *279*(5350), 509–514.

8   RINNERTHALER, G., B. GEIGER, and J.V. SMALL, Contact forma-
     tion during fibroblast locomotion: involvement of membrane ruf-
     fles and microtubules. *J Cell Biol,* **1988**, *106*(3), 747–760.

9   KAVERINA, I., K. ROTTNER, and J.V. SMALL, Targeting, capture,
     and stabilization of microtubules at early focal adhesions. *J Cell
     Biol,* **1998**, *142*(1), 181–190.

10 KAVERINA, I., O. KRYLYSHKINA, and J.V. SMALL, Microtubule tar-
     geting of substrate contacts promotes their relaxation and disso-
     ciation. *J Cell Biol,* **1999**, *146*(5), 1033–1044.

11 KRYLYSHKINA, O., et al., Nanometer targeting of microtubules to
     focal adhesions. *J Cell Biol,* **2003**.

12 BURRIDGE, K. and M. CHRZANOWSKA-WODNICKA, Focal adhe-
     sions, contractility, and signaling. *Annu Rev Cell Dev Biol,* **1996**,
     12, 463–518.

13 BERSHADSKY, A., et al., Involvement of microtubules in the con-
     trol of adhesion-dependent signal transduction. *Curr Biol,* **1996**,
     6(10), 1279–1289.

14 RIVELINE, D., et al., Focal contacts as mechanosensors: externally
     applied local mechanical force induces growth of focal contacts
     by an mDia1-dependent and ROCK-independent mechanism. *J
     Cell Biol,* **2001**, *153*(6), 1175–1186.

15 DANOWSKI, B.A., Fibroblast contractility and actin organization
     are stimulated by microtubule inhibitors. *J Cell Sci,* **1989**, *93*
     (Pt 2), 255–266.

16 ENOMOTO, T., Microtubule disruption induces the formation of
     actin stress fibers and focal adhesions in cultured cells: possible
     involvement of the rho signal cascade. *Cell Struct Funct,* **1996**,
     21(5), 317–326.

17 KAVERINA, I., et al., Enforced polarisation and locomotion of fi-
     broblasts lacking microtubules. *Curr Biol,* **2000**, *10*(12), 739–742.

18 HIROKAWA, N., Kinesin and dynein superfamily proteins and the
     mechanism of organelle transport. *Science,* **1998**, *279*(5350), 519–
     526.

19 KARCHER, R.L., S.W. DEACON, and V.I. GELFAND, Motor-cargo in-
     teractions: the key to transport specificity. *Trends Cell Biol,* **2002**,
     12(1), 21–27.

20 RODIONOV, V.I., et al., Microtubule-dependent control of cell
     shape and pseudopodial activity is inhibited by the antibody to
     kinesin motor domain. *J Cell Biol,* **1993**, *123*(6 Pt 2), 1811–1820.

21 KAVERINA, I.N., et al., Kinesin-associated transport is involved in
     the regulation of cell adhesion. *Cell Biol Int,* **1997**, *21*(4), 229–
     236.

22 KRYLYSHKINA, O., et al., Modulation of substrate adhesion dy-
     namics via microtubule targeting requires kinesin-1. *J Cell Biol,*
     2002, *156*(2), 349–360.

23 VERHEY, K.J., et al., Cargo of kinesin identified as JIP scaffolding
     proteins and associated signaling molecules. *J Cell Biol,* **2001**,
     152(5), 959–970.

24 SCHUYLER, S.C. and D. PELLMAN, Microtubule "plus-end-tracking
     proteins": The end is just the beginning. *Cell,* **2001**, *105*(4), 421–
     424.

25 GALJART, N. and F. PEREZ, A plus-end raft to control microtubule
     dynamics and function. *Curr Opin Cell Biol,* **2003**, *15*(1), 48–53.

26 PEREZ, F., et al., CLIP-170 highlights growing microtubule ends in vivo. *Cell*, **1999**, *96*(4), 517–527.

27 KAVERINA, I., O. KRYLYSHKINA, and J. V. SMALL, Regulation of substrate adhesion dynamics during cell motility. *Int J Biochem Cell Biol*, **2000**, *34*(7), 746–761.

28 SMALL, J. V. and I. KAVERINA, Microtubules meet substrate adhesions to arrange cell polarity. *Curr Opin Cell Biol*, **2003**, *15*(1), 40–47.

29 KAWASAKI, Y., et al., Asef, a link between the tumor suppressor APC and G-protein signaling. *Science*, **2000**, *289*(5482), 1194–1197.

30 FUKATA, M., et al., Rac1 and Cdc42 capture microtubules through IQGAP1 and CLIP-170. *Cell*, **2002**, *109*(7), 873–885.

31 SANDER, E. E., et al., Rac downregulates Rho activity: reciprocal balance between both GTPases determines cellular morphology and migratory behavior. *J Cell Biol*, **1999**, *147*(5), 1009–1022.

32 REN, Y., et al., Cloning and characterization of GEF-H1, a microtubule-associated guanine nucleotide exchange factor for Rac and Rho GTPases. *J Biol Chem*, **1998**, *273*(52), 34954–34960.

33 GLAVEN, J. A., et al., The Dbl-related protein, Lfc, localizes to microtubules and mediates the activation of Rac signaling pathways in cells. *J Biol Chem*, **1999**, *274*(4), 2279–2285.

34 VAN HORCK, F. P., et al., Characterization of p190RhoGEF, a RhoA-specific guanine nucleotide exchange factor that interacts with microtubules. *J Biol Chem*, **2001**, *276*(7), 4948–4956.

35 KRENDEL, M., F. T. ZENKE, and G. M. BOKOCH, Nucleotide exchange factor GEF-H1 mediates cross-talk between microtubules and the actin cytoskeleton. *Nat Cell Biol*, **2002**, *4*(4), 294–301.

36 VIGNAL, E., et al., Kinectin is a key effector of RhoG microtubule-dependent cellular activity. *Mol Cell Biol*, **2001**, *21*(23), 8022–8034.

37 MIMORI-KIYOSUE, Y., N. SHIINA, and S. TSUKITA, Adenomatous polyposis coli (APC) protein moves along microtubules and concentrates at their growing ends in epithelial cells. *J Cell Biol*, **2000**, *148*(3), 505–518.

38 JIMBO, T., et al., Identification of a link between the tumour suppressor APC and the kinesin superfamily. *Nat Cell Biol*, **2002**, *4*(4), 323–327.

39 KOMAROVA, Y. A., et al., Cytoplasmic linker proteins promote microtubule rescue in vivo. *J Cell Biol*, **2002**, *159*(4), 589–599.

40 CASSIMERIS, L., The oncoprotein 18/stathmin family of microtubule destabilizers. *Curr Opin Cell Biol*, **2002**, *14*(1), 18–24.

41 BAGRODIA, S. and R. A. CERIONE, Pak to the future. *Trends Cell Biol*, **1999**, *9*(9), 350–355.

42 DAUB, H., et al., Rac/Cdc42 and p65PAK regulate the microtubule-destabilizing protein stathmin through phosphorylation at serine 16. *J Biol Chem*, **2001**, *276*(3), 1677–1680.

43 KAVERINA, I., KRYLYSHKINA, O., BENINGO, K., ANDERSON, K., WANG, Y.-L., SMALL, J. V. Tensile stress stimulates microtubule outgrowth in living cells. *J. Cell Sci.*, **2002**, *115*, 2283–2291.

44 YIN, H., et al., Myosin V orientates the mitotic spindle in yeast. *Nature*, **2000**, *406*(6799), 1013–1015.

45 Hwang, E., et al., Spindle orientation in Saccharomyces cerevisiae depends on the transport of microtubule ends along polarized actin cables. *J Cell Biol,* **2003**, *161*(3), 483–488.

46 Goode, B.L., D.G. Drubin, and G. Barnes, Functional cooperation between the microtubule and actin cytoskeletons. *Curr Opin Cell Biol,* **2000**, *12*(1), 63–71.

3

Mechanisms of Eukaryotic Chemotaxis

Chris Janetopoulos, Yu Long, and Peter Devreotes

Summary

Chemotaxis is a fundamental cellular response that plays a central role in health and disease. Experiments in amoebae and neutrophils have shown that local accumulations of $PI(3,4,5)P_3$ mediate the ability of cells to migrate directionally. Random pseudopod extension, phagocytosis and other spontaneous events also lead to $PI(3,4,5)P_3$ synthesis and mark the sites of new actin filled projections. During directional sensing, it is a discreet step in the signaling pathway, downstream of G protein activation but upstream of the accumulation of $PI(3,4,5)P_3$, that sets up the initial asymmetry. The enzymes PI3K and PTEN that regulate the local levels of the $PI(3,4,5)P_3$ move to the anterior and posterior of the cell. Quantitative measurements have shown that the localization is dependent on the relative chemo-attractant gradient, and the final distribution of $PI(3,4,5)P_3$ is amplified, even without the presence of an actin cytoskeleton. Multi-stimulus inputs on latrunculin treated cells can lead to bimodal responses, which can be extinguished and redistributed by shifting the concentration gradients. These observations suggest that a local excitation–global inhibition model can account for the localization of PI3K and PTEN and thereby explain directional sensing.

3.1
Chemotaxis is a Fundamental Cellular Response

While all cells have the capacity to sense extracellular signals, many can also determine the direction of the signal source and respond by changing shape or moving towards or away from the highest concentration. For example, in embryogenesis, morphogen gradients determine patterns of differentiation in tissues and during chemotaxis, a chemical gradient provides a directional cue that organizes cell movement [1–3]. This ability to sense direction plays a central role not only in development, but in

Cell Migration. Edited by Doris Wedlich
Copyright © 2005 WILEY-VCH Verlag GmbH & Co. KGaA, Weinheim
ISBN: 3-527-30587-4

immunity and tissue homeostasis in adults [4–7]. It is also a key to a variety of pathological situations involving cell migration [8–12].

By investigating simple organisms such as the social amoeba Dictyostelium discoideum, researchers are discovering how cells sense gradients of external stimuli [13–15]. The general principles of chemotaxis in mammalian cells are remarkably similar to those in this genetically well-behaved model system [16–18]. During their growth phase, D. discoideum amoebae use chemotaxis to find food and nutrients. When starved, the cells develop the capacity to polarize and migrate directionally towards secreted 3′,5′-cyclic adenosine monophosphate (cAMP). This extracellular signaling molecule is detected by four serpentine receptors, designated cAR1-cAR4, coupled to a unique heterotrimeric G protein [19]. A similar strategy is used in mammalian leukocytes where twenty types of receptors for "chemokines" principally couple to the G protein designated G_i [20, 21]. Other similarities with mammalian systems include stimulus-induced transient increases in phosphoinositides (PIs), cAMP, cGMP, IP_3 and Ca^{2+} and rearrangements in the cytoskeleton [16, 22]. Phosphatidylinositol tris phosphate, $PI(3,4,5)P_3$, is a key intermediate in directional sensing in D. discoideum amoebae and mammalian leukocytes [23–34].

Chemotaxis involves directional sensing, polarity, and the periodic extension of pseudopodia. Directional sensing is the process that detects asymmetric extracellular signals and generates an internal amplified response [14, 15]. Even when cells are exposed to very shallow gradients of chemoattractant, signaling molecules accumulate at the membrane and in the cortex adjacent to the higher concentration. This localized activation can be visualized, for example, with proteins containing a PH domain fused to green fluorescent protein (GFP). As described below, this amplified response can occur in a cell that is unpolarized and immobilized. Polarization refers to the cell's ability to elongate and display a defined anterior and posterior [14]. Molecules associated with the front of a cell include actin and actin-binding proteins Scar, WASP, filopodin, cofilin, and coronin while molecules associated with the trailing edge include myosin II and cortexillin [35–39]. Polarization helps focus the polymerization of the actin cytoskeleton at the leading edge. This likely promotes efficient movement towards a chemo-attractant source and prevents the cell from straying in the wrong direction during moments when the gradient is weak. Because of this, the cell has polarized sensitivity, the sensing of the gradient must occur within a restricted area at the very front of the cell. This also prevents the cell from reacting rapidly when gradients are rapidly shifted. Periodic pseudopod extension is required for motility and occurs spontaneously even in the absence of receptor stimulation or functional G proteins. Cells such as Dictyostelium amoeba and mammalian leukocytes extend pseudopodia at approximately 60-second intervals. The pseudopodia are either retracted or attach to a substrate and propel the cell forward.

3.2
Directional Sensing Occurs Downstream of G Protein Activation
and Upstream of the Accumulation of PI(3,4,5)P₃

Chemo-attractant receptors and G protein α-, β-, and γ-subunits are uniformly localized along the cell membrane and receptor occupancy closely resembles the external gradient of chemo-attractant (Fig. 3.1) [40–43]. Furthermore, the kinetics of G protein dissociation suggests that it mirrors receptor occupancy [44]. The first point in the signaling pathway where a strong spatially localized response is observed is the accumulation of PI(3,4,5)P₃ at the cell anterior. Local accumulation of PI(3,4,5)P₃ was first inferred from studies in D. discoideum by visualization with the GFP-tagged PH domain of the cytosolic regulator of adenylyl cyclase (CRAC) [45]. Similar localizations of other PH domains to the front of cells have been subsequently been shown to occur in D. discoideum and leukocytes exposed to gradients of chemoattractants [46, 47]. In cells exposed to an increment of chemoattractant, PI(3,4,5)P₃ increases are transient (Fig. 3.1, top). PH domain association with the membrane increases within 5 s and then declines to a plateau phase during the next several minutes. Direct measurements of total PI(3,4,5)P₃ levels in D. discoideum and mammalian leukocytes parallel the membrane binding of the PH domains. In cells lacking functional G proteins, stimuli do not elevate PI(3,4,5)P₃ levels [47, 48]. On the other hand, inhibitors of the cytoskeleton, such as latrunculin A, do not interfere. Taken together, these observations pinpoint that gradient detection occurs between G protein activation and accumulation of PI(3,4,5)P₃ [15, 44].

The distributions of PI3Ks and PTEN in D. discoideum have further revealed the importance of PI(3,4,5)P₃ in directional sensing [49, 50]. In naive cells, a small portion of PTEN is bound to the plasma membrane while PI3K is in the cytoplasm. Uniform application of chemo-attractant results in the movement of PI3Ks to the membrane while PTEN rapidly dissociates. Over the next few minutes, PI3Ks return to the cytosol and PTEN reassociates with the membrane even while receptors are still occupied. When a cell is placed in a chemo-attractant gradient, PI3Ks are recruited to the front of the cell and PTEN re-localizes to the back (Fig. 3.1, bottom). This spatial distribution is more evident in polarized versus unpolarized cells.

The movements of PI3K and PTEN and the changes in PI(3,4,5)P₃ ₂ levels suggest that the enzyme activities are reciprocally regulated during the response. In fact, these binding events likely regulate the enzyme activity since the substrates and products are found on the inner face of the membrane. PI3K is activated immediately once chemo-attractant is added [51]. Cell lysates examined within 5 s of addition of a stimulus show rapid incorporation of ^{32}P-γ-ATP into ^{32}P-PI(3,4,5)P₃. This activation is transient and peaks in 20–30 s. PI3K activity in lysates of cells

Unpolarized cell exposed to uniform stimulus:

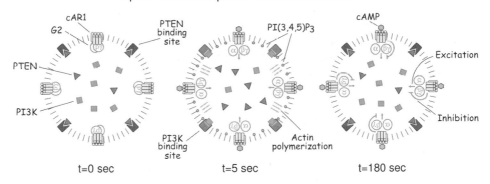

Unpolarized cell
exposed to a gradient: Polarized cell
exposed to a gradient:

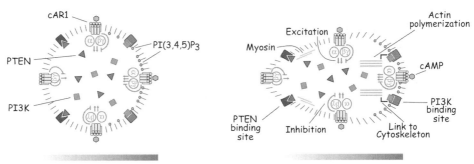

Fig. 3.1 Cartoon depicting the distribution and activity of signaling molecules in unpolarized and polarized cells during temporal and spatial stimulation. Chemo-attractant receptors (cAR1), receptor occupancy, associated and dissociated (activated) G protein $\alpha\beta\gamma$-subunits, excitation, inhibition, PI3Ks, PTEN, PI3K and PTEN binding sites, PI(3,4,5)P$_3$, F-actin, and myosin are indicated. Upper diagram illustrates an unpolarized cell at times before and during stimulation. In resting cells, PTEN is bound to the membrane and PI3Ks are in the cytosol (t=0 s). An increase in receptor occupancy by chemo-attractant (orange hexagons) triggers, through the heterotrimeric G proteins, a rapid increase in excitation (green arrow) which leads to binding of PI3Ks (light blue squares) to their binding sites (dark blue squares) at the membrane and causes PTEN (maroon triangles) to dissociate from binding sites (purple squares) at the membrane (t=5 s). The combined effect causes a large increase in PI(3,4,5)P$_3$ (green lollipops). At longer times, inhibition (red line) increases and eventually balances excitation. PI3Ks return to the cytosol, PTEN returns to the membrane, and PI(3,4,5)P$_3$ returns to prestimulus levels (t=180 s). Lower diagram shows the response of a cell treated with latrunculin A (left) and of a polarized cell (right) in a spatial gradient. The appearance of the polarized cell would be similar in a uniform concentration of attractant (see text and Fig. 3.5). The "global" inhibition (red line) is equal at the two ends of the cell. "Local" excitation is slightly higher at the front causing the binding of PI3K to and the loss of PTEN from the membrane at the front. This leads to a large steady-state accumulation of PI(3,4,5)P$_3$ selectively at the front, and in untreated cells, actin polymerization polarization and directed motility. (This figure also appears with the color plates.) (Figure reprinted from [14] with permission from the American Society of Biochemistry and Molecular Biology.)

pretreated for 30 s or more returns to a plateau level which is slightly elevated above the pre-stimulus value. This demonstrates that receptor-mediated activation of PI3K contributes to the large increases in PI(3,4,5)P$_3$ following stimulation. It is likely that the reciprocal loss of PTEN from the membrane further enhances the response, while the return of the enzyme to the membrane helps return PI(3,4,5)P$_3$ to basal levels. Regulation of PI3K activity is normal in pten$^-$ cells, yet increases in PI(3,4,5)P$_3$ are higher and prolonged, providing further evidence for an important role of PTEN in regulating the response [51]. This coordinated regulation of synthesis and degradation of PI(3,4,5)P$_3$ provides a vigorous system that is resistant to perturbation [52]. Since changes in both enzymes contribute to the accumulation of the phosphoinositide, inhibition of either synthesis or degradation alone may do little to impair the response.

3.3
Input–Output Relationships Reveal Gradient Amplification in Polarized and Unpolarized Cells

To characterize the contributions of the enzymes and their relationship to the response upon stimulation, quantitative measurements of cAMP, PI3K-GFP, PTEN-GFP, and PH-GFP to temporal and spatial stimuli have been performed [53]. In a shallow chemo-attractant gradient, PI(3,4,5)P$_3$ is present on the front, with none detectable on the membrane at the rear of the cell (Fig. 3.2A). The response starts to appear about midway along the cell and then increases about three times more steeply than the external chemo-attractant gradient. Data from the same cell exposed to gradients of different mean levels demonstrates how the same absolute concentrations can give vastly different responses (Fig. 3.2 B). These characteristics, indicative of significant amplification, were found whether or not cells were treated with latrunculin. PI3K shows a profile similar to that of PI(3,4,5)P$_3$, but with less amplification (Fig. 3.3A). In contrast, PTEN binding is inversely correlated with receptor occupancy, and decreases proportionally along the length of the cell (Fig. 3.3 B). These data provide evidence that localized PI3K-induced synthesis and degradation of the lipid at the rear and sides of the cell act together to focus the PI(3,4,5)P$_3$ response. The differently shaped profiles of the two enzymes suggest that there are different mechanisms regulating their binding to the membrane.

Experiments have been performed to further elucidate the sensing mechanism by presenting cells with multiple stimulus inputs. Latrunculin-treated cells when exposed to two very steep gradients from a pair of micropipettes, respond on both ends (Fig. 3.4). PI3K and PH domains move to the poles (Fig. 3.4A, C) while PTEN-GFP moves to the midline

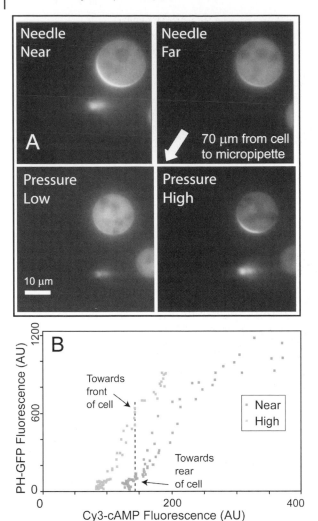

Fig. 3.2 Response to chemo-attractant gradients of varying steepness and absolute concentrations. A) Cells were exposed to steep gradients (needle near) and shallow gradients (needle far). The mid-point concentration of chemo-attractant was varied by changing the pressure in the micropipette (pressure high and low). B) Shown are the Cy3-cAMP and PH-GFP fluorescence levels for the same cell in two different gradients. Following the dotted line it is clear that the same Cy3-cAMP concentration elicited vastly different PH-GFP responses. (This figure also appears with the color plates.) (Figure reprinted from [53] with permission from the National Academy of Sciences of the United States of America.)

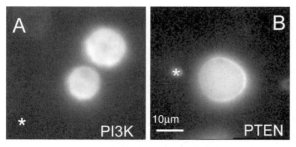

Fig. 3.3 Response of PI 3′ enzyme. Response of latrunculin-treated cells expressing PI3K-GFP (A) or PTEN GFP
(B) when exposed to a cAMP gradient. (Figure reprinted
from [53] with permission from the National Academy of
Sciences of the United States of America.)

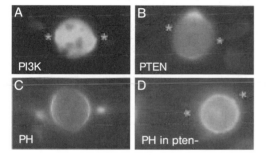

Fig. 3.4 Response to multiple gradients. Two micropipettes were brought in close proximity to latrunculin-
treated cells creating cAMP profiles with two sharp gradients on either side. A) PI3K-GFP localized to both ends
of the cells. B) PTEN-GFP re-localized to the cell membrane at the point of lowest cAMP concentration. C)
Cells expressing PH-GFP responded at both ends similar
to that of PI3K-GFP. D) pten⁻ cells expressing PH-GFP
are incapable of responding with two sharp crescents as
in panel C. (This figure also appears with the color
plates.) (Figure reprinted from [53] with permission from
the National Academy of Sciences of the United States of
America.)

of a cell (Fig. 3.4 B). Cells lacking PTEN failed to display two $PI(3,4,5)P_3$
crescents since each response was much broader than that in wild-type
cells (Fig. 3.4 D and [50]). With polarized cells, it was difficult to obtain a
stable bipolar response, since the cells tended to choose one micropipette
and move towards it [53].

The complementary creation of the binding sites for PI3K and PTEN
likely produces the initial asymmetry in signaling that leads to directional migration. The movements and regulation of PI3K and PTEN can

be explained by a local excitation–global inhibition (LEGI) model (Fig. 3.1). It is proposed that a balance between an excitation and an inhibition process regulates the membrane binding and activity of the enzymes. Thus, each enzyme responds to changes in receptor occupancy and adapt when this occupancy is held constant. For PI3K, excitation mirrors local levels of receptor occupancy and leads to the recruitment and activation of the enzyme while global inhibition, determined by the average receptor occupancy, counteracts these effects. For PTEN, the situation is reversed and local excitation decreases its association with the membrane while global inhibition restores membrane binding. The model provides a basis for understanding many of the observed responses under various stimulus paradigms. It accounts for transient responses, to uniform increments and steady-state responses to gradients as well as responses to combinations of temporal and spatial stimuli. It is also consistent with the cell's ability to respond to gradients with a wide range of midpoint concentrations.

3.4
Increase in Local PI(3,4,5)P$_3$ Precedes Actin Polymerization Responses

A number of observations suggest that PI(3,4,5)P$_3$ is an important component for directing where and when sites of actin-filled projections form in a variety of cellular responses. In wild-type cells, chemo-attractant stimulation typically triggers a biphasic increase in PI3K activation, PI(3,4,5)P$_3$ levels, and acting polymerization. The stimulus-induced accumulation of PI(3,4,5)P$_3$ as visualized by binding of PH domains to the membrane invariably colocalizes with sites of new actin filament formation [24, 30, 50]. Attempts to reduce PI(3,4,5)P$_3$ and thereby block actin polymerization triggered by chemo-attractant yielded differing results. Inhibitors of P13K and gene disruptions, which knock down chemoattractant-elicited increases in PI(3,4,5)P$_3$ by over 90%, eliminate the second phase of actin polymerization but do not affect the initial rapid phase. It may be that the first phase of actin polymerization is independent of or requires only very slight increases in these PIs. Increasing the PI(3,4,5)P$_3$ levels by genetically disrupting PTEN had the most dramatic effect on the second phase of the response. In pten$^-$ cells, where the PI(3,4,5)P$_3$ response was dramatically prolonged and broadened, the second phase of actin polymerization is much larger and extended as compared to wild-type cells.

Interestingly, the PH domains label the surface membranes of pseudopodia, ruffles, filopods, macropinosomes, phagosomes, and sites of cell-to-cell contact (Fig. 3.5) [14, 55–59]. These events can be elicited and occur spontaneously in the absence of functional G proteins, thus implying there are also G protein-independent activators that regulate P13K and PTEN binding sites. D. discoideum cells lacking P13Ks or treated with

Pseudopods

Ruffles

Filopods

Macropinosomes

Phagosomes

Cell-cell contact

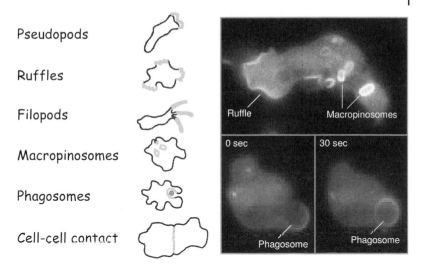

Fig. 3.5 Cells expressing PH$_{Crac}$-GFP show localization of this marker for PI(3,4,5)P$_3$ and/or PI(3,4)P$_2$ on a variety of membrane structures. The cartoons on the left illustrate the role that PI(3,4,5)P$_3$ and/or PI(3,4)P$_2$ (gray labeling) plays in a number of events such as pseudopodia extension, membrane ruffling, filopod extension, macropinocytosis, phagocytosis, and cell–cell contact that require cytoskeletal remodeling. Examples on the right include PH$_{Crac}$-GFP localization on ruffles, which mediate random movements of D. discoideum cells and on macropinosomes (top panel). Bottom two frames show a phagocytic cup during phagocytosis of a yeast cell by growing cells. PH$_{Crac}$-GFP signal at the membrane appears with the initial encounter of the yeast cell and usually disappears from the phagosome within minutes once complete engulfment has occurred. (Figure reprinted from [14] with permission from the American Society of Biochemistry and Molecular Biology.)

PI3K inhibitors display profound defects in ruffling, macropinocytosis, and phagocytosis (unpublished observations). Disruption of PTEN also interferes with these events [50].

3.5
Positive Feedback and the Actin Cytoskeleton
May Stabilize Directional Sensing and Establish Polarity

It is interesting that each of the phenomena involved in chemotaxis, directional sensing, polarity, and periodic pseudopod extension are associated with local accumulation of PI(3,4,5)P$_3$. It is clear that directional sensing

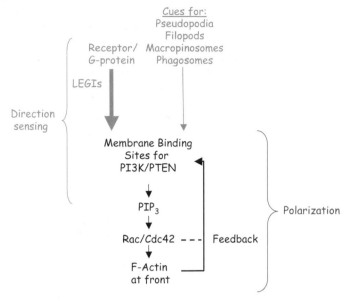

Fig. 3.6 Speculative model indicating that positive feedback and the actin cytoskeleton stabilize directional sensing and establish polarity. Also see Fig. 3.1, bottom for cartoon representations of unpolarized and polarized cells. Elements and events critical for directional sensing and polarization are bracketed. During chemotaxis, directional sensing is initiated by binding of chemo-attractant to the receptor. Occupied receptors activate the heterotrimeric G protein $\alpha\beta\gamma$-subunits. Local excitation results in the transient binding and activation of PI3Ks and transient loss of PTEN from the plasma membrane. This leads to the production and local accumulation of $PI(3,4,5)P_3$. This $PI(3,4,5)P_3$, possibly acting through exchange factors and the small G proteins Rac and cdc42 determines sites of new actin polymerization. Cells that undergo morphological changes and polarize redistribute many cytoskeletal components and further stabilize the distribution of signaling molecules such as PI3K and PTEN. The polarization of PI3K and PTEN (and other signaling molecules) confer a leading edge which is more sensitive to chemo-attractant than the rear of the cell. We speculate that leading edge events such as assembly of actin and actin binding proteins stabilize PI3K while posteriorly localized events such as recruitment to the cell cortex of Pak A and myosin II participate in the localization of PTEN. Myosin recruitment involves chemo-attractant mediated changes in cGMP [61]. Disruption of the actomyosin cytoskeleton by treating cells with drugs such as latrunculin A leads to a loss of this polarized sensitivity and makes cells equally sensitive along their length.

does not require the cytoskeleton and is complete with the localization of $PI(3,4,5)P_3$, upstream of actin polymerization. Polarity and periodic pseudopod extension, however, depend critically on reorganization of the cytoskeleton. The establishment of polarization and the generation of pseudopodia must involve the regulated interaction of many of the same components used for directional sensing (see Fig. 3.1, bottom and Fig. 3.6). Thus, PI3K, PTEN, and $PI(3,4,5)P_3$, together with the cytoskeleton, bring about these events. A likely role of the cytoskeleton may be to stabilize the asymmetric distribution of important signaling players, such as the enzymes, so that the associations of PI3K and PTEN with the membrane at the front and back of a polarized cell, respectively, are reinforced by interactions with components of the cytoskeleton. Since $PI(3,4,5)P_3$ stimulates actin polymerization, elements of the anterior cytoskeleton may help stabilize the interaction of PI3K with the membrane, and a positive feedback loop may result and reinforce the initial asymmetry (Fig. 3.6). Similarly, a relationship of PTEN with components of the cytoskeleton like myosin II and Pak A, which localize to the back when cells are in a chemo-attractant gradient, might create a second feedback loop at the rear [14, 60, 61]. Given sensitive feedback loops, small changes of chemo-attractant would allow the cell to acquire a polarized morphology and sensitivity even in the presence of a uniform concentration.

3.6
References

1   BIBER, K., ZUURMAN, M. W., DIJKSTRA, I. M., and BODDEKE, H. W., *Curr. Opin. Pharmacol.* **2002** *2*, 63–68.
2   FERNANDIS, A. Z. and GANJU, R. K., *Sci. STKE* **2001** *91*, PE1.
3   RUBEL, E. W. and CRAMER, K. S., *J. Comp. Neurol.* **2002** *448*, 1–5.
4   CARLOS, T. M., *J. Leuko. Biol.* **2001** *70*, 171–184.
5   LALOR, P. F., SHIELDS, P., GRANT, A., and ADAMS, D. H., *Immunol. Cell Biol.* **2002** *80*, 52–64.
6   PATEL, D. D. and HAYNES, B. F., *Curr. Dir. Autoimmun.* **2001** *3*, 133–167.
7   THELEN, M., *Nat. Immunol.* **2001** *2*, 129–134.
8   CONDEELIS, J. S., WYCKOFF, J. B., BAILLY, M., PESTELL, R., LAWRENCE, D., BACKER, J., and SEGALL, J. E., *Semin. Cancer. Biol.* **2001** *11*, 119–128.
9   GANGUR, V., BIRMINGHAM, N. P., and THANESVORAKUL, S., *Vet. Immunol. Immunopathol.* **2002** *86*, 127–136.
10 MOORE, M. A., *Bioessays* **2001** *8*, 674–676.
11 ARYA, M., PATEL, H. R., and WILLIAMSON, M., *Curr. Med. Res. Opin.* **2003** *19* (6), 557–564.
12 MURPHY, P. M., *N. Engl. J. Med.* **2001** *345*, 833–835.
13 CHUNG, C. Y., FUNAMOTO, S., and FIRTEL, R. A., *Trends Biochem. Sci.* **2001** *26*, 557–566.
14 DEVREOTES, P. N. and JANETOPOULOS, C., *J. Biol. Chem.* **2003** *278*(3), 20445–2448.

15 PARENT, C. and DEVREOTES, P.N., *Science* 1999 *284*, 765–770.
16 DEVREOTES, P.N. and ZIGMOND, S.H., *Annu. Rev. Cell Biol.* 1988 *4*, 649–686.
17 STEPHENS, L., ELLSON, C., and HAWKINS, P., *Curr. Opin. Cell Biol.* 2002 *14*, 203–213.
18 VAN ES, S. and DEVREOTES, P.N., *Cell. Mol. Life Sci.* 1999 *55*, 1341–1351.
19 PARENT, C. and DEVREOTES, P.N., *Annu. Rev. Biochem.* 1996 *65*, 411–440.
20 KLINKER, J.F., WENZEL-SEIFERT, K., and SEIFERT, R., *Gen. Pharmacol.* 1996 *27*, 33–54.
21 MURPHY, P.M., *Annu. Rev. Immunol.* 1994 *12*, 593–633.
22 MAGHAZACHI, A.A., *Int. J. Biochem. Cell. Biol.* 2000 *32*, 931–943.
23 CURNOCK, A.P., LOGAN, M.K., and WARD, S.G., *Immunology* 2002 *105*, 125–136.
24 FUNAMATO, S., MILAN, K., MEILI, R., and FIRTEL, R., *J. Cell Biol.* 2001 *153*, 795–810.
25 LI, Z., JIANG, H., XIE, W., ZHANG, Z., SMRCKA, A.V., and WU, D., *Science* 2000 *287*, 1046–1049.
26 NIGGLI, V. and KELLER, H., *Eur. J. Pharmacol.* 1997 *335*, 43–52.
27 RICKERT, P., WEINER, O.D., WANG, F., BOURNE, H.R., and SERVANT, G., *Trends Cell Biol.* 2000 *10*, 466–473.
28 HANNIGAN, M., ZHAN, L., LI, Z., AI, Y., WU, D., and HUANG, C.K., *Proc. Natl. Acad. Sci. USA* 2002 *99*, 3603–3608.
29 HIRSCH, E., KATANAEV, V.L., GARLANDA, C., AZZOLINO, O., PIROLA, L., SILENGO, L., SOZZANI, S., MANTOVANI, A., ALTRUDA, F., and WYMANN, M.P., *Science* 2000 *287*, 1049–1053.
30 INSALL, R.H. and WEINER, O.D., *Dev. Cell* 2001 *1*, 743–747.
31 SASAKI, T., IRIE-SASAKI, J., JONES, R.G., OLIVEIRA-DOS-SANTOS, A.J., STANFORD, W.L., BOLON, B., WAKEHAM, A., ITIE, A., BOUCHARD, D., KOZIERADZKI, I., JOZA, N., MAK, T.W., OHASHI, P.S., SUZUKI, A., and PENNIGNER J.M., *Science* 2000 *287*, 1046–1049.
32 SOTSIOS, Y. and WARD, S.G., *Mol. Biol. Cell* 2000 *177*, 217–235.
33 WANG, F., HERZMARK, P., WEINER, O.D., SRINIVASAN, S., SERVANT, G., and BOURNE, H.R., *Nat. Cell Biol.* 2002 *4*, 513–518.
34 WYMANN, M.P., SOZZANI, S., ALTRUDA, F., MANTOVANI, A., and HIRSCH, E., *Immunol. Today* 2000 *21*, 260–264.
35 AIZAWA, H., SUTOH, K., TSUBUKI, S., KAWASHIMA, S., ISHII, A., and YAHARA, I., *J. Biol. Chem.* 1995 *270*, 10923–10932.
36 CLOW, P.A. and MCNALLY, J.G., *Mol. Biol. Cell* 1999 *10*, 1309–1323.
37 GERISCH, G., ALBRECHT, R., DE HOSTOS, E., WALLRAFF, E., HEIZER, C., KREITMEIER, M., and MULLER-TAUBENBERGER, A., *Symp. Soc. Exp. Biol.* 1993 *47*, 297–315.
38 MISHIMA, M. and NISHIDA, E., *J. Cell Sci.* 1999 *112*, 2833–2842.
39 WEBER, I., NEUJAHR, R., DU, A., KOHLER, J, FAIX, J., and GERISCH G., *Curr. Biol.* 2000 *10*, 501–506.
40 JIN, T., ZHANG, N., LONG, Y., PARENT, C.A., and DEVREOTES, P.N., *Science* 2000 *287*, 1034–1036.
41 SERVANT, G., WEINER, O.D., NEPTUNE, E.R., SEDAT, J.W., and BOURNE, H.R., *Mol. Biol. Cell* 1999 *10*, 1163–1178.
42 XIAO, Z., ZHANG, N., MURPHY, D.B., and DEVREOTES, P.N., *J. Cell Biol.* 1997 *139*, 365–374.

43 Ueda, M., Sako, Y., Tanaka, T., Devreotes, P., and Yanagida, T., *Science* **2001** *294*, 864–867.

44 Janetopoulos, C., Jin, T., and Devreotes, P.N., *Science* **2001** *291*, 2408–2411.

45 Parent, C., Blacklock, B., Froelich, W., Murphy, D., and Devreotes, P.N., *Cell* **1998** *95*, 81–91.

46 Meile, R., Ellsworth, C., Lee, S., Reddy, T.B., Ma, H., and Firtel, R.A., *EMBO J.* **1999** *18*, 2092–2105.

47 Servant, G., Weiner, O.D., Herzmark, P., Balla, T., Sedat, J.W., and Bourne, H.R., *Science* **2000** *287*, 1037–1040.

48 Lilly, P.J. and Devreotes, P.N., *J. Cell Biol.* **1995** *129*, 1659–1665.

49 Funamoto, S., Meili, R., Lee, S., Parry, L., and Firtel, R.A., *Cell* **2002** *109*, 611–623.

50 Iijima, M. and Devreotes, P.N., *Cell* **2002** *109*, 599–610.

51 Huang, E., Iijima, M., Funamoto, S., Firtel, R., and Devreotes, P.N., *Mol. Biol. Cell* **2002** *14*(5), 1913–1922.

52 Iijima, M., Huang, E., and Devreotes, P.N., *Dev. Cell* **2002** *3*, 469–478.

53 Janetopoulos, C., Ma, L., Devreotes, P.N., and Iglesias, P., *PNAS* **2004** *101(24)*, 8951–8956.

54 Chen, L., Janetopoulos, C., Huang, E., Iijima, M., Borleis, J., and Devreotes, P., *Mol. Biol. Cell* **2003** *14*(12), 5028–5037.

55 Wennstrom, S., Hawkins, P., Cooke, F., Hara, K., Yonezawa, K., Kasuga, M., Jackson, T., Claesson-Welsh, L., and Stephens, L., *Curr. Biol.* **1994** *4(5)*, 385–393.

56 Diakonova, M., Bokoch, G., and Swanson, J.A., *Mol. Biol. Cell* **2002** *13*, 402–411.

57 Amyere, M., Mettlen, M., Van, D., Platek, A., Payrastre, B., Veithen, A., and Courtoy, P.J., *Int. J. Med. Microbiol.* **2002** *291*, 487–494.

58 Costello, P.S., Gallagher, M., and Cantrell, D., *Nat. Immunol.* **2002** *3*, 1082–1089.

59 Harriague, J. and Bismuth, G., *Nat. Immunol.* **2002** *3*, 1090–1096.

60 Berlot, C.H., Devreotes, P.N., and Spudich, J.A., *J. Biol. Chem.* **1987** *262*, 3918–3926.

61 Chung, S. and Firtel, R.A., *J. Cell Biol.* **1999** *14*, 559–576.

62 Bosgraaf, L., Russcher, H., Smith, J.L., Wessels, D., Soll, D.R., and Van Haastert, P.J., *EMBO J.* **2002** *21*(17), 4560–4570.

4
Dual Location Proteins:
Communication Between Cell Adhesions and the Nucleus

Erin G. Cline and W. James Nelson

4.1
Introduction

Multicellular organisms are composed of many different cell types that become organized during development into distinct tissues and organs. One important process required for these cell rearrangements is cell adhesion, both to the extracellular matrix and to adjacent cells [1]. The timing of these adhesions and the types of adhesion specified by different proteins are critical to the correct formation and subsequent function of cells in tissues and organs. While it is generally considered that these cell adhesions play structural roles in the organization of cells, there is increasing evidence that many of the proteins involved also regulate signaling pathways that connect cell adhesions to the regulation of gene expression in the nucleus.

Cells are connected to the substratum through integrin-mediated focal adhesions. Upon engagement with extracellular matrix proteins, integrins recruit structural proteins, including zyxin and paxillin, which link the transmembrane integrins to the actin cytoskeleton [2]. Cell–cell adhesion is mediated by a variety of membrane proteins, including classical cadherins, claudins/occludin, and desmosomal cadherins. Classical cadherins are required to initiate cell–cell contacts, and other adhesion protein complexes subsequently assemble specific junctions required for controlling paracellular diffusion (tight junction) and maintaining the structural continuum of the epithelium (desmosomes) [3–5]. Cadherins on opposing cells bind members of the catenin family, comprised of armadillo repeat proteins and a-catenin, that bind the protein complex to actin filaments. Desmosomes connect cells along their lateral membranes through desmosomal cadherins. Accessory proteins, including plakoglobin and plakophilin, bound to the cytoplasmic face of these cadherins connect desmosomes to the intermediate filament cytoskeleton [6]. Tight junctions are located at the apex of the lateral membranes of contacting cells. The transmembrane proteins occludin and claudin form interdigitating strands between cells and bind accessory proteins, including ZO-1 and ZO-2, that connect to the actin cytoskeleton [7].

Cell Migration. Edited by Doris Wedlich
Copyright © 2005 WILEY-VCH Verlag GmbH & Co. KGaA, Weinheim
ISBN: 3-527-30587-4

In addition to its proline-rich and LIM domains, zyxin contains a short leucine-rich sequence that has been shown to be a functional nuclear export sequence (NES). The NES sequence has been shown to be the necessary domain to keep zyxin from accumulating in the nucleus at steady state. However, endogenous zyxin has been shown to shuttle between the nucleus and the cytoplasm [12]. Both the proline-rich regions and the LIM domains of zyxin can independently target the protein to the nucleus although a classical nuclear localization signal (NLS) is not present in either of these domains [13].

The function of zyxin in the nucleus is unknown. Efforts to induce nuclear accumulation of zyxin by manipulating culture conditions or inducting various cellular pathways have failed. Treatment of cells with leptomycin B (an inhibitor of CRM1-mediated nuclear export) causes heterogeneous nuclear accumulation of zyxin, suggesting that import of zyxin into a cell nucleus is determined by individual cell physiologies [13]. Although full-length zyxin fused to the Gal4-DNA binding domain exhibits transactivation activity in yeast, it does not have transactivation activity alone in mammalian cells. In mammalian cells transcription activation requires a binding partner [6]E6 (E6 protein from low risk HPV type 6), a protein which when overexpressed can cause the redistribution of zyxin from focal adhesions to the nucleus. These findings suggest that normally another protein may bind zyxin in response to an extracellular cue and facilitate its translocation to the nucleus, thus allowing zyxin to stimulate transcription [14].

4.2.1.2 **LPP (Lipoma-Prefered Partner)**

The *LPP* (Lipoma-Prefered Partner) gene is the preferred translocation partner of the *HMGIC* gene in a subclass of human benign mesenchymal tumors known as lipomas [15]. LPP is found in focal adhesions and at sites of cell–cell contact. Like zyxin, LPP has a proline-rich N-terminal domain followed by three LIM domains at its C terminus. Also, the N-terminal domain of LPP binds VASP (*V*asodilator *S*timulated *P*hosphoprotein) and α-actinin [15–17].

LPP has an NES that is partially conserved with that of zyxin, though it has been found that another leucine-rich stretch of amino acids acts as the NES in LPP. The protein accumulates in the nucleus in response to treatment with leptomycin B, though this accumulation is more homogeneous than zyxin [16]. There is no classical NLS in LPP and the domain necessary for nuclear import has not been elucidated [18]. A nuclear function has not yet been found for LPP, but in a GAL4-based transactivation assay LPP can enhance transcription, and this enhancement is increased when the NES is deleted [16].

4.2.1.3 Trip6

Trip6 has been identified as a protein that interacts with several proteins in yeast two-hybrid screens including the thyroid hormone receptor, PTP-BL, RIL, and v-REL, most likely through its LIM domains [19]. Trip6 is found in focal adhesions and overexpression of Trip6 results in decreased cell migration [20].

Trip6 contains an NES conserved with that of LPP, and treatment with leptomycin B causes Trip6 accumulation in the nucleus. The Trip6 NES also directs transport of v-REL from the nucleus. As with zyxin, no NLS is apparent in the sequence of Trip6, but both the N-terminal and C-terminal domains have nuclear targeting capabilities. Again, no nuclear function has been found, but GAL4-based transactivation assays reveal transactivation domains in both the N and C termini. Interestingly, the N-terminal transactivation domain overlaps with the NES [21], suggesting that an interplay between engagement in transcription and nuclear export could be responsible for regulating Trip6 nuclear localization.

4.2.1.4 WTIP (Wilms Tumor protein 1 Interaction Protein)

WTIP (Wilms Tumor protein 1 Interaction Protein) was found in a screen for proteins that interact with WT1, a transcription factor essential for normal nephrogenesis and maintenance of podocyte differentiation. The interaction of these two proteins is through the LIM domains of WTIP. WTIP localizes to sites of cell–cell contact in cells at steady state. The N terminus of the protein diverges somewhat from other zyxin family members, but does retain the proline-rich regions. In addition, there is a PDZ binding domain at the C terminus of WTIP.

There is an NES in the N-terminal domain and WTIP accumulates in the nucleus upon leptomycin B treatment. Deletion of the NES represses WT1-dependent transcription. WTIP may normally localize to podocyte cell–cell junctions, but upon injury translocate to the nucleus where it exerts its repressive effect on WT1, thus allowing the dedifferentiation of the podocyte phenotype that is characteristic of the disease state [22].

4.2.1.5 Ajuba

Ajuba is recruited to sites of cell–cell contact in embryonal carcinoma cells and is also found at adherens junctions containing E-cadherin in primary keratinocytes [23, 24]. Recruitment of ajuba to E-cadherin adhesions in primary keratinocytes relies on its binding to a-catenin and it was found that ajuba contributes to cell–cell junction formation and/or stabilization [23].

Proliferation and differentiation of embryonal carcinoma cells seem to depend on ajuba localization. While overexpression of full-length ajuba or just its N-terminal domain (both cytoplasmically localized) enhances proliferation, overexpression of a construct of the LIM domains, which localizes to the nucleus, suppresses proliferation and causes spontaneous endodermal differentiation. Furthermore, ajuba accumulates in the nucleus in response to induction of endodermal differentiation with all-*trans* retinoic acid. There is a functional NES in the N terminus of ajuba and endogenous protein is accumulated in the nucleus upon treatment with leptomycin B, indicating that ajuba shuttles between the cytosol/cell–cell contacts and the nucleus. Like other members of the zyxin protein family ajuba does not contain a classical NLS, though truncation experiments indicate that nuclear targeting is mediated by the LIM domains [24].

4.2.2
Paxillin Subfamily

The paxillin subfamily of LIM domain proteins has four contiguous LIM domains at the C terminus and common LD motifs at the N terminus [25]. LD domains are protein–protein interaction domains and their presence in multiple copies suggests that these proteins may act as adaptors for multi-protein complexes. The members of the paxillin subfamily are paxillin, Hic-5, leupaxin, and paxB [26, 27], and they all localize to focal adhesion complexes. Paxillin and Hic-5 are also found in the nucleus and hence may be involved in transferring signals from the extracellular matrix to changes in gene expression.

4.2.2.1 Paxillin
Paxillin localizes to focal adhesions and is important in embryonic development and cell spreading and motility. At focal adhesions, protein–protein interaction domains in paxillin promote binding to several different proteins, many of which are involved in actin cytoskeleton organization. Vinculin and actopaxin both bind to paxillin through its N-terminal LD motifs [28]. In turn, vinculin can bind actin and talin and may function to connect paxillin to the cytoskeleton. Actopaxin also binds actin and functions in cell spreading and adhesion [29]. Several ARF-GAPs (GTPase activating proteins for the ARF family of GTP-binding proteins) have also been shown to bind the LD motifs of paxillin. This interaction may regulate cell motility [30–32]. Kinases, including FAK (focal adhesion kinase), ILK (integrin linked kinase) and CAKβ, also bind to the LD motifs of paxillin. Some of these interactions may be important for paxillin tyrosine phosphorylation following integrin engagement [28], which creates a binding site for SH2 domain containing signaling proteins. Papilloma virus 16 E6 protein binds to an LD domain of paxillin adding it to zyxin as a focal adhesion dual location protein targeted by this oncoprotein [33, 34].

At steady state paxillin is found in focal adhesions. The LIM domains are responsible for this localization [35]. However, addition of leptomycin B can cause paxillin accumulation in the nucleus [36]. Neither an NLS nor NES is apparent in the paxillin amino acid sequence, but the signal for paxillin nuclear localization may involve tyrosine phosphorylation [37]. A function for paxillin in the nucleus is suggested by the finding that it associates with two proteins involved in mRNA production and transport [36]. One of these proteins is poly(A) binding protein 1 (PABP1) which contains an RNA recognition motif that binds to the poly(A)-tail of mRNAs. PABP1 is involved in the biogenesis of mRNA in the nucleus and shuttles to the cytoplasm where it is part of the translation initiation complex. PABP1 protein associates directly with the N terminus of paxillin. The other protein is PSF (polypyrimidine tract binding (PTB) protein-associated splicing factor) which binds PTB, a protein involved in mRNA splicing. PTB in turn can bind raver1, a protein with three RNA recognition motifs that shuttles between the cytoplasm and RNA metabolism sites in the nucleus, and is also a ligand for focal adhesion proteins vinculin and α-actinin [38].

4.2.2.2 **Hic-5**

Hic-5 was originally cloned as a gene whose expression was increased by treatment of cells with TGFβ1 or hydrogen peroxide [39]. Hic-5 overexpression causes senescence-like phenotypes [40]. Hic-5 binds vinculin, FAK, and CAKβ through its N-terminal domain [41, 42]. Hic-5, like paxillin, is localized to focal adhesions through its LIM domains, but it is not tyrosine phosphorylated in response to integrin engagement. In fact, overexpression of Hic-5 can inhibit paxillin phosphorylation. By interfering with phosphorylation of paxillin, Hic-5 can prevent SH2 binding sites from being formed, and hence antagonize downstream signaling by paxillin. This could be responsible for decreased cell proliferation and migration that are characteristic of the Hic-5-induced senescent phenotype [42, 43].

The LIM domains of Hic-5 are capable of localizing to the nucleus, though they do not contain a classical NLS. There is, however, an NES in the N-terminal domain, and Hic-5 accumulates in the nuclei of cells treated with leptomycin B. The NES of Hic-5 is special compared to that of other LIM family dual localization proteins in that it harbors two cysteines that are similar to the Yap/Pap-type of NES in yeast. The Yap/Pap proteins are transcription factors involved in oxidative stress response and they accumulate in the nucleus in response to oxidants. Similarly, Hic-5 accumulates in the nucleus when cells are exposed to oxidants [27]. The LIM domains of Hic-5 were found to have DNA-binding capabilities, a feature not found in the other LIM family dual location proteins [44]. Hic-5 also exhibits transactivation activity in a GAL4-transactivation assay and potentiates the activation of reporter genes by steroid receptors when transfected into mammalian cells [45]. Additionally, Hic-5 has been found to activate the transcription of c-*fos* [27, 46].

4.3
MAGUK Protein Family

The MAGUKs (*m*embrane-*a*ssociated *gu*anylate *k*inase homologs) are a family of proteins that act as molecular scaffolds for signaling pathway components at the plasma membrane of animal cells. They are localized to, and required for the formation of a variety of cell junctions including synaptic and neuromuscular junctions, tight junctions in mammalian cells, and septate junctions in *Drosophila* cells. These proteins have a distinct domain structure that consists of (from N to C terminus): 1) one to three PDZ domains; 2) an SH3 domain; and 3) a domain homologous to the mammalian and yeast cytosolic guanylate kinase (GUK), which in MAGUKs does not retain any kinase activity. All of these domains are involved in protein–protein interactions. In addition, some MAGUKs have additional protein–protein interaction domains that define subfamilies within the MAGUKs. The Dlg-like MAGUKs consist of proteins with a domain structure like *Drosophila* lethal (1) discs large (Dlg) – three PDZ domains, and SH3 and GUK domains. Some also contain a HOOK domain that binds cytoskeletal protein Band 4.1. The p55-like MAGUKs contain a single PDZ, SH3 and GUK domains, and some also have a HOOK domain. The ZO-1-like MAGUKs have three PDZ domains, SH3 and GUK domains, and a C-terminal proline-rich extension. Finally, the Lin-2-like MAGUKs have one PDZ domain, SH3 and GUK domains, and an N terminus similar to calcium/calmodulin-dependent protein kinase (CaM kinase) followed by a Veli/Lin-7 binding region [47]. It is these last two subfamilies of MAGUKs that contain proteins that have been found to have dual localization. ZO-1 and ZO-2 are part of the ZO-1 family, and CASK is part of the Lin-2 family.

4.3.1
ZO-1

In mammalian epithelial and endothelial cells ZO-1 is associated with the cytoplasmic face of tight junctions. ZO-1 binds to the cytoplasmic tail of the tight junction integral membrane proteins occludin and claudin and acts as a molecular adaptor, linking the tight junction to the actin cytoskeleton [7]. In subconfluent cultures of cells, ZO-1 localizes to the nucleus and to tight junctions. These localizations are also observed upon wounding confluent cell monolayers. Sequence inspection reveals two possible NLSs within ZO-1 [48]. No nuclear function for ZO-1 has been found, but its interaction with another protein may explain the dual localization of ZO-1. ZONAB (ZO-1-associated nucleic acid-binding protein) binds the SH3 domain of ZO-1 and localizes to both the tight junction and nucleus.

ZONAB binds promoters containing an inverted CCAAT box and has been shown to regulate ErbB2 expression. During cell growth ZONAB levels are high while ZO-1 levels are low, and ErbB2 expression is repressed. As cells reach confluency and decrease proliferation the level of ZO-1 increases, ZONAB is sequestered at the membrane, and expression of ErbB2 is de-repressed [49]. ZONAB also has an effect on CDK4 (cyclin-dependent kinase 4) localization. CDK4 associates with D-type cyclins to facilitate exit from the G1 phase of the cell cycle [50]. During proliferative stages of cell growth when ZONAB levels are high in the nucleus, CDK4 is also localized to the nucleus and this localization is necessary for proliferation. Changes in the ZONAB level in the nucleus have a parallel effect on CDK4 nuclear localization [51]. It should be noted that an association of ZO-1 with ZONAB has not been found in the nucleus.

4.3.2
ZO-2

ZO-2 is another MAGUK protein that associates with the cytoplasmic face of tight junctions to connect the actin cytoskeleton to the membrane-spanning proteins occludin and claudin. As with ZO-1, ZO-2 contains a putative NLS, though this cannot be the only signal for nuclear localization because deletion of this sequence still allows translocation of mutant ZO-2 to the nucleus [52]. ZO-2 also contains two NESs, and leptomycin B treatment causes nuclear accumulation of the protein. ZO-2 is present in the nuclei of both sparse and confluent cell cultures, but to a greater extent in the former. In wounding experiments, ZO-2 accumulates in the nuclei of cells at the wound edge and this does not require new protein synthesis. This accumulation is also dependent on an intact actin cytoskeleton [53].

ZO-2 in the nucleus may be involved in regulation of gene transcription. Firstly, ZO-2 is localized in discreet speckles in the nucleus that partially overlap with spliceosomes [53]. Secondly, yeast two-hybrid and co-immunoprecipitation results show a direct interaction of ZO-2 with Scaffold Attachment Factor B (SAF-B), a DNA-binding protein that is known to concentrate in spliceosome speckles. SAF-B is thought to be involved in transcriptional regulation [52]. Lastly, ZO-2 interacts with the transcription factors Fos, Jun, and C/EBP. This interaction is found in the nucleus and, like the interaction of ZO-1 with the transcription factor ZONAB, ZO-2 also interacts with these transcription factors at the tight junction. Overexpression of ZO-2 downregulates gene expression from AP1 (Fos and Jun) promoters in a dose dependent manner [54].

4.4.1.2 **Plakoglobin**

Whether plakoglobin is a true dual location protein is unclear. Although plakoglobin shares significant homology in sequence and domain organization with β-catenin, it seems that the two proteins have divergent functions. While both bind to cadherin family adhesion proteins (see above), there are significant differences in the effects of deletion of the genes on embryonic development. While deletion of β-catenin is lethal shortly after implantation due to a developmental signaling defect (see above), deletion of the plakoglobin gene leads to structural abnormalities in tissue integrity including heart muscle, which leads to embryonic death at E12– 16, and the epidermis, which results in skin blistering in embryos that survive until later in gestation [73]. A conclusion from these phenotypes is that plakoglobin plays an essential structural role in the organization of adherens and desmosomal junctions, rather than in regulating gene transcription.

On the other hand, plakoglobin can translocate to the nucleus, bind TCF/LEF, and inhibit TCF/LEF transcriptional activity [74, 75]. Plakoglobin is also accumulated in response to Wnt signals [76, 77] and injection of either β-catenin or plakoglobin mRNAs into *Xenopus* embryos results in the same phenotype – axis duplication [78, 79]. Whether plakoglobin exerts its effects in these cases directly or acts only to modulate β-catenin's function is an area of debate [80, 81]. Other evidence points to plakoglobin as a growth suppressor. Overexpression of plakoglobin in highly transformed cells decreases their tumorigenicity [82–85] and overexpression in adult tissues suppresses epithelial proliferation and hair growth [86].

4.4.2
p120 Armadillo Repeat Subfamily

The p120 subfamily of armadillo repeat proteins consists of p120, ARVCF, and the plakophillins [58]. Through their armadillo domains p120 and ARVCF bind to the juxtamembrane region of classical cadherins, a region important for cadherin clustering, cell–cell adhesion, and motility. p120 and ARVCF may play roles in these functions [87]. Plakophilins bind through their N-terminal head domain to desmosomal cadherins desmoglein and desmocolin and function to cluster these cadherins and connect them to the intermediate filament network [58]. p120, ARVCF, and the plakophilins exist as multiple splice variants, each with slightly different binding profiles and nuclear targeting capabilities [87– 92]. The expression of these different variants of p120 subfamily members allows fine tuning of the nature of cell–cell adhesions and possibly the nuclear signals generated by the dual location of these proteins. All members of the p120 subfamily contain a conserved basic sequence in their armadillo domains that resembles an NLS. However, this putative

NLS is not essential for nuclear targeting. It is thought that, as with β-catenin, the armadillo domains of the p120 subfamily may directly facilitate nuclear import (see above) [87, 88, 92].

4.4.2.1 **p120**

p120 was originally identified as a Src tyrosine kinase substrate whose phosphorylation correlated by mutational analysis with a transformed phenotype [93]. The N-terminal sequence that contains the tyrosine phosphorylation targets are required for the nuclear localization of p120 [94]. The ability of p120 to localize to the nucleus is also dependent on the splice variant. Isoforms containing the so-called "exon B" are excluded from the nucleus. This exon contains a classical leucine-rich NES that is capable of directing the nuclear export of ovalbumin. Leptomycin B can induce the nuclear accumulation of p120 containing exon B, but only in cells lacking E-cadherin expression. In E-cadherin expressing cells, p120 remains sequestered at cell-cell contacts. Removing E-cadherin from the membrane with phorbol ester results in a redirection of p120 to the nucleus [90, 94, 95].

Different splice variants of p120 have different nuclear shuttling capabilities and it is interesting to note that none of these splice variants of p120 lose their ability to bind cadherins. Only the nuclear localization is affected. Overexpression of p120 isoforms capable of targeting the nucleus causes a plasma membrane branching phenotype reminiscent of filopodia and dendrites seen in fibroblasts [94, 96]. A possible mechanism by which nuclear p120 is able to effect this branching phenotype is through its interaction with Kaiso, a BTB/POZ (Broad complex, Tramtrak, Bric a brac/Pox virus and zinc finger) transcription factor that requires either cytosine methylation at its DNA binding sequence or the consensus sequence CTGCNA [97–99]. p120 inhibits Kaiso binding to both of these target DNA sequences and thus may modulate Kaiso transcriptional activity required for expression of proteins involved in branching [99].

4.4.2.2 **ARVCF**

ARVCF (armadillo repeat gene deleted in Velo-cardio-facial syndrome) is a candidate gene for the related developmental abnormalities Velo-cardio-facial syndrome and DiGeorge syndrome [100]. ARVCF is the closest homolog of p120 and competes with p120 for binding to the juxtamembrane region of cadherins [88, 89, 101]. Like p120, ARVCF is phosphorylated [88]. The C terminus of ARVCF contains a putative NES and the protein is sensitive to leptomycin B treatment [101].

Despite their similarities, ARVCF and p120 probably serve different functions in the cell. Multiple cell types show ARVCF expression to be at least ten-fold less than that of p120, making it unlikely that ARVCF func-

would be a second-order or indirect sequestration by the adhesion plaque. This scenario is especially appealing for those proteins that have a strong NES and a short dwell time in the nucleus. It is possible that these proteins function to keep a protein out of the nucleus and stored at a cell adhesion until the appropriate extracellular signal causes the transcriptional regulator cargo to be released. When the signal ceases the dual location protein could then be important in removing the transcriptional regulator from the nucleus. Though not a dual location protein from adhesion plaques, an example of this scenario involves IκBα. IκBα regulates the transcriptional activity of NF-kB by retaining NF-kB in the cytoplasm. Upon stimulation of the cell, IκBα is degraded and NF-kB is released and translocated to the nucleus where it regulates transcription. However, newly synthesized IκBα then enters the nucleus and binds to NF-kB, inhibiting its transcriptional activation and escorting it out of the nucleus. This mechanism allows the signal from the stimulus to the cell to be ended [109]. Effects seen in experiments that accumulate dual location proteins in the nucleus could be due to the action of their cargoes. For example, deletion of an NES could trap a dual location protein and its export cargo in the nucleus. The unregulated effect of this cargo protein, not the dual location protein, on transcription could then be responsible for the phenotype observed.

4.6.2
mRNA Localization

Dual location proteins may chaperone mRNAs out of the nucleus, and possibly to the adhesion site at which the dual location protein also resides, thus leading to localized protein translation at that site. Cytoplasmic RNA localization is a mechanism conserved throughout evolution for creating cellular asymmetries. RNA is specifically localized in *Drosophila*, *Xenopus*, ascidian, zebrafish, and echinoderm oocytes and embryos, as well as in a variety of developing and differentiated polarized cells from yeast to mammals [110, 111].

Paxillin binds to two proteins involved in mRNA biogenesis and transport – PABP1 and PSF. Paxillin and PABP1 co-localize at the leading lamella during cell migration [36]. This could be significant given the finding that integrin engagement, which is known to recruit paxillin, also induces a rapid relocation of mRNA and ribosomes to the forming focal adhesions [112]. An attractive hypothesis is that before focal adhesion formation paxillin exists in the cytoplasm in complex with transcripts that will be needed upon integrin engagement. When focal adhesions form, paxillin is recruited not only to perform its structural role and possible downstream signaling role, but also to bring in transcripts for proteins that are needed on a timescale too fast for *de novo* transcription in response to substrate adhesion.

Adhesion sites not only provide anchorage points between cells and with their substratum, but also are specific for different regions of the plasma membrane. Thus, targeting an mRNA to focal adhesions also targets the mRNA to the basal area of the cell where a specific transcript might be needed. Likewise, targeting an mRNA to the tight junction leads to its localization at the apex of the lateral membrane, whereas targeting to adherens junctions or desmosomes leads to localization at the lateral membrane. Therefore, dual location proteins need not only be involved in the localization of transcripts for proteins involved in adhesion. Instead they could localize transcripts for proteins with some other function that needs to be localized to a specific region of the cell. For example, a cytoplasmic binding partner for a receptor that is targeted to the apical membrane could have its transcript localized to the tight junction so that the protein is translated near the apical side of the cell.

4.6.3
Scaffolding

Some dual location proteins may have evolved as scaffolding proteins that the cell has adapted for multiple uses. In this scenario the dual location of a protein does not actually function to connect adhesion to gene expression. Many of the proteins reviewed here have multiple protein–protein interaction domains that allow them to build up combinatorial complexes with interacting proteins. This explanation of dual location seems most plausible for proteins that are constitutively located in both the nucleus and at cell adhesions. Proteins that are not found in the nucleus at steady state seem most likely to redistribute in response to a specific signal and thus to have a nuclear function somehow related to that signal.

Symplekin is always present in the nuclei of all cells. Symplekin also localizes to tight junctions in epithelial cells. The pre-mRNA polyadenylation machinery consists of six factors that can be purified to homogeneity (CPSF, CstF, CFI, CFII, PAP, and RNA polymerase II). Symplekin is not found in the purified forms of any of these factors. However, it does co-fractionate with CPSF, CFI, and CFII during early purification steps and can bind directly to a component of CstF when this protein is free of the assembled complex [107]. Symplekin could function to bring the polyadenylation machinery together and/or it could be stabilizing the machinery once it is assembled. Alternatively, symplekin could be linking the polyadenylation machinery to another nuclear process. The function of symplekin at the tight junction is not known but perhaps it is to act as a scaffold for several proteins or protein complexes that contribute to the function of the tight junction, just as it is involved in organizing several factors of the polyadenylation machinery.

4.7
Conclusion

We have highlighted a class of proteins called "dual location" proteins. Though there are many other proteins that translocate from the cell membrane to the nucleus – steroid hormone receptors for example – we have limited discussion to those proteins that are part of adhesion plaques between cells or between cells and the substratum by virtue or their direct or indirect interaction with the transmembrane proteins that form these adhesions. These proteins are interesting because they represent a mechanism by which the adhesion status of a cell can be translated to the gene expression machinery directly.

Several families of proteins have demonstrated dual location to both cell adhesions and the nucleus – LIM domain proteins, MAGUK family proteins, and armadillo repeat proteins. There are likely to be more proteins to be discovered and as they are found perhaps the differences and similarities between them will give some clues to their functions. So far, those families discovered all exhibit multiple protein–protein interaction domains. At adhesion plaques these domains allow proteins to interact with transmembrane proteins, the cytoskeleton, and other proteins involved in establishing and maintaining adhesion. As more is learned it will be interesting to see how these protein–protein interaction domains are utilized in nuclear processes and whether common themes emerge.

The results reviewed above were obtained in a variety of cell types. It is possible to extrapolate from one cell type to another to a point because the basic structure of cell adhesions is the same in many instances (for example a focal adhesion in a fibroblast is similar to a focal adhesion in an epithelial cell, at least structurally). However, care must be taken because there is cross-talk between the different cell adhesions within a cell. Given that there are several types of junctions (focal adhesions, tight junctions, adherens junctions, desmosomes) and the fact that each of these will have a slightly different function or character in different cell types, accounting for cross-talk between the adhesions results in enormous complexity. The dual location of proteins in these adhesions may be interdependent just as the adhesions are, and hence a full accounting of one dual location protein will require an understanding of it in the context of the others. The field of dual location proteins will therefore benefit greatly when each of the dual location proteins known (and those yet to be found) is characterized in a variety of cell types and cellular contexts.

4.8
Acknowledgments

EGC is supported by NIH NIGMS pre-doctoral training program GM07276. Work in the laboratory of WJN is supported by NIH GM35227.

4.9
References

1 Trinkaus, J. P. In *Organogenesis* (Dehaan, R. I., Urspring, H., Eds.) Holt, Rinehart and Winston: New York, 1965, 55–104.
2 Ridley, A. J., Schwartz, M. A., Burridge, K., Firtel, R. A., Ginsberg, M. H., Borisy, G., Parsons, J. T., and Horwitz, A. R., *Science* 2003, *302*, 1704–1709.
3 Gumbiner, B. M., *Cell* 1996, *84*, 345–357.
4 Jamora, C. and Fuchs, E., *Nat Cell Biol* 2002, *4*, E101–108.
5 Takeichi, M., *Science* 1991, *251*, 1451–1455.
6 Green, K. J. and Gaudry, C. A., *Nat Rev Mol Cell Biol* 2000, *1*, 208–216.
7 Anderson, J. M., van Itallie, C. M., and Fanning, A. S., *Curr Opin Cell Biol* 2004, *16*, 140–145.
8 Bach, I., *Mech Dev* 2000, *91*, 5–17.
9 Beckerle, M. C., *Bioessays* 1997, *19*, 949–957.
10 Crawford, A. W. and Beckerle, M. C., *J Biol Chem* 1991, *266*, 5847–5853.
11 Drees, B. E., Andrews, K. M., and Beckerle, M. C., *J Cell Biol* 1999, *147*, 1549–1560.
12 Nix, D. A. and Beckerle, M. C., *J Cell Biol* 1997, *138*, 1139–1147.
13 Nix, D. A., Fradelizi, J., Bockholt, S., Menichi, B., Louvard, D., Friederich, E., and Beckerle, M. C., *J Biol Chem* 2001, *276*, 34759–34767.
14 Degenhardt, Y. Y. and Silverstein, S., *J Virol* 2001, *75*, 11791–11802.
15 Petit, M. M., Mols, R., Schoenmakers, E. F., Mandahl, N., and Van de Ven, W. J., *Genomics* 1996, *36*, 118–129.
16 Petit, M. M., Fradelizi, J., Golsteyn, R. M., Ayoubi, T. A., Menichi, B., Louvard, D., Van de Ven, W. J., and Friederich, E., *Mol Biol Cell* 2000, *11*, 117–129.
17 Li, B., Zhuang, L., Reinhard, M., and Trueb, B., *J Cell Sci* 2003, *116*, 1359–1366.
18 Petit, M. M., Meulemans, S. M., and Van de Ven, W. J., *J Biol Chem* 2003, *278*, 2157–2168.
19 Wang, Y., Dooher, J. E., Koedood Zhao, M., and Gilmore, T. D., *Gene* 1999, *234*, 403–409.
20 Yi, J., Kloeker, S., Jensen, C. C., Bockholt, S., Honda, H., Hirai, H., and Beckerle, M. C., *J Biol Chem* 2002, *277*, 9580–9589.
21 Wang, Y., Gilmore, T. D., *Biochim Biophys Acta* 2001, *1538*, 260–272.

22 SRICHAI, M. B., KONIECZKOWSKI, M., PADIYAR, A., KONIECZKOWS-KI, D. J., MUKHERJEE, A., HAYDEN, P. S., KAMAT, S., EL-MEANAWY, M. A., KHAN, S., MUNDEL, P., LEE, S. B., BRUGGEMAN, L. A., SCHELLING, J. R., and SEDOR, J. R., *J Biol Chem* **2004**.

23 MARIE, H., PRATT, S. J., BETSON, M., EPPLE, H., KITTLER, J. T., MEEK, L., MOSS, S. J., TROYANOVSKY, S., ATTWELL, D., LONGMORE, G. D., and BRAGA, V. M., *J Biol Chem* **2003**, *278*, 1220–1228.

24 KANUNGO, J., PRATT, S. J., MARIE, H., and LONGMORE, G. D., *Cell, M. B.* **2000**, *11*, 3299–3313.

25 BROWN, M. C., CURTIS, M. S., and TURNER, C. E., *Nat Struct Biol* **1998**, *5*, 677–678.

26 LIPSKY, B. P., BEALS, C. R., and STAUNTON, D. E., *J Biol Chem* **1998**, *273*, 11709–11713.

27 SHIBANUMA, M., KIM-KANEYAMA, J. R., ISHINO, K., SAKAMOTO, N., HISHIKI, T., YAMAGUCHI, K., MORI, K., MASHIMO, J., and NOSE, K., *Mol Biol Cell* **2003**, *14*, 1158–11571.

28 SCHALLER, M. D., *Oncogene* **2001**, *20*, 6459–6472.

29 NIKOLOPOULOS, S. N. and TURNER, C. E., *J Cell Biol* **2000**, *151*, 1435–1448.

30 KONDO, A., HASHIMOTO, S., YANO, H., NAGAYAMA, K., MAZAKI, Y., and SABE, H., *Mol Biol Cell* **2000**, *11*, 1315–1327.

31 MAZAKI, Y., HASHIMOTO, S., OKAWA, K., TSUBOUCHI, A., NAKA-MURA, K., YAGI, R., YANO, H., KONDO, A., IWAMATSU, A., MIZO-GUCHI, A., and SABE, H., *Mol Biol Cell* **2001**, *12*, 645–662.

32 ZHAO, Z. S., MANSER, E., LOO, T. H., and LIM, L., *Mol Cell Biol* **2000**, *20*, 6354–6363.

33 TONG, X. and HOWLEY, P. M., *Proc Natl Acad Sci USA* **1997**, *94*, 4412–4417.

34 VANDE POL, S. B., BROWN, M. C., and TURNER, C. E., *Oncogene* **1998**, *16*, 43–52.

35 BROWN, M. C., PERROTTA, J. A., and TURNER, C. E., *J Cell Biol* **1996**, *135*, 1109–1123.

36 WOODS, A. J., ROBERTS, M. S., CHOUDHARY, J., BARRY, S. T., MAZA-KI, Y., SABE, H., MORLEY, S. J., CRITCHLEY, D. R., and NORMAN, J. C., *J Biol Chem* **2002**, *277*, 6428–64237.

37 OGAWA, M., HIRAOKA, Y., and AISO, S., *Biochem Biophys Res Commun* **2003**, *304*, 676–683.

38 HUTTELMAIER, S., ILLENBERGER, S., GROSHEVA, I., RUDIGER, M., SINGER, R. H., and JOCKUSCH, B. M., *J Cell Biol* **2001**, *155*, 775–786.

39 SHIBANUMA, M., MASHIMO, J., KUROKI, T., and NOSE, K., *J Biol Chem* **1994**, *269*, 26767–26774.

40 SHIBANUMA, M., MOCHIZUKI, E., MANIWA, R., MASHIMO, J., NISHIYA, N., IMAI, S., TAKANO, T., OSHIMURA, M., and NOSE, K., *Mol Cell Biol* **1997**, *17*, 1224–1235.

41 MATSUYA, M., SASAKI, H., AOTO, H., MITAKA, T., NAGURA, K., OHBA, T., ISHINO, M., TAKAHASHI, S., SUZUKI, R., and SASAKI, T., *J Biol Chem* **1998**, *273*, 1003–1014.

42 FUJITA, H., KAMIGUCHI, K., CHO, D., SHIBANUMA, M., MORIMO-TO, C., and TACHIBANA, K., *J Biol Chem* **1998**, *273*, 26516–26521.

43 NISHIYA, N., TACHIBANA, K., SHIBANUMA, M., MASHIMO, J. I., and NOSE, K., *Mol Cell Biol* **2001**, *21*, 5332–5345.

44 NISHIYA, N., SABE, H., NOSE, K., and SHIBANUMA, M., *Nucleic Acids Res* **1998**, *26*, 4267–4273.

45 YANG, L., GUERRERO, J., HONG, H., DeFRANCO, D.B., and STALL-
    CUP, M.R., *Mol Biol Cell* **2000**, *11*, 2007–2018.
46 KIM-KANEYAMA, J., SHIBANUMA, M., and NOSE, K., *Biochem Bio-
    phys Res Commun* **2002**, *299*, 360–365.
47 DIMITRATOS, S.D., WOODS, D.F., STATHAKIS, D.G., and BRYANT,
    P.J., *Bioessays* **1999**, *21*, 912–921.
48 GOTTARDI, C.J., ARPIN, M., FANNING, A.S., and LOUVARD, D.,
    Proc Natl Acad Sci USA **1996**, *93*, 10779–10784.
49 BALDA, M.S. and MATTER, K., *EMBO J* **2000**, *19*, 2024–2033.
50 SHERR, C.J., *Trends Biochem Sci* **1995**, *20*, 187–190.
51 BALDA, M.S., GARRETT, M.D., and MATTER, K., *J Cell Biol* **2003**,
    160, 423–432.
52 TRAWEGER, A., FUCHS, R., KRIZBAI, I.A., WEIGER, T.M., BAUER,
    H.C. and BAUER, H., *J Biol Chem* **2003**, *278*, 2692 2700.
53 ISLAS, S., VEGA, J., PONCE, L., and GONZALEZ-MARISCAL, L., *Exp
    Cell Res* **2002**, *274*, 138–148.
54 BETANZOS, A., HUERTA, M., LOPEZ-BAYGHEN, E., AZUARA, E.,
    AMERENA, J., and GONZALEZ-MARISCAL, L., *Exp Cell Res* **2004**,
    292, 51–66.
55 COHEN, A.R., WOODS, D.F., MARFATIA, S.M., WALTHER, Z.,
    CHISHII, A.H., ANDERSON, J.M., and WOOD, D.F., *J Cell Biol*
    1998, *142*, 129–138.
56 HSUEH, Y.P., WANG, T.F., YANG, F.C., and SHENG, M., *Nature*
    2000, *404*, 298–302.
57 STRAIGHT, S.W., KARNAK, D., BORG, J.P., KAMBEROV, E., DARE,
    H., MARGOLIS, B., and WADE, J.B., *Am J Physiol Renal Physiol*
    2000, *278*, F464–475.
58 HATZFELD, M., *Int Rev Cytol* **1999**, *186*, 179–224.
59 HUBER, A.H., NELSON, W.J., and WEIS, W.I., *Cell* **1997**, *90*, 871–882.
60 NELSON, W.J. and NUSSE, R., *Science* **2004**, *303*, 1483–1487.
61 HECHT, A. and KEMLER, R., *EMBO Rep* **2000**, *1*, 24–28.
62 HAEGEL, H., LARUE, L., OHSUGI, M., FEDOROV, L., HERREN-
    KNECHT, K., and KEMLER, R., *Development* **1995**, *121*, 3529–3537.
63 YOKOYA, F., IMAMOTO, N., TACHIBANA, T., and YONEDA, Y. *Mol
    Biol Cell* **1999**, *10*, 1119–1131.
64 FAGOTTO, F., GLUCK, U., and GUMBINER, B.M., *Curr Biol* **1998**, *8*,
    181–190.
65 KUTAY, U., IZAURRALDE, E., BISCHOFF, F.R., MATTAJ, I.W., and
    GORLICH, D., *EMBO J* **1997**, *16*, 1153–1163.
66 MALIK, H.S., EICKBUSH, T.H., and GOLDFARB, D.S., *Proc Natl
    Acad Sci USA* **1997**, *94*, 13738–13742.
67 HENDERSON, B.R., *Nat Cell Biol* **2000**, *2*, 653–660.
68 NEUFELD, K.L., ZHANG, F., CULLEN, B.R., and WHITE, R.L.,
    EMBO Rep **2000**, *1*, 519–523.
69 HENDERSON, B.R. and FAGOTTO, F., *EMBO Rep* **2002**, *3*, 834–839.
70 GOTTARDI, C.J., WONG, E., and GUMBINER, B.M., *J Cell Biol*
    2001, *153*, 1049–1060.
71 FAGOTTO, F., FUNAYAMA, N., GLUCK, U., and GUMBINER, B.M., *J
    Cell Biol* **1996**, *132*, 1105–1114.
72 HUBER, O., KORN, R., McLAUGHLIN, J., OHSUGI, M., HERRMANN,
    B.G., and KEMLER, R., *Mech Dev* **1996**, *59*, 3–10.
73 BIERKAMP, C., SCHWARZ, H., HUBER, O., and KEMLER, R., *Devel-
    opment* **1999**, *126*, 371–381.

II

Classical Examples of Cell Migration in Development

time of their internalization and their localization. The epiblast will form ectodermal and neuroectodermal tissues.

Cell intercalations along the mediolateral axis of the gastrula mediate convergence and extension movements within the epiblast and hypoblast. By this rearrangement, cells move towards the dorsal side of the gastrula and, at the same time, are dispersed along the anterior-posterior axis of the embryo eventually leading to the formation of the embryonic body axis at the end of gastrulation.

These different cell movements take place in a highly coordinated spatial and temporal manner indicating that they must be controlled by a complex interplay of morphogenetic and inductive events. In this chapter, we will attempt to summarize the progress achieved towards understanding the cellular and molecular mechanisms underlying the different types of cell behaviors during gastrulation in zebrafish. We will first focus on the spreading of the blastoderm around the yolk cell during epiboly. We will then describe how the most marginal blastodermal cells move inwards to form the hypoblast cell layer. Finally, we will discuss convergence and extension movements, which will lead to the anterior-posterior axis formation. Furthermore, we will highlight some recent advances shedding light on the signaling pathways and cell-cell interaction molecules that regulate gastrulation movements.

5.2
Cellular and Molecular Mechanisms

After fertilization of the oocyte, the chorion surrounding the zygote swells and non-yolky cytoplasm streams towards the animal pole of the zygote segregating the blastodisc from the yolk cell. This is followed by the cleavage period when the blastodisc undergoes synchronous and rapid cell divisions. During those initial cleavages, the cleavage furrow extends through the blastodisc but not the yolk cell, which is characteristic for a meroblastic cell division pattern. When the embryo reaches the 64-cell stage, division planes become more stochastic and many cells start to lose their contact with the underlying yolk. Cells located at the most superficial locations will eventually form the enveloping layer (EVL), a thin sheet of cells that covers the blastomeres, while cells at deeper positions (normally called the deep cell layer) will give rise to all other embryonic tissues (Kimmel, 1995).

The blastula period begins at the 128-cell stage and lasts until gastrulation starts. During this period, the embryo enters the midblastula transition (MBT), the yolk syncytial layer (YSL) is formed, and epiboly begins. The YSL forms by marginal blastomeres releasing their cytoplasm and nuclei into the underlying yolk cell, to which they are connected. As the embryo enters MBT, zygotic gene transcription starts and the deep cells become more motile (Kane and Kimmel, 1993).

5.2.1
Epiboly

5.2.1.1 **Epiboly Movements**

At late blastula stage, the embryo consists of three basic tissues: the blastoderm, which consists of the EVL and the deep cell layer; the underlying yolk cell, and the YSL, which surrounds the animal portion of the yolk cell. This is also the time when epiboly begins. Epiboly is the process by which the blastoderm thins and spreads over the yolk cell until it has completely engulfed the yolk at the end of gastrulation. During epiboly, deep blastomeres typically become more loosely associated with each other. They round up, exert bleb-like protrusions and stream outwards towards the surface where they intermix with more superficial blastomeres in a process called radial intercalation (Fig. 5.1A, D, G) (Keller, 1980, Warga, 1990). Radial cell intercalations are restricted to the deep cell layer, while the EVL remains a monolayer that will eventually give rise to the periderm (Kimmel et al., 1990b; Wilson et al., 1993).

Interestingly, deep blastomeres at the margin, which will later form the mesoderm and endoderm, do not intermix as actively as the central ones that are fated to become ectodermal tissues, suggesting that the difference in the mixing behavior between these regions is already an expression of their different identities (Helde et al., 1994; Kimmel et al., 1991).

At the same time as radial cell intercalations occur, the yolk cell domes and occupies the space of the deep blastomeres. Furthermore, a portion of the YSL nuclei, the internal YSL nuclei, begins to move towards the animal pole of the embryo while the rest of the YSL nuclei, the external YSL, undergoes epiboly. In *Fundulus heteroclitus*, a related teleost, these movements of the YSL are independent of the blastomeres and autonomous to the yolk cell (Betchaku and Trinkaus, 1978). In contrast, *Fundulus* blastodermal cells do not undergo epiboly movements unless they are attached to the underlying YSL (Trinkaus, 1951).

It has been suggested that the doming movement of the yolk towards the animal pole exerts the force that drives the deep blastomeres to move outwards and undergo radial cell intercalations (Wilson et al., 1995). Consistent with its function as a motor in this process, the yolk cell contains microtubules and microfilaments in the proper orientation to generate such movements. In fact, treatments that interfere with these structures also interfere with epiboly (Solnica-Krezel and Driever, 1994; Strähle and Jesuthasan, 1993).

The role of the EVL in epiboly is still unclear. The EVL is attached through tight junctions with the yolk cell at the blastoderm margin and, based on observations in *Fundulus*, is passively pulled towards the vegetal pole by the YSL. It is conceivable that the movement of the EVL controls the epiboly movements of the underlying blastodermal cells by simply providing sufficient space for the blastoderm to move towards the vegetal pole of the gastrula.

5.2.1.2 Epiboly Mutants

Five mutations that arrest epiboly were found in the first Tübingen and Boston screens for zebrafish mutants affecting embryonic development: *volcano* (*vol*) (Solnica-Krezel et al., 1996), *half-baked* (*hab*), *avalanche* (*ava*), *lawine* (*law*), and *weg* (Kane et al., 1996). These mutations map to a single locus, and they are considered alleles of the zebrafish cadherin-1 gene (D. Kane, personal communication). The *hab* phenotype has been characterized most extensively. Heterozygous embryos for *hab* display a partial dominant phenotype consisting of an enlarged hatching gland, while *hab* homozygous mutants display a recessive phenotype that results in the arrest of the vegetal movement of the deep blastodermal cells. However, the morphogenetic behavior of the yolk cell nuclei and the EVL appears to be unaffected, indicating that these tissues complete epiboly independent of the deep cells (Kane et al., 1996).

5.2.2
Internalization

5.2.2.1 Internalization Movements

When the blastoderm reaches the equator of the embryo (50% epiboly), the first marginal blastodermal cells start to undergo internalization movements, which will lead to the formation of the primary germ layers, ectoderm, mesoderm and endoderm (Fig. 5.1 B, E, H). This internalization marks the onset of gastrulation characterized by a transient pause in epiboly movements and a concomitant thickening of the marginal blastoderm rim, which is called the germ ring (Kimmel, 1995; Warga, 1990).

At the germ ring, deep cells in marginal regions first move towards the yolk cell, then sharply change direction and move towards both the overlying cell layer and the animal pole. This leads to the formation of the two germ layers within the germ ring: the outer epiblast layer and the inner hypoblast layer (Warga et al., 1990; Nelsen, 1953). The epiblast is a multilayer of cells of nearly uniform thickness, but eventually forms a single-cell layered pseudostratified epithelium (Papan and Campos-Ortega, 1994). The hypoblast is located between the epiblast and the YSL. Epiblast and hypoblast cells move in different directions, with early internalizing hypoblast cells migrating towards the animal pole and epiblast cells undergoing epiboly movements. Hypoblast cells internalizing at later stages of gastrulation do not move towards the animal pole, but instead, move together with the overlying epiblast towards the vegetal pole. When gastrulation ends, cells remaining in the epiblast correspond to the definitive ectoderm and will give rise to tissues such as the epidermis, the central nervous system, neural crest, and sensory placodes. In contrast, early internalizing hypoblast cells will form anterior mesendodermal tissues, while later internalizing cells will become posterior mesendoderm (D'Amico and Cooper, 1997; Kane and Adams, 2002).

As a consequence of the inward movement of prospective mesendodermal cells, a fissure (also named 'Brachet's cleft') forms between the hypoblast and the epiblast (Ballar, 1981). Hypoblast and epiblast cells do not usually cross this border, although hypoblast cells, when transplanted into the epiblast, can re-enter the hypoblast by crossing Brachet's cleft without impediment (Ho and Kimmel, 1993).

In the prospective dorsal region of the embryo, blastodermal cells accumulate locally along the germ ring and form the so-called embryonic shield. The embryonic shield is the zebrafish equivalent of the amphibian Spemann organizer region (Ho, 1992b). Cells internalizing at the shield form axial mesendodermal tissues such as the prechordal plate and notochord. Prospective prechordal plate cells are tightly clustered together and move as a coherent sheet of cells, while internalizing notochordal progenitors are less tightly associated (D'Amico and Cooper, 1997; Kane and Adams, 2002).

Interestingly, accompanying the movement of hypoblast cells that have internalized early, the internal YSL nuclei also move towards the animal pole. This movement seems to be independent from the overlying hypoblast cells, because in mutants lacking hypoblast internalization, the movement of these nuclei towards the animal pole remains unchanged (D'Amico and Cooper, 2001; Kane and Adams, 2002).

5.2.2.2 Internalization Mutants

One of the signaling pathways that operates in mesendodermal specification involves the TGF-β (Transforming Growth Factor β) like signaling molecule Nodal (Schier and Talbot, 2001). Two *nodal*-related genes have been found in zebrafish to control cell fate specification during gastrulation, *cyclops* (*cyc*; Rebagliati et al., 1998; Sampath et al., 1998) and *squint* (*sqt*; Feldman et al., 1998). In $cyc^{-/-}/sqt^{-/-}$ double mutant embryos, as well as in mutants that are defective in receiving Nodal signals such as maternal-zygotic *one-eyed-pinhead* (mzoep) mutants, induction of mesendodermal marker genes and internalization of mesendodermal cell progenitors are strongly reduced (Feldman et al., 1998; Gritsman et al., 1999).

A direct role of Nodal signaling in mesendodermal cell internalization has been demonstrated by studies showing that cells transplanted from the germ ring margin of wild-type embryos into the blastoderm margin of mzoep$^{-/-}$ embryos could individually internalize and express mesendodermal specific genes; this indicates that Nodal signals are sufficient to induce mesendodermal cell fate specification and internalization (Car-

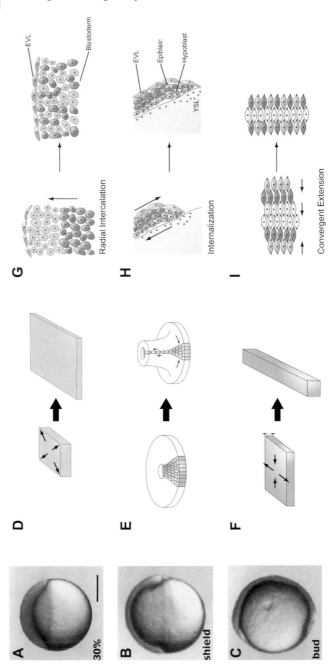

many-Rampey and Schier, 2001). Similar observations have been made in experiments where cells expressing a constitutively active form of the presumptive Nodal receptor Taram-A were transplanted into the animal pole of wild-type embryos and moved directly to the yolk surface, thereby mimicking mesendodermal cell internalization movements (David and Rosa, 2001).

5.2.3
Convergence and Extension

5.2.3.1 **Convergence and Extension Movements**
Convergence and extension (CE) refers to the overall process of mediolateral narrowing and anterior-posterior elongation of embryonic tissues, thereby defining the dorso-ventral and rostro-caudal embryonic axes (Fig. 5.1 C, F, I) Although epiboly and mesendodermal cell internalization are largely completed by the end of the gastrula period, CE movements continue to shape the embryonic body during segmentation.

Much of our knowledge about CE movements comes from studies in *Xenopus*. There, both mesodermal and ectodermal cells elongate mediolaterally and, while moving to the dorsal side of the gastrula, undergo mediolateral cell intercalations that lead to a concomitant extension of the forming body axis along its anterior-posterior extent. Oriented bipolar protrusive activities in mesodermal cells and directed monopolar protrusive activities in ectodermal cells mediate CE movements in *Xenopus* (Elul and Keller, 2000; Shih and Keller, 1992).

◄ ───

Fig. 5.1   Cell movements during zebrafish gastrulation. (A–C) Differential Interference Contrast (DIC) images of wild type zebrafish embryos at 30% epiboly (A), shield stage (B), and bud stage (C). In these and the following views, the animal pole is on top and dorsal to the right, with the exception of (I) which shows a dorsal view. (D–F) Schematic drawings illustrating the tissue rearrangements at the stages shown in (A–C). During epiboly (D), the tissue thins and spreads towards the vegetal pole. At shield stage (E), cells at the germ ring internalize giving rise to the external epiblast cell layer, which continues undergoing epiboly, and the internal hypoblast, which moves towards the animal pole of the embryo. During convergence and extension (F), cells move towards the dorsal side of the embryo (convergence) and lengthen the forming body axis along its anterior-posterior extent (extension). (G–I) Schematic representations of the main cellular movements during zebrafish gastrulation. During epiboly (G), deep blastomeres (in red) intermix with more superficial blastomeres (in orange) in a process called radial intercalation. During internalization (H), hypoblast cells (in orange) move towards the animal pole of the embryo, in a direction opposite to that of the overlying epiblast (in red). During convergence extension (I), mediolateral intercalations of cells lead to the extension of the tissue in an anterior-posterior direction. EVL (Enveloping Layer). (This figure also appears with the color plates.)

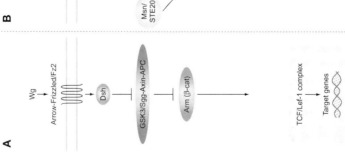

While studies in *Xenopus* point to the small GTPase Cdc42 as a downstream mediator of Wnt11 function (Djiane et al., 2000), direct evidence for the involvement of small GTPases in zebrafish CE is still lacking. However, in mammalian tissue culture and in the *Drosophila* PCP pathway, Rho-associated kinase-2 (Rok-2) mediates some aspects of Rho signaling and is implicated in zebrafish CE movements (Marlow et al., 2002). Over-expression of a dominant negative form of Rok-2 in wild-type embryos phenocopies aspects of both the *slb* and *kny* mutant phenotype, while over-expression of wild-type Rok-2 in a *slb* mutant background can rescue the *slb* mutant phenotype, suggesting that Rok-2 functions downstream of Slb/Wnt11 in regulating CE movements during gastrulation (Fig. 5.2 C) (Marlow et al., 2002).

The cellular basis of how all these genes regulate gastrulation movements has only begun to be elucidated. In zebrafish *slb/wnt11* mutant embryos, CE movements of anterior and posterior mesendodermal and neuroectodermal cells are reduced, leading to a transiently shortened body axis at the end of gastrulation and a partial fusion of the eyes at later developmental stages (Heisenberg et al., 1996; Heisenberg and Nusslein-Volhard, 1997; Heisenberg et al., 2000). Moreover, three-dimensional reconstruction and motion analysis of individual cells show that Slb/Wnt11 is required for the directionality and velocity of hypoblast cell movements in the forming germ ring at the onset of gastrulation. At the same time, *slb* hypoblast cells exhibit defects in the orientation of cellular processes along their individual movement directions, indicating that Wnt11 mediates cell movements by synchronizing cell polarization with the direction of cell migration at the onset of gastrulation (Ulrich et al., 2003). This is similar to observations made both in *Xenopus* and zebrafish showing that interfering with non-canonical Wnt signaling causes defects in both CE movements and mesodermal cell elongation. This indicates that mediolateral cell elongation is a prerequisite for normal CE movements during both *Xenopus* and zebrafish gastrulation (Jessen et al., 2002; Marlow et al., 2002; Topczewski et al., 2001; Wallingford et al., 2000).

Fig. 5.2 The Wnt/PCP pathway. A) A simplified version of the canonical Wnt/β catenin pathway for comparison. B), C) The Wnt/PCP cascades regulating planar cell polarity (PCP) in *Drosophila* (B) and convergence extension movements in zebrafish (C). C) Like the Wnt/β-catenin signaling, the Wnt/PCP pathway involves the cytoplasmic signal transduction protein Dishevelled. Dishevelled regulates the activity of members of the Rho family of small GTPases. RhoA, which is linked to Dishevelled by Daam 1, activates the effector kinase Rok, which in turn regulates the actin cytoskeleton. Additionally, Dishevelled can also activate the small GTPase Cdc42 and further downstream the JNK signaling pathway, which regulates transcription of target genes. D) The Frizzled receptors can also stimulate calcium influx and the activation of the calcium-sensitive kinases PKC and CamKII, which determine cellular adhesion. (This figure also appears with the color plates.)

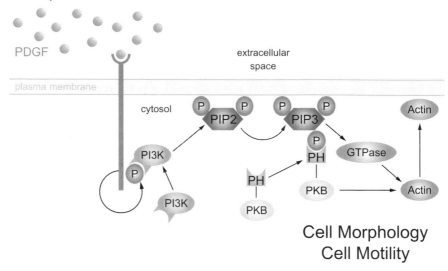

Fig. 5.3 Schematic diagram of the PI3Kinase signaling pathway during zebrafish gastrulation. PDGF (Platelet Derived Growth Factor) binds and activates its receptor. The receptor becomes phosphorylated and binds phosphoinositol-3-kinase (PI3K), which then converts Phosphoinositide-biphosphate (PI2P) to Phosphoinositide-triphosphate (PIP3) at the plasma membrane. Protein Kinase B (PKB) binds to PIP3 via its Pleckstrin Homology (PH) domain, triggering the localized accumulation of Actin and subsequent polarization, process formation and cell motility. (This figure also appears with the color plates.)

(PDGF) signaling pathways, where they are needed for the induction and morphogenesis of the mesoderm (Ataliotis et al., 1995; Montero et al., 2003; Symes and Mercola, 1996). Furthermore, in zebrafish, PDGF/PI3K signaling is required for cell polarization and process formation of mesendodermal cells at early stages of gastrulation. Similar to observations made in mammalian neutrophils and *Dictyostelium*, PI3Ks appear to polarize mesendodermal cells by asymmetrically localizing PKB to the plasma membrane at their leading edge. Interestingly, although blocking PI3K activity results in reduced polarization and process formation in mesendodermal cells of the zebrafish gastrula, the overall movement direction of mesendodermal cells remains unaffected; this suggests that cell polarization and net movement direction can be uncoupled in those cells (Montero et al., 2003).

d) Eph receptors and ephrins Eph receptors constitute a large family of tyrosine kinase receptors and, together with their plasma membrane-bound Ephrin ligands, play important roles for tissue morphogenesis during development and adulthood. These include the regulation of vascular development, tissue-border formation, axon guidance, synaptic plasticity and cell migration (Holder and Klein, 1999). Since both Eph receptors and Ephrin ligands are membrane bound, their binding and activation requires direct cell-cell interactions. The influence of Eph/Ephrin activation on cell behavior differs depending on the cell type, but can be generally attributed to cell adhesion and repulsion (Mellitzer et al., 2000).

Eph receptors are divided into two major classes: Eph-A and Eph-B. In general, Eph-A receptors bind A-ephrins that are anchored to the plasma membrane by a Glycosyl-Phosphatidyl-Inositol (GPI) linkage while Eph-B receptors bind B-ephrins that have transmembrane and short highly conserved cytoplasmic domains (Flanagan and Vanderhaeghen, 1998) (Fig. 5.4). Binding of Ephrins to their Eph receptors promotes clustering of receptors and propagation of the signal in both the ligand- and receptor-bearing cell (Bruckner et al., 1997; Davy et al., 1999; Davy and Robbins, 2000).

In zebrafish development, binding between Eph receptors and their Ephrin ligands has been shown to generate a mutual exclusion effect on adjacent cells allowing for the formation of rhombomere and somite boundaries (Durbin et al., 1998; Mellitzer et al., 1999; Xu et al., 1999). Several members of the Ephrin and Eph families are expressed during zebrafish gastrulation (Chan et al., 2001; Oates et al., 1999). Over-expression of a soluble dominant negative form of Ephrin-B2b has been shown to cause defects in notochord and prechordal plate morphogenesis at the end of gastrulation (Chan et al., 2001). Similarly, over-expression of dominant negative forms of human Eph-A3 and Eph-A5 in zebrafish embryos led to abnormal notochord, somite and brain formation. These effects were preceded by apparent cell migration defects throughout gastrulation, without alteration of cell fates (Oates et al., 1999). It is therefore likely that Ephrin-Eph interactions play important roles in regulating zebrafish gastrulation movements. However, the precise molecular and cellular mechanisms by which Ephrins and Ephs function during gastrulation are still unclear.

e) Robo/Slit Slits are secretory proteins that were originally identified in *Drosophila* to regulate axonal guidance (Brose et al., 1999). The molecular target for Slit is a repulsive guidance transmembrane receptor known as the Roundabout (Robo) receptor (Brose et al., 1999; Li et al., 1999; Rothberg et al., 1998). Slit/Robo signaling in *Drosophila* has been shown not only to prevent axons from crossing the midline, but also to function both as a repellent and attractant during mesodermal cell migration (Kidd et al., 1999).

In zebrafish, two *slit* homologs (*slit1* and *slit2*) and three *robo* homologs (*robo-1/2/3*) have been identified that are expressed locally in axial mesendodermal tissues (*slit-1/2*) or uniformly (*robo-1/2/3*) during gastrula-

blast and hypoblast cells different from each other, and what prevents these two populations of cells from mixing?

One explanation is that these two populations of cells actively repell each other. Eph receptors and their Ephrin ligands are prime candidates to mediate this process. In zebrafish, gastrulating cells expressing Eph-B type receptors do not intermix with cells expressing the corresponding Ephrin-B ligands, suggesting that de-adhesion and/or repulsion between gastrulating cells can be mediated by Ephrin-Eph signaling (Mellitzer et al., 1999). Furthermore, interference with Ephrin-Eph signaling during gastrulation perturbs multiple morphogenetic processes such as mesendodermal cell internalization and CE within both the epiblast and hypoblast (Chan et al., 2001; Oates et al., 1999). However, the specific mechanisms by which Ephrin-Eph signaling controls gastrulation movements are still unclear. It has been shown that, depending on the cell type involved, Ephrin-Eph signaling can play a role not only in cell repulsion but also in cell adhesion (Holmberg and Frisen, 2002; Wilkinson, 2003; Zimmer et al., 2003). For example, in *Xenopus* gastrulation, the overexpression of activated EphA4 or Ephrin-B1 cause a loss of cell adhesion in the *Xenopus* blastula, suggesting that Ephrin-Eph signaling is needed for cell adhesion rather than repulsion within the gastrula (Jones et al., 1998; Winning et al., 1996). It is therefore conceivable that Ephrin-Eph signaling might play a role in controlling cellular adhesion within and between the epiblast and hypoblast. To test this idea, it will be necessary to closely monitor the behavior of epiblast and hypoblast in relation to different states of Ephrin-Eph signaling on a single cell basis.

A second possibility is that epiblast and hypoblast cells acquire different cohesive properties. In support of this notion it has been shown that at the onset of gastrulation, epiblast and hypoblast cells exhibit differential expression of Cadherin proteins with hypoblast cells expressing higher levels than epiblast cells (Fig. 5.5) (Juan Antonio Montero and C. P. H., unpublished data). Similarly, in *Xenopus*, a variety of cell adhesion molecules, including Cadherins, Protocadherins, Syndecans, and Integrins, are expressed in gastrulating cells and are involved in mediating tissue morphogenesis at different stages of gastrulation (Marsden and DeSimone, 2003; Davidson et al., 2002; Shi and Keller, 1992).

But how are the activity and the expression of adhesion molecules regulated during gastrulation? A number of recent studies suggest that localized extracellular signals can regulate cell adhesion at early developmental stages, although the underlying molecular mechanisms are not yet fully understood. Such modulation of cell adhesion can be achieved through signaling molecules such as Activin, but also through adhesion mediated cell-cell and cell-matrix communication (Marsden and DeSimone, 2003; Ramos et al., 1996). In *Xenopus*, normal levels of C-cadherin expression are required for Integrin-dependent FN matrix assembly, which in turn is essential for normal radial cell intercalations and CE

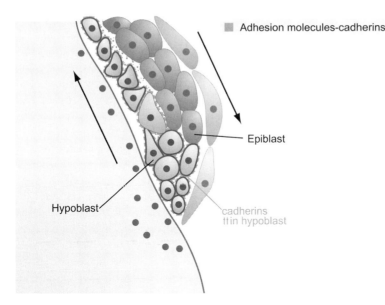

Adhesion molecules-cadherins

Epiblast

Hypoblast

cadherins
↑↑ in hypoblast

Fig. 5.5 Schematic diagram illustrating the (potential) role of differential cell adhesion in germ layer formation and scparation at the onset of gastrulation. During zebrafish gastrulation, hypoblast cells (in orange) internalize and migrate towards the animal pole of the embryo (represented by the left arrow) using the overlying epiblast cells (in red) as a substrate. The epiblast cells continue undergoing epiboly (represented by the right arrow). The levels of adhesion molecules, such as Cadherins are higher in hypoblast cells (green outline in hypoblast cells) than in epiblast cells. This differential expression of adhesion molecules (Cadherins) between epiblast and hypoblast cells triggers the sorting of hypoblast from epiblast cells and might stabilize the separation of the newly formed hypoblast and epiblast cell layers. The enveloping cell layer is represented in blue. The green dots show the yolk syncytial layer and the yellow background represents the yolk cell. (This figure also appears with the color plates.)

movements (Winklbauer, 1998). Conversely, Integrin activation alters C-cadherin activity and thereby promotes cellular motility during gastrulation (Marsden and DeSimone, 2003), indicating that there is extensive cross-talk between these different types of adhesion molecules during gastrulation.

Interestingly, it has also been shown that Integrin signaling is required for the localization of Dsh to the plasma membrane, and that this localization coincides with the activation of the non-canonical Wnt signaling pathway (Marsden and DeSimone, 2001; see also section II.2.C.a). This observation suggests that Integrin and the non-canonical Wnt signaling pathways cooperate in regulating CE movements during gastrulation.

However, the molecular mechanisms by which Cadherin and/or Integrin adhesive activity are controlled remain to be elucidated. One possible explanation is that the dynamic turnover of adhesion molecules at the plasma membrane, e.g. through regulated endocytosis and/or secretion, controls the stability and availability of these molecules and thereby determines the rate by which adhesion is built up and disassembled between cells and their substrate. Such a model is also supported by recent findings showing that Clathrin-mediated endocytosis of C-cadherins is needed for proper CE movements in *Xenopus* (Jarrett et al., 2002). Alternative possibilities for regulating the activity of adhesion molecules include phosphorylation, ubiquitination followed by degradation and the binding of regulatory proteins such as certain Catenins that link Cadherins to the cytoskeleton.

5.3.2
Cellular Polarity

Both epiblast and hypoblast cells show very dynamic changes in their individual cell morphology during the course of gastrulation. However, in contrast to epithelial cells that are polarized along their apical-basolateral axis, no stable polarization can be seen in zebrafish gastrulating cells. Instead, hypoblast cells exhibit some transient elongation along the direction of their individual cell movement. But are these hypoblast cells clearly polarized on a molecular level during gastrulation?

Different components of the non-canonical Wnt signaling pathway have been implicated in the regulation of cell polarization at different stages of *Xenopus* and zebrafish gastrulation (Jessen et al., 2002; Marlow et al., 2002; Topczewski et al., 2001; Ulrich et al., 2003; Wallingford et al., 2000). However, none of these components show a convincing polarized distribution in gastrulating cells nor have they been shown to actively polarize those cells. Furthermore, PI3Ks are required for cellular process formation and cell polarization in mesendodermal (prechordal plate) progenitors at the onset of zebrafish gastrulation by localizing PKB to the leading edge of those cells (Montero et al., 2003). However, the involvement and intracellular localization of potential downstream and upstream interaction partners of PKB, such as P21 Associated Kinase (PAK) and the small GTPases RhoA, Rac and Cdc42, remain to be determined.

Likely candidate processes that lead to the asymmetric distribution of intracellular components are the directed endocytosis, secretion and recycling of such molecules. Indeed, experiments in which the endocytic and recycling pathways are blocked in prechordal plate cell progenitors indicate that those pathways are required for cell polarization and process formation (I.C. and C.P.H., unpublished results). However, whether they function in a permissive or instructive manner to polarize these cells is still unknown.

Interestingly, although when mesendodermal cell polarization is defective, e.g. when PI3K function is blocked in those cells, the overall direction of mesendodermal cell migration appears to be largely unaffected. This raises the question of what mechanisms regulate directed cell migration of mesendodermal cells at the onset of gastrulation. One likely explanation is that mesendodermal progenitors are pushed towards the animal pole of the gastrula by newly internalizing cells at the germ ring margin and that it is the rate of ingression around the germ ring that determines the direction in which these cells are moving. In this scenario, the cell autonomous movement of internalized mesendodermal progenitors would be undirected and the overall movement direction of these cells would be determined cell-non-autonomously through interactions with surrounding cells. This idea is also supported by studies showing that single wild-type mesendodermal cells transplanted into the germ ring margin of *mzoep* mutants that otherwise lack any mesendodermal cell ingression could individually internalize but do not move anteriorly towards the animal pole. This suggests that this movement requires the simultaneous ingression of other mesendodermal cells (Carmany-Rampey and Schier, 2001).

Independent of whether cell polarization and directed cell migration of gastrulating cells is cell-autonomously or non-autonomously regulated, the molecular nature of this regulation is still unclear. Experiments in both zebrafish and *Xenopus* show that PDGF can function upstream of PI3K to regulate mesendodermal cell movement and morphology at the onset of gastrulation (Montero et al., 2003; Symes and Mercola, 1996). Furthermore, the finding that the introduction of beads coated with PDGF-A in cultured mesendodermal cells causes cells next to the bead to extend processes towards the bead indicates that PDGF has the potential to polarize those cells (Montero et al., 2003). However, PDGF-A is ubiquitously expressed during zebrafish gastrulation, which argues against the idea that local differences in the concentration of external PDGF directly polarize mesendodermal cells (Liu et al., 2002).

Another signaling pathway, which potentially regulates mesendodermal cell polarization and directed cell migration during gastrulation is the *Ja*-nus *K*inase family of *S*ignal *T*ransducers and *A*ctivators of *T*ranscription (JAK/STAT) pathway. *Stat3* in zebrafish has been shown to be required cell-autonomously for the anterior movement of hypoblast cells and non-autonomously for the dorsal convergence of paraxial mesodermal cells (Yamashita et al., 2002). This, together with recent findings that the chemokine receptor CXCR4, a known regulator of JAK/STAT signaling (Vila-Coro et al., 1999), and its proposed ligand SDF-1 are needed for the migration of zebrafish and mouse primordial germ cells during gastrulation

(Molyneaux et al., 2003; Raz, 2003), points at the intriguing possibility that a yet unidentified chemokine ligand signals through the JAK/STAT pathway to also polarize and direct the movement of mesendodermal cells during gastrulation.

5.3.3
Cellular Adhesion

At the onset of gastrulation, internalizing hypoblast cells move anteriorly towards the animal pole of the embryo using the overlying epiblast as a substrate to migrate (Juan Antonio Montero and C.P.H., unpublished data). At the same time, the epiblast undergoes epiboly movements in the opposite direction. This means that the epiblast and hypoblast cell layers move on top of each other but in opposite directions raising the question of how mechanistically net movement between these two cell layers is generated.

One possibility is that hypoblast and epiblast cells move directly on top of each other. As mentioned above (Section 5.3.1), hypoblast and epiblast cells express differential levels of Cadherins at the onset of zebrafish gastrulation with higher levels of expression within hypoblast cells (Juan Antonio Montero and C.P.H., unpublished data). This differential distribution of adhesion molecules might both restrict the intermixing of cells between these layers and promote the movement of hypoblast cells on the epiblast surface (Fig. 5.6).

One requirement for restricting the intermixing between both cell types is that the adhesion between cells within each cell layers must be stronger than that between the two different cell layers. To promote the movement of mesendodermal cells along the epiblast interface, one would expect higher adhesion between epiblast cells than between hypoblast cells, which consequently would lead to enhanced spreading of the hypoblast cells on the epiblast surface. As cadherin expression is higher in the hypoblast than in the epiblast, spreading of the hypoblast on the epiblast appears unlikely. This indicates that other mechanisms than spreading must contribute to the extension of the hypoblast on the epiblast. One such mechanism might be the localized internalization of new mesendodermal progenitors at the germ ring margin combined with the convergence of more lateral mesendodermal progenitors towards the shield. This would then push axial mesendodermal progenitors progressively towards the animal pole.

The hypothesis that cell-cell interactions between neighboring cells are required to convert the forces produced by the cytoskeleton into actual cell motility has been best studied for the movement of border cells during *Drosophila* oogenesis (Montell, 2001). The border cells in the *Drosophila* ovary originate within the anterior extent of an epithelium of cells known as follicle cells that form the egg chamber. As border cells initiate

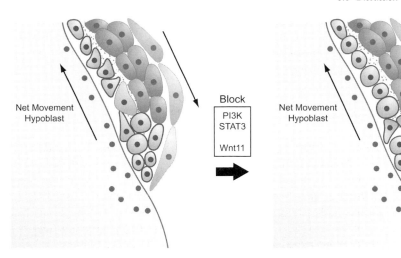

Net Movement
Hypoblast

Block
PI3K
STAT3
Wnt11

Net Movement
Hypoblast

Epiblast

Hypoblast
lost of
Polarization

Fig. 5.6 Schematic diagram illustrating the (potential) role of cell polarization and directed cell migration for germ layer formation at the onset of gastrulation. At 50% epiboly, cells begin to internalize forming the internal hypoblast cell layer (in orange), which is placed below the external epiblast cell layer (in red). Epiblast cells undergo epiboly (represented by the right arrow). Hypoblast cells are elongated, and their processes are oriented towards the overlying epiblast. Hypoblast cells might use either the overlying epiblast or the extracellular matrix (e.g. FN – small green dots) between the epiblast and hypoblast as a substrate for their migration towards the animal pole (represented by the left arrow). Several signaling pathways have been shown to be involved in cell polarization and/or cell process formation/ orientation during zebrafish gastrulation, such as the non-canonical Wnt11-, the PI3K-, and the STAT3-signaling pathways. Blocking any of these signaling pathways results in the loss of hypoblast cell polarity (note 'round' hypoblast cells in the scheme to the right) and defects in cell motility (for details see text). The enveloping cell layer is represented in blue. The green dots show the yolk syncytial layer and the yellow background represents the yolk cell.
(This figure also appears with the color plates.)

migration, they undergo a morphological change that resembles an epithelial to mesenchymal transition. However, border cells do not lose all epithelial characteristics. They remain tightly attached to each other via junctions that contain E-cadherins (Montell, 2001). Interestingly, E-cadherin expression remains higher in border cells than in the surrounding follicle cells and this expression appears to be required for normal border cell migration. This is similar to the situation in zebrafish, where prechordal plate progenitors that actively migrate towards the animal pole, exhibit higher expression levels of Cadherins than the overlying epiblast cells they move on (Juan-Antonio Montero and C.P.H.; unpublished data). It is therefore tempting to speculate that E-cadherin might have a similar role in controlling the movements of zebrafish prechordal plate cells and *Drosophila* border cells.

But what could be the cellular mechanisms by which Cadherin-mediated cell-cell adhesion regulates the movement of the epiblast and hypoblast cell layers on top of each other? Presumably, the role of Cadherins during this process is first to maintain the integrity of specific groups of cells, and second to provide sufficient traction to allow cells to migrate on their substrate. However, in order to migrate, cells expressing high levels of cadherin require an efficient turnover of adhesion complexes. In stationary epithelial cells, adherence junctions are very stable, whereas in cells that lack stable adhesions, Cadherins are recycled through the endocytic pathway at a rapid rate (Le et al., 1999). It is therefore conceivable that during zebrafish gastrulation, cells at the border between hypoblast and epiblast must very efficiently recycle their cadherin adhesion complexes in order to guarantee transient adhesion.

Another possibility is that epiblast and hypoblast cells do not move by directly contacting each other. Instead, cells at the border between the two cell layers might secrete extracellular matrix proteins such as FN, which is then being used by both cell types as a substrate on which to move. In fact, FN has been implicated in cellular movements in a variety of vertebrate embryos. In *Xenopus*, FN is required for radial cell intercalations during epiboly, the movement of mesodermal cells after involution and CE movements at later stages of gastrulation (Marsden and DeSimone, 2001; Winklbauer, 1998). The function of FN appears to be mediated by Integrin β_1 receptors, which in turn can activate C-cadherin at the plasma membrane (Marsden and DeSimone, 2003). Similar to the situation in *Xenopus*, the preliminary analysis of zebrafish maternal-zygotic *natter/fibronectin1* (*MZnat*) mutants suggests that FN also plays an important role in the regulation of gastrulation movements in zebrafish (Trinh and Stainier, 2004). Future studies will have to address common and divergent features of FN function during *Xenopus* and zebrafish gastrulation.

5.3.4
Molecular Topics: Non-canonical Wnt Signaling

In *Drosophila*, many of the genes that are required for the establishment of Planar Cell Polarity (PCP) within an epithelium, are either components of the Wingless (Wg) signal transduction pathway, such as *fz* and *dsh*, or they genetically interact with this pathway, such as *flamingo* (*fmi*) and *van gogh/strabismus* (*vang/stbm*) (Mlodzik, 2002; Strutt, 2002). The Wg signaling pathway establishing PCP in *Drosophila* does not include downstream members of the canonical Wnt signaling pathway such as Axin, GSK-3 and β-catenin but rather signals via small GTPases including RhoA, Rac and Cdc42 and Jun N-terminal kinases (JNKs), known regulators of the cytoskeleton (Adler, 2002; Axelrod and McMeil, 2002; Mlodzik, 1999).

It has been shown that zebrafish and *Xenopus* homologues of many components of the *Drosophila* non-canonical Wnt signaling show defects in CE movements, providing evidence that a similar non-canonical Wnt/PCP pathway might operate in vertebrates to control cellular movements and polarization during gastrulation (Tada et al., 2002; Wallingford et al., 2002). This leads to the intriguing hypothesis that the molecular and cellular mechanisms controlling (planar) cell polarization are conserved between *Drosophila* and vertebrates. Although this is an attractive hypothesis, there is still no compelling experimental evidence supporting it.

A series of studies in zebrafish and *Xenopus* addressed the question if cell polarization during gastrulation is controlled by components of the non-canonical Wnt signaling pathway and if the cell polarization defect might account for the cell movement phenotype observed. It has been noted that the elongation of mesendodermal cells along the mediolateral axis of the gastrula is dependent on non-canonical Wnt signaling and that a reduction in the degree of mediolateral elongation of these cells is accompanied by defects in the velocity and direction of individual cell migration. Although these results indicate that non-canonical Wnt signaling is required for mesendodermal cell elongation, evidence for a direct role of the pathway in this process is still lacking. Furthermore, a mechanistic link between cell polarization and directed cell migration has not yet been demonstrated.

In *Drosophila*, the polarization of epithelial cells is dependent on the polarized distribution of non-canonical Wnt/PCP components. In contrast, so far no satisfying evidence exists for asymmetric distribution of Wnt/PCP components in cells during vertebrate gastrulation. In some studies, Dsh has been shown to localize at the lateral ends of strongly elongated mesendodermal cells. However, it is not clear wether this accumulation is due to asymmetrical localization of Dsh or a mere accumulation of plasma membrane at the ends of those cells (Kinoshita et al., 2003; Wallingford et al., 2000).

Similarly, it is still unclear if cell elongation is a valid representation of cell polarization and if non-canonical Wnt/PCP signaling controls cellular movements through its effect on cellular elongation/polarization. In the *Drosophila* wing epithelium, the proximo-distal polarization of epithelial cells becomes apparent by the distal outgrowth of a wing hair in those cells. In contrast, although gastrulating mesendodermal cells are elongated along their mediolateral extent, no clear cellular and/or molecular difference between the medial and lateral ends has been observed, suggesting that these cells are elongated but not necessarily polarized. Furthermore, the elongation defect in mesendodermal cells of mutant

KIMMEL, C. B., KANE, D. A., and Ho, R. K. (1991). Lineage specification during early embryonic development of the zebrafish. In *"Cell-Cell Interaction in Early Development"*. GERHART, J. (ed). New York: Wiley-Liss, Inc., 203–225.

KIMMEL, C. B., WARGA, R. M., and SCHILLING, T. F. (1990b). Origin and organization of the zebrafish fate map. *Development* 108, 581–594.

KINOSHITA, N., IIOKA, H., MIYAKOSHI, A., and UENO, N. (2003). PKC delta is essential for Dishevelled function in a noncanonical Wnt pathway that regulates *Xenopus* convergent extension movements. *Genes Dev.* 17, 1663–1676.

KUHL, M., SHELDAHL, L. C., PARK, M., MILLER, J. R., and MOON, R. T. (2000). The Wnt/Ca2+ pathway: a new vertebrate Wnt signaling pathway takes shape. *Trends Genet.* 16, 279–283.

LE, T. L., YAP, A. S., and STOW, J. L. (1999). Recycling of E-cadherin: a potential mechanism for regulating cadherin dynamics. *J. Cell Biol.* 146, 219–232.

LEE, C. H., and GUMBINER, B. M. (1995). Disruption of gastrulation movements in Xenopus by a dominant-negative mutant for C-cadherin. *Dev Biol.* 171, 363–373.

LEE, J., RAY, R., and CHIEN, C. (2001). Cloning and expression of three zebrafish roundabout homologs suggest roles in axon guidance and cell migration. *Dev Dyn* 221, 216–230.

LELE, Z., FOLCHERT, A., CONCHA, M., RAUCH, G. J., GEISLER, R., ROSA, F. M., and WILSON, S. W. (2002). Parachute/n-cadherin is required for morphogenesis and maintained integrity of the zebrafish neural tube. *Development* 129, 3281–3294.

LEVINE, E., LEE, C. H., KINTNER, C., and GUMBINER, B. M. (1994). Selective disruption of E-cadherin function in early *Xenopus* embryos by a dominant negative mutant. *Development* 120, 901–909.

LI, H. S., CHEN, J. H., WU, W., FAGALY, T., ZHOU, L., YUAN, W., DUPUIS, S., JIANG, Z. H., NASH, W., GICK, C., ORNITZ, D. M., WU, J. Y., and RAO, Y. (1999). Vertebrate slit, a secreted ligand for the transmembrane protein roundabout, is a repellent for olfactory bulb axons. *Cell* 96, 807–818.

LIU, L., KORZH, V., BALASUBRAMANIYAN, N. V., EKKER, M., and GE, R. (2002). Platelet-derived growth factor A (pdgf-a) expression during zebrafish embryonic development. *Dev. Genes Evol.* 212, 298–301.

MARLOW, F., TOPCZEWSKI, J., SEPICH, D. S., and SOLNICA-KREZEL, L. (2002). Zebrafish Rho kinase 2 acts downstream of Wnt11 to mediate cell polarity and effective convergence and extension movements. *Curr. Biol.* 12, 876–884.

MARSDEN, M., and DESIMONE, D. W. (2001). Regulation of cell polarity, radial intercalation and epiboly in Xenopus: novel roles for integrin and fibronectin. *Development* 128, 3635–3647.

MARSDEN, M., and DESIMONE, D. W. (2003). integrin-EMC interactions regulate cadherin-dependent cell adhesion and are required for convergent extension in Xenopus. *Curr Biol.* 13, 1182–1191.

MELLITZER, G., XU, Q., and WILKINSON, D. G. (1999). Eph receptors and ephrins restrict cell intermingling and communication. *Nature* 400, 77–81.

MELLITZER, G., XU, Q., and WILKINSON, D. G. (2000). Control of cell behaviour by signalling through Eph receptors and ephrins. *Curr Opin Neurobiol.* 10, 400–408.

MILLER, J.R. (2003). The Wnts. *Genome Biol.* 3(1):REVIEWS3001.

MLODZIK, M. (1999). Planar polarity in the *Drosophila* eye: a multifaceted view of signaling specificity and cross-talk. *EMBO J.* 18, 6873–6879.

MLODZIK, M. (2002). Planar cell polarization: do the same mechanisms regulate *Drosophila* tissue polarity and vertebrate gastrulation? *Trends Genet.* 18, 564–571.

MOLYNEAUX, K.A., ZINSZNER, H., KUNWAR, P.S., SCHAIBLE, K., STEBLER, J., SUNSHINE, M.J., W., O.B., RAZ, E., LITTMAN, D., WYLIE, C., and LEHMANN, R. (2003). The chemokine SDF1/CXCL12 and its receptor CXCR4 regulate mouse germ cell migration and survival. *Development* 130, 4279–4286.

MONTELL, D.J. (2001). Command and control: regulatory pathways controlling invasive behavior of the border cells. *Mech Dev.* 105, 19–25.

MONTERO, J.A., KILIAN, B., CHAN, J., BAYLISS, P.E., and HEISENBERG, C.P. (2003). Phosphoinositide 3-kinase is required for process outgrowth and cell polarization of gastrulating mesendodermal cells. *Curr. Biol.* 13(15), 1279–1289.

NELSEN, O.E. (1953). *Comparative Embryology of the Vertebrates.* New York: McGraw-Hill Book Inc.

OATES, A.C., LACKMANN, M., POWER, M.A., BRENNAN, C., DOWN, L.M., DO, C., EVANS, B., HOLDER, N., and BOYD, A.W. (1999). An early development role for eph-ephrin interaction during vertebrate gastrulation. *Mech Dev.* 83, 77–94.

PAPAN, C., and CAMPOS-ORTEGA, J. (1994). On the formation of the neural keel and neural tube in the zebrafish *Danio* (*Brachydanio*). *Rouxs Arch. Dev. Biol.* 203, 178–186.

RAMOS, J.W., WHITTAKER, C.A., and DeSIMONE, D.W. (1996). integrin-dependent adhesive activity is spatially controlled by inductive signals at gastrulation. *Development* 122, 2873–2883.

RAZ, E. (2003). Primordial germ-cell development: the zebrafish perspective. *Nat Rev Genetic.* 4, 690–700.

REBAGLIATI, M.R., TOYAMA, R., HAFFTER, P., and DAWID, I.B. (1998). *cyclops* encodes a *nodal*-related factor involved in midline signaling. *Proc. Natl. Acad. Sci. USA* 18, 9932–9937.

ROTHBERG, J.M., HARTLEY, D.A., WALTHER, Z., and ARTAVANIS-TSAKONAS, S. (1998). slit: and EGF-homologous locus of D. melanogaster involved in the development of the embryonic central nervous system. *Cell* 55, 1047–1059.

SAMPATH, K., RUBINSTEIN, A.I., CHENG, A.M., LIANG, J.O., FEKANY, K., SOLNICA-KREZEL, L., KORZH, V., HALPERN, M.E., and WRIGHT, C.V. (1998). Induction of the zebrafish ventral brain and floorplate requires *cylcops/nodal* signaling. *Nature* 10, 185–189.

SANG-YEOB, Y., LITTLE, M.H., YAMADA, T., MIYASHITA, T., HALLORAN, M.C., KUWADA, J.Y., HUH, T., and OKAMOTO, H. (2001). Overexpression of a Slit homologue impairs convergent extension of the mesoderm and causes cyclopia in embryonic zebrafish. *Dev Biol.* 230, 1–17.

SCHIER, A.F., and TALBOT, W.S. (2001). Nodal signaling and the zebrafish organizer. *Int J Dev Biol.* 45, 289–297.

SCHOENWOLF, G.C., and SMITH, J.L. (2000). Gastrulation and early mesodermal patterning in vertebrates. *Methods Mol Biol.* 135, 113–125.

SCHWARZ-ROMOND, T., ASBRAND, C., BAKKERS, J., KUHL, M., SCHAEFFER, H.-J., HUELSKEN, J., BEHRENS, J., HAMMERSCHMIDT, M., and BIRCHMEIER, W. (2002). The ankyrin repeat protein Diversin recruits Casein Kinase Ie to the β-catenin degradation complex and acts in both canonical Wnt and Wnt/JNK signaling. *Genes Dev.* 16, 2073–2084.

SHIH, J., and KELLER, R. E. (1992). Cell motility driving mediolateral intercalation in explants of *Xenopus laevis*. *Development* 116, 901–914.

SOLNICA-KREZEL, L., and DRIEVER, W. (1994). Microtubule arrays of the zebrafish yolk cell: organization and funciton during epiboly. *Development* 120, 2443–2455.

SOLNICA-KREZEL, L., STEMPLE, D. L., MOUNTCASTLE-SHAH, E., RANGINI, Z., NEUHAUSS, S. C., MALICKI, J., SCHIER, A. F., STAINIER, D. Y., ZWARTKRUIS, F., ABDELILAH, S., and DRIEVER, W. (1996). Mutations affecting cell fates and cellular rearrangements during gastrulation in zebrafish. *Development* 123, 67–80.

STRÄHLE, U., and JESUTHASAN, S. (1993). Ultraviolet irradiation impairs epiboly in zebrafish embryos: evidence for a microtubule-dependent mechanism of epiboly. *Development* 119, 909–919.

STRUTT, D. I. (2002). The asymmetric subcellular localization of components of the planar polarity pathway. *Cell Dev Biol.* 13, 225–231.

SUMANAS, S., and EKKER, S. C. (2001). Xenopus frizzled-7 morphant displays defects in dorsoventral patterning and convergent extension movements during gastrulation. *Genesis* 30, 119–122.

SYMES, K., and MERCOLA, M. (1996). Embryonic mesoderm cells spreadd in response to platelet-derived growth factor and signaling by phosphatidylinositol 3-kinase. *Proc. Natl. Acad. Sci. USA* 93, 9641–9644.

TADA, M., CONCHA, M., and HEISENBERG, C. P. (2002). Non-canonical Wnt signalling and regulation of gastrulation movements. *Semin Cell Dev Biol.* 13, 251–260.

TADA, M., and SMITH, J. C. (2000). *Xwnt11* is a target of Xenopus Brachyury: regulation of gastrulation movements via Dishevelled, but not through the canonical Wnt pathway. *Development* 127, 2227–2238.

TEPASS, U., TROUNG, K., GODT, D., IKURA, M., and PEIFER, M. (2000). Cadherins in embryonic and neural morphogenesis. *Nat Rev Mol Cell Biol.* 1, 91–100.

TOPCZEWSKI, J., SEPICH, D. S., MYERS, D. C., WALKER, C., AMORES, A., LELE, Z., HAMMERSCHMIDT, M., POSTLETHWAIT, J., and SOLNICA-KREZEL, L. (2001). The zebrafish glypican knypek controls cell polarity during gastrulation movements of convergent extension. *Dev Cell.* 1, 251–264.

TRINH, L. A., and STAINIER, D. Y. (2004). Fibronectin regulates epithelial organization during myocardial migration in zebrafish. *Dev Cell.* 6, 371–382.

TRINKAUS, J. P. (1951). A study of the mechanism of epiboly in the egg of *Fundulus heteroclitus*. *J Exp Zool* 118, 269–320.

ULRICH, F., CONCHA, M. L., HEID, P. J., VOSS, E., WITZEL, S., ROEHL, H., TADA, M., WILSON, S. W., ADAMS, R. J., SOLL, D. R., and HEISENBERG, C. P. (2003). Slb/Wnt11 controls hypoblast cell migration and morphogenesis at the onset of zebrafish gastrulation. *Development* 130, 5375–5384.

UNGAR, A.R., KELLY, G.M., and MOON, R.T. (1995). Wnt4 affects morphogenesis when misexpressed in the zebrafish embryo. *Mech Dev.* 52, 153–164.

VILA-CORO, A.J., RODRIGUEZ-FRADE, J.M., MARTIN DE ANA, A., MORENO-ORTIZ, M.C., MARTINEZ, A.C., and MELLADO, M. (1999). The chemokine SDF-1alpha triggers CXCR4 receptor dimerization and activates the JAK/STAT pathway. *FASEB J.* 13, 1699–1710.

WACKER, S., GRIMM, K., JOOS, T., and WINKLBAUER, R. (2000). Development and control of tissue separation at gastrulation in *Xenopus. Dev Biol.* 224, 428–439.

WALLINGFORD, J.B., FRASER, S.E., and HARLAND, R.M. (2002). Convergent extension: the molecular control of polarized cell movement during embryonic development. *Dev Cell.* 2, 695–706.

WALLINGFORD, J.B., ROWNING, B.A., VOGELI, K.M., ROTHBACHER, U., FRASER, S.E., and HARLAND, R.M. (2000). Dishevelled controls cell polarity during Xenopus gastrulation. *Nature* 405, 81–85.

WARGA, R.M., and KIMMEL, C.B. (1990). Cell movements during epiboly and gastrulation in zebrafish. *Development* 108, 569–580.

WILKINSON, D.G. (2003). How atraction turns to repulsion. *Nat Cell Biol.* 5, 851–853.

WILSON, E.T., CRETEKOS, C.J., and HELDE, K.A. (1995). Cell mixing during early epiboly in the zebrafish embryo. *Dev Genet.* 17, 6–15.

WILSON, E.T., HELDE, K.A., and GRUNWALD, D.J. (1993). Something's fishing here-rethinking cell movement and cell fate in the zebrafish embryo. *TIG* 9, 348–352.

WINKLBAUER, R. (1998). Conditions for fibronectin fibril formation in the early Xenopus embryo. *Dev Dyn* 212, 335–345.

WINKLBAUER, R., MEDINA, A., SWAIN, R.K., and STEINBEISSER, H. (2001). Frizzled-7 signalling controls tissue separation during *Xenopus* gastrulation. *Nature* 413, 856–860.

WINNING, R.S., SCALES, J.B., and SARGENT, T.D. (1996). Disruption of cell adhesion in Xenopus embryos by Pagliaccio, an Eph-class receptor tyrosine kinase. *Dev Biol.* 179, 309–319.

XU, Q., MELLITZER, G., ROBINSON, V., and WILKINSON, D.G. (1999). In vivo cell sorting in complementary segmental domains mediated by Eph receptors and ephrins. *Nature* 399, 267–271.

YAMAMOTO, A., AMACHER, S.L., KIM, S.H., GEISSERT, D., KIMMEL, C.B., and DE ROBERTIS, E. (1998). Zebrafish paraxial protocadherin is a downstream target of spadetail involved in morphogenesis of gastrula mesoderm. *Development* 125, 3389–3397.

YAMASHITA, S., MIYAGI, C., CARMANY-RAMPEY, A., SHIMIZU, T., FUJII, R., SCHIER, A.F., and HIRANO, T. (2002). Stat3 controls cell movements during zebrafish gastrulation. *Dev Cell.* 2, 363–375.

ZHONG, Y., BRIEHER, W.M., and GUMBINER, B.M. (1999). Analysis of C-cadherind regulation during tissue morphogenesis with an activating antibody. *J. Cell Biol.* 144, 351–359.

ZIMMER, M., PALMER, A., KOHLER, J., and KLEIN, R. (2003). EphB-ephrinB bi-directional endocytosis terminates adhesion allowing contact mediated repulsion. *Nat Cell Biol.* 5, 869–878.

6

Aspects of the Origin and Function of Cell Mediolateral Intercalation Behavior (MIB) During Convergent Extension

Ray Keller

6.1
Introduction

In most if not all cases, convergent extension of tissues during embryonic development involves an oriented intercalation of cells, which produces a tissue that is narrower along one axis (convergence) and longer along the perpendicular axis (extension) (Fig. 6.1 A–D). Convergences of this type usually also produce a certain amount of thickening of the tissue because not all the convergence goes into length, but some is transformed into increased thickness (Keller, 2002; Keller et al., 2000) (Fig. 6.1 A, B). In an active, force-producing convergent extension, one in which the resident cells are generating the forces for tissue deformation, convergence can be viewed as the active process, with the extension and thickening being the outputs, or results, of cell repacking in a system in which cell volume is conserved (Keller et al., 2003). Volume may not be conserved in other systems, such as the amniotes, for example, which show cell division and growth during the gastrula stage. In these systems, oriented cell division and directed growth may occur (see Keller et al., 2003).

In the amphibian, the dorsal axial and paraxial mesodermal cells undergo a stereotyped change in polarity and shape prior to mediolateral intercalation. The cells are initially pleiomorphic and show protrusive activity in all directions in the plane of the tissue (Fig. 6.1 C). They then form large filo-lamelliform protrusions on their medial and lateral ends, both between adjacent cells (Davidson et al., 2004; Shih and Keller, 1992 a), and on the underlying and overlying interfaces with endoderm and neural ectoderm, respectively, interfaces which contain the extracellular matrix molecule, fibronectin (Davidson et al., 2004; Marsden and DeSimone, 2001) (Fig. 6.1 D). Probably as a result of this oriented protrusive activity, these cells then become elongated mediolaterally and flattened in their anterior–posterior dimension. They then intercalate between one another along the mediolateral axis to form a longer, narrower array (Shih and Keller, 1992 a, b) (Fig. 6.1 D). Scanning electron microscopy

Cell Migration. Edited by Doris Wedlich
Copyright © 2005 WILEY-VCH Verlag GmbH & Co. KGaA, Weinheim
ISBN: 3-527-30587-4

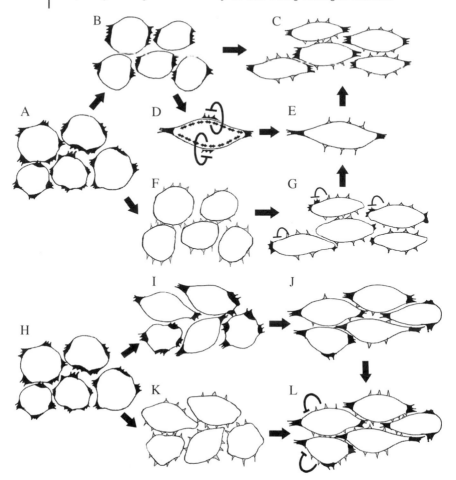

Fig. 6.2 Diagrams illustrate possible events in the emergence of MIB and the elongated, bipolar parallel arrays of cells characterizing intercalation in the mesoderm of the amphibian. The dark protrusions indicate the filo-lamelliform protrusions that are eventually localized at the medial and lateral ends of the cells, and the grey protrusions represent the filiform contacts found at the elongated anterior and posterior sides of the cells. See the text for an explanation.

axis, prior to pulling the cells between one another (Fig. 6.2 B, C). In this scenario, the elongation and alignment along the mediolateral axis would result directly from the initial polarized expression of filo-lamelliform protrusions medially and laterally. Tangential tension along the elongated sides of the cells would be expected to inhibit filo-lamelliform protrusive activity there (Kolega, 1986) (Fig. 6.2 D), and thus constitute a mechanical feedback inhibition that would reinforce the primary polarity. The small filiform protrusions on these elongated anterior and posterior sides could be a default state, occurring in absence of the primary filo-lamelliform protrusions, or they could be specified as a separate polarization event (Fig. 6.2 E).

Alternatively, the primary polarization event could be the specification of the small filiform protrusive activity at the anterior and posterior sides (Fig. 6.2 A, F). These protrusions could potentially set up a specialized adhesion or mode of contact interaction that would maximize the contact along these anterior and posterior surfaces, thus leading to the flattening of the cells on one another and thus to mediolateral elongation (Fig. 6.2 F, G). The presence of this mode of protrusive activity, or adhesion, could inhibit the tractive filo-lamelliform protrusive activity on the elongated sides, thereby confining this activity to the medial and lateral ends (Fig. 6.2 C, E). Alternatively, the filo-lamelliform activity could be specified as a separate aspect of polarization. In this scenario, restriction of this protrusive activity to the mediolateral zones would be a secondary process, and result in specification of mediolateral traction as a secondary event, following the primary polarization in the anterior–posterior axis.

In an alternate scenario, cell polarization could be independent of orientation. The cells could become bipolar, either by forming the large filo-lamelliform protrusions at the narrow ends as the primary event in polarization (Fig. 6.2 H, I), or by formation of the small contacts at the elongated sides as the primary event (Fig. 6.2 H–K). The polarized but unoriented cells would then become aligned, either by mutual traction (Fig. 6.2 I–J) or by selective apposition of the elongated sides (Fig. 6.2 K, L). Thus far the dyamnics of cell shape change and intercalation observed in the amphibian mesoderm favors the scenario in which the polarization directly accounts for the mediolateral orientation and alignment (Fig. 6.2 A–C, E) rather than the primary polarization of the cells occurring first without orientation, followed by a secondary orientation of the polarized cells (Fig. 6.2 H–J, L) (Shih and Keller, 1992 b).

The pathways leading to the various aspects of polarization, and the relationships between them, are not understood although evidence offering new insights is emerging (Tahinci and Symes, 2003). Therefore it is not known whether the large lamelliform protrusions or the small filiform protrusions are the initial event in polarization, whether both are involved simultaneously, and result from the same primary polarization process, or whether one is specified and the other is a permissive or default state.

6.3
Is the Elongation of the Cells due to External Traction or Polarized Cell Adhesion, or to Internal Cytoskeletal Events?

The cell traction/cell substrate model postulates that the mediolateral elongation is due to traction exerted in the medial and lateral directions, thus stretching the cell along this axis (Fig. 6.2 B, C), but there is no direct evidence that this occurs. The cells could also become polarized in their adhesions such that the anterior and posterior sides have a high affinity for one another, and flatten on one another (Fig. 6.2 F, G). Or perhaps both these mechanisms could contribute to elongation. Thirdly, the elongation may not be due to external cell interactions (traction) at all, but instead it might be generated internally by the cytoskeleton, independent of traction on a substrate. The fact that these cells lose their characteristic shape when isolated as individual cells does not rule out an internally-generated shape. Contact cues may signal the cytoskeletal machinery to polarize and elongate the cell by internally generated forces, independent of traction, as distinguished from using a substrate for polarized traction. Such a mechanism would be dependent on cell contact-encoded information, or signaling, other than that embodied in adhesion and traction. Currently, there is no evidence in favor of such a substrate-independent mechanism, nor is there evidence ruling it out.

6.4
Possible Mechanisms of Secondary Alignment and Orientation

Whether by external traction, or by internally generated forces, the cells could become bipolar, but un-oriented (Fig. 6.2 I, K), and then secondarily align parallel to one another and parallel to the mediolateral axis of the embryo (Fig. 6.2 J, L). Such a secondary alignment could occur as a result of cell interactions in which their elongate shape, polarized traction, and contact behavior would play roles. Such interactions can not only align weakly elongated cells, but can also reinforce the elongated morphology of the cells in a positive feed back between shape, motility, and initial elongation. Weakly polarized human lung fibroblasts in a monolayer culture become bipolar and strongly polarized as they become confluent (Elsdale, 1968, 1973; Elsdale and Wasoff, 1976). These cells tend to adhere along their sides and tend move along one another's surfaces, even before they become crowded, and on crowding, they form highly organized bipolar arrays in which cells encounter one another at lower angles or move parallel to one another. At lower densities they tend to move at high angles with respect to one another and show a higher frequency of collisions (high angle encounters). Thus a strongly bipolar and morphologically elongated cell, very similar to the intercalating mesodermal cells, emerges from these parallel contact interactions (Elsdale and Wasoff, 1976). Contact interactions of elongated objects results in their alignment, similar to the alignment resulting from shaking a number of pencils in a box, for example, and similar density-dependent interactions may align, or contribute to alignment of intercalating cells. For example, if cells elongate due to intrinsic forces of the cytoskeleton, and that process is not globally constrained, the cells would elongate in all directions. But as the density of elongated cells increases, the cells would have a tendency to align parallel to one another, due to these contact interactions (Fig. 6.2 H–L). However, if experimental manipulations result in isolated cells elongating, they could only interact with unelongated cells, and they might not align parallel to the mediolateral axis, or one another, thus hiding the density dependent relationship between elongation and alignment, and leading to the conclusion that they are independently specified processes.

Likewise, tractional elongation of cells might normally be un-oriented initially, when there are few cells generating traction, but as the density of elongating cells increases, the cells may have a tendency to pull one another into alignment (Fig. 6.2 I, J). It is unclear how local alignment fields could be integrated with other local alignment fields that do not necessarily match, to finally generate a global, mediolaterally organized pattern (see Elsdale and Wasoff, 1976). In the case of MIB, this mechanism may function in the local context to reinforce alignment, rather than in a global sense, the latter being controlled by directional signals (see Section 6.5). The onset of MIB suggests that elongation and alignment are closely coupled to the process of polarization, as the cells appear to elongate in parallel arrays that are mediolaterally aligned at the outset (Shih and Keller, 1992b), but the local contact interactions between already elongated cells and ones about to do so, may play a role.

6.5
Global Patterning: the Progressive Expression of MIB

Explants of the involuting marginal zone of *Xenopus* show that in the presumptive deep mesoderm, MIB occurs in an anterior-to-posterior (black arrows, Fig. 6.3 A, B) and a lateral-to-medial progression (white arrows, Fig. 6.3 A, B) (Shih and Keller, 1992b). The pattern of the progression is easier to visualize if mapped on to the marginal zone of an explant that does not extend (Fig. 6.3 B, C). Normally, the progressive pattern of expression results in a zone of convergence proceeding posteriorly, which results in convergent extension of the explant (Fig. 6.3 D). In the embryo, the advance of the posterior progression of MIB lies at lip of the blastopore and results in highly anisotropic (dorsally biased) constriction of the blastopore in the embryo (Fig. 6.3 E).

In the posteriorly progressing zone of MIB onset, elongated and aligned cells lie adjacent to ones that have not yet responded to the signals organizing MIB, or are in the process of responding to these signals

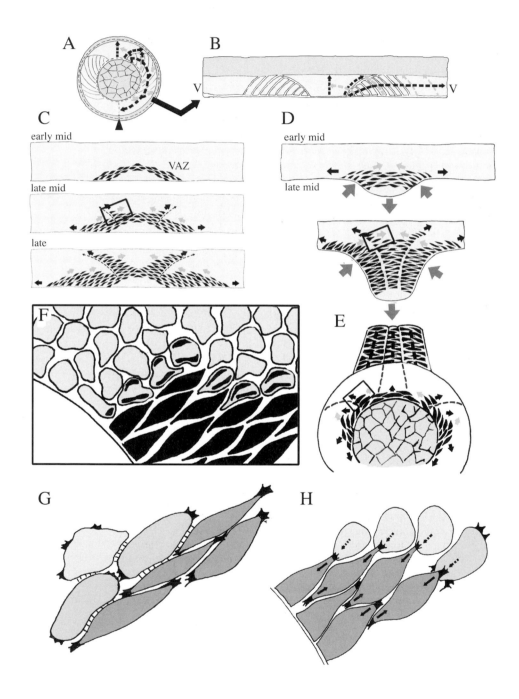

(Fig. 6.3 F). Recent, very exciting evidence suggests that a signal (or signals) proceeding from anterior to posterior (black arrows, Fig. 6.3) sets up local differences in anterior–posterior positional values, and perhaps local differences in adhesion or contact behavior, which then mediate MIB (Ninomiya et al., 2004). The lateral-to-medial progression may result from different rates of progress of this anterior-to-posterior signal in lateral and medial regions (Fig. 6.3 F). In these experiments, induction of animal cap explants with different concentrations of activin results in differences in expression of the genes chordin and brachyury, which are normally expressed in a graded fashion along the anterior–posterior axis, and convergent extension by mediolateral intercalation occurs only in regions of apposition of tissues with differing levels of induction (Ninomiya et al., 2004). This study is very important as it shows that variation in anterior–posterior positional values is essential for inducing convergent extension by cell intercalation. Interfering with *Xenopus* Dishevelled function does not block establishment of anterior–posterior polarity but does block convergent extension, showing that Dishevelled, and the PCP pathway probably acts downstream or in parallel to the primary anterior–posterior polarity pathway in organizing polarized cell behavior and convergent extension. Another important finding is the fact that in *Drosophila* germ band convergent extension, the epithelial cells are likewise dependent on anterior–posterior patterning, and they localize nonmuscle myosin II to the anterior and posterior sides of the cells and Bazooka (Par3) to the medial and lateral ends (Zallen and Wieschaus, 2004). Moreover, Frizzled and Dishevelled, two components of the PCP pathway, are not essential for polarization of these cells and for germband extension (Zallen and Wieschaus, 2004). These findings indicate that in flies and frogs anterior–posterior patterning is the primary process in the polarizing the cells prior to mediolateral intercalation, and that patterning pathways other than the PCP pathway play a role, most probably the

◄──────────────────────────────────────

Fig. 6.3 A diagram of the vegetal pole of an early gastrula, with dorsal uppermost (A), and the isolated marginal zone in an explant form (B) shows the direction of the presumptive anterior–posterior axis (black arrows) and the presumptive lateral to medial axis (grey arrows). The presumptive notochordal (n) and somitic (s) mesoderm is indicated. MIB proceeds along these axes from anterior to posterior (black arrows) and from lateral to medial (grey arrows) in explants that are mechanically restrained in their movements (C), in ones that are allowed to converge and extend (D), and in the embryo (E). The local context of the posterior and medial progress of MIB is shown (F). Local anterior–posterior interactions reinforcing cell polarization could include specialized contact-mediated alignment behavior on anterior and posterior surfaces of cells that have undergone MIB (dark grey) and those in the process of responding (light grey) (G). Local mediolateral interactions could include tension mediated polarization of cells as the polarized protrusive activity of cells undergoing MIB (dark grey) exert traction and tension on those that are in the process (light grey) (H).

primary role, in the cell and tissue polarization underlying convergent extension. Although it is now clear that the anterior–posterior axis is the important signaling dimension in organizing cell intercalation and convergent extension in both the frog and the fly, there are different gaps in our knowledge of the two; in the case of the frog, the polarized behavior of the cells (MIB) has been described, but no corresponding polarized localization of molecular components of polarization has been described, and in the fly, molecular localizations have been described but no polarized cell behavior has been described. It is important to note that the process of polarizing, and then orienting the polarized cells with regard to the tissue axes, makes the underlying pathway more appropriately named a planar tissue polarity (PTP) pathway than a planar cell polarity (PCP) pathway (Adler, 2002).

6.6
Local Contact-mediated Processes of Polarization

In time lapse recordings, the progression of the overt signs of polarization, such as mediolateral elongation and alignment, and the associated bipolar protrusive activity, proceed posteriorly cell by cell, such that cells that express overt polarization lie next to those that have not (Fig. 6.3 F). It is in this context that such factors as differences in adhesion, local contact behavior, and tension-mediated responses could occur as secondary processes that would reinforce the primary polarity. For example, cells in the process of responding might develop specialized interactions or adhesion on the anterior and posterior surfaces, and these interactions could function to template the alignment of the elongating cells (Fig. 6.3 G). Likewise, the emerging tractional forces on more medial cells could put these cells under tension, which might contribute to their polarization (Fig. 6.3 H). Tension can differentially activate the rho family GTPases regulating the cytoskeleton and protrusive activity (Katsumi et al., 2002), and tension along the cantilevered sides of elongated cells inhibits protrusive activity there (Kolega, 1986).

6.7
Bipolarity Versus Monopolarity: The Importance of Being Decisive

Comparison of various forms of the "bipolar" mode of intercalation with the "monopolar" mode expressed in the neural tissue point to the balance of protrusive activity as a major element in the mechanism of intercalation. As described above, the bipolar mode of cell intercalation, first described in the notochordal and somitic mesoderm, consists of protrusive activity statistically biased towards the medial and lateral sectors of the cell, thus forming a bowtie like distribution in rose diagrams (Keller et al., 1989; Shih and Keller, 1992a). We postulated that balanced traction of these cells on one another leads to shortening of mediolateral arrays of cells by intercalation. The effective mediolateral arrays are constantly remodeled as cells intercalate, the end result being the placement of the tissue under mediolateral, internal tension. However, this mechanism requires that the cell maintain balance in traction at the two ends a large part of the time. If the cell becomes unbalanced for any length of time, with one end exerting substantially more traction than the other, then the cell will tend to move that way. Random temporal unbalance of traction, in the medial and lateral directions would be expected to result in local "migrations" of individual cells between one another, in opposite directions, which would produce much neighbor change but little remodeling of the shape of the tissue.

These expectations are supported in the observations of the bipolar behavior displayed by the deep neural cells when isolated alone, without underlying somitic and notochordal mesoderm (Elul et al., 1997). Under these conditions, the deep neural cells show a bipolar protrusive activity, similar to that seen in the mesoderm, but in the neural tissue it is considerably less effective in producing convergent extension (Elul et al., 1997). As individuals, the cells tend to make large excursions medially and laterally, presumably as the balance of tractions varies from medial to lateral, and back again, and as a result, the cells exchange neighbors frequently, but much less organized intercalation occurs, and little convergence and extension occurs compared to the balanced bipolarity of the deep mesoderm. The hypothesis emerging from these observations is that protrusive activity, and most important, the resulting traction, must be constantly balanced at the two, tractive (medial and lateral) ends of the cells for an effective "bipolar" mode of intercalation. Statistical balance over long time periods, but with transient bias in first one direction and then the other, is not sufficient, as it leads to the radical medial and lateral excursions, and promiscuous, unproductive neighbor exchange. More detailed analysis correlating protrusive activity, and traction, with cell displacement and neighbor change should be done to further explore the relationship between balanced traction and productive intercalation; however, traction is not easy to measure (see Section 6.9).

The native mode of intercalation in the neural tissue supports the idea that either the cells must either have protrusive activity that is bipolar and constantly balanced, or is constantly biased in one direction. The "bipolar" mode of intercalation is displayed only when the deep neural tissue is isolated *without* the underlying mesoderm at the late mid gastrula stage. When isolated *with* the underlying mesoderm, the notoplate develops over the notochord and the neural plate develops over the somitic mesoderm, and under these conditions the lateral neural plate cells show protrusive activity strongly biased toward the midline, often referred to in our laboratory as the "monopolar" mode (Elul and Keller, 2000). It should be kept in mind that it is not really a "monopolar" mode but a medially biased mode. The monopolar mode depends on the presence of the midline tissues of notochord and notoplate, and when these are removed and the two halves of the neural plate abutted to make a midlineless explant, the cells again become bipolar in their protrusive activity (Ezin et al., 2002). Thus the underlying somitic mesoderm is not sufficient to support the monopolar mode. An ectopic midline placed in lateral abutment to such a midlineless explant results in induction of monopolar cells throughout the midlineless component and those cells near the ectopic midline orient their "monopolarity" towards the ectopic midline, facts which suggest that the midline is the source of the "monopolarizing" and orienting signals.

6.8
Specialization of Cell Polarity and Intercalation Behavior

It is clear that there are a number of variants of the "bipolar" mode of cell intercalation. The definitive deep mesodermal bipolar mode is used alone in formation of the initial vegetal alignment zone that initiates convergent extension in the mesoderm (Lane and Keller, 1997; Shih and Keller, 1992b). However, this biopolar mode is quickly supplemented with a monopolar mode at the notochordal–somitic mesodermal boundary, as this boundary forms (Fig. 6.3C). As the bipolar cells contact the boundary, the end making contact spreads on the boundary and rarely, if ever, leaves the boundary thereafter, and is thus "captured" there; this "boundary capture" mechanism of elongation proposed as a major mechanism of tissue elongation in classic work on urodele neural/noto-plate morphogenesis (Jacobson and Gordon, 1976; Jacobson and Moury, 1995). The bipolar mode continues to function to converge the non-boundary regions of both notochordal and somitic mesoderm while the monopolar, boundary capture mode functions to pull more bipolar cells into the boundary on its notochord side; the behavior on the somitic side of the boundary has not been characterized.

In the neural tissue, the bipolar mode is likewise modified, or perhaps replaced entirely. As described above, the midline tissue, consisting of notochord, and that region of the neural plate in contact with it, the notoplate (Jacobson and Gordon, 1976), induce the "monopolar" medially biased behavior (Elul and Keller, 2000; Ezin et al., 2002); whether the notochord or notoplate alone will suffice is not known. The monopolar mode is probably the definitive neural mode of neural cell intercalation, and the "bipolar" mode may be used little or not at all. The bipolar mode may be an emergent behavior in the explants; that is, these behaviors are uncovered when signaling is disrupted in making the explant without a

midline. Since both the midline notoplate, as defined by specialized notoplate cell behavior, and the monopolar mode develop very early (in the
late gastrula stage), at best the bipolar mode might operate in the embryo for only a very short time (stage 10.5–11.5). The bipolar neural
mode of intercalation may represent an ancient mechanism that at one
time may have been used alone, as an effective, balanced bipolar mode,
similar to the current mesodermal mode. It may have been retained only
as a necessary underpinning for the definitive monopolar mode, and in
this role, it has lost its effectiveness as a stand-alone mechanism. As
such, it may only be expressed when uncovered by manipulating tissue
interactions (Elul and Keller, 2000). It may be that during evolution, definitive mechanisms of shaping tissues are often cobbled together using
many components of morphogenic machines that may once have been
stand-alone mechanisms, but are no longer, and that as regulatory cir-
cuits and tissue interactions are manipulated, these atavistic elements
will reveal themselves.

6.9
Traction is the Key, But it Has Not been Measured

The key element in evaluating these theories of how local cell behavior
translates into macroscopic tissue distortion is traction, and traction is
something that has not been measured in this context. A local "reporter"
is needed to determine whether, in fact, the bipolar and monopolar protrusive activities actually produce traction, and if so, on what. Protrusive
activity is often an indicator of traction on substrates (Harris et al., 1980),
but not necessarily all protrusions exert traction, not to mention that the
magnitude and direction of tractional forces are important. Moreover, on
what is traction being exerted? Cells could either exert traction directly
on one another's surfaces (the original cell substrate/cell traction model),
or they could exert traction indirectly on one another by intervening
tethers of extracellular matrix, or they could also exert directional traction
on an external matrix in the case of the monopolar behavior. It remains
to be seen which of these, or which combination of these is operating in
a given cell intercalation mechanism, both within amphibians, and in
the many other cases of convergent extension by cell intercalation among
both epithelial and nonepithelial (mesenchymal) cells.

6.10
References

ADLER, P. (2002). Planar signaling and morphogenesis in *Drosophila*. *Dev. Cell* **2**, 5–8.

ADLER, P. N., CHARLTON, J., and PARK, W. J. (1994). The Drosophila tissue polarity gene inturned functions prior to wing hair morphogenesis in the regulation of hair polarity and number. *Genetics* **137**, 829–36.

ADLER, P. N., KRASNOW, R. E., and LIU, J. (1997). Tissue polarity points from cells that have higher Frizzled levels towards cells that have lower Frizzled levels. *Curr Biol* **7**, 940–9.

DARKEN, R. S., SCOLA, A. M., RAKEMAN, A. S., DAS, G., MLODZIK, M., and WILSON, P. A. (2002). The planar polarity gene strabismus regulates convergent extension movements in Xenopus. *EMBO J* **21**, 976–85.

DAVIDSON, L., MARSDEN, M., KELLER, R., and DeSIMONE, D. (2004). *Dev. Cell*, in preparation.

DJIANE, A., RIOU, J., UMBHAUER, M., BOUCAUT, J., and SHI, D. (2000). Role of frizzled 7 in the regulation of convergent extension movements during gastrulation in *Xenopus laevis*. *Development* **127**, 3091–3100.

ELSDALE, T. (1968). Parallel orientation of fibroblasts in vitro. *Exp. Cell Res.* **51**, 439–450.

ELSDALE, T. (1973). The generation and maintenance of parallel arrays in cultures of diploid fibroblasts. In *"Biology of Fibroblasts"* (E. KURLONEN and J. PIKKARAINEN, Eds.), pp. 41–58. London, Academic Press.

ELSDALE, T., and WASOFF, F. (1976). Fibroblast cultures and dermatoglyphics: the topology of two planar patterns. *Wilhelm Roux's Archives* **180**, 121–147.

ELUL, T., and KELLER, R. (2000). Monopolar protrusive activity: a new morphogenic cell behavior in the neural plate dependent on vertical interactions with the mesoderm in *Xenopus*. *Dev Biol* **224**, 3–19.

ELUL, T., KOEHL, M. A., and KELLER, R. (1997). Cellular mechanism underlying neural convergent extension in *Xenopus laevis* embryos. *Dev Biol* (Orlando) **191**, 243–58.

EZIN, M., SKOGLUND, P., and KELLER, R. (2002). The midline (notochord and notoplate) patterns cell motility underlying convergence and extension of the *Xenopus* neural plate. *Dev Biol*. in press.

GLICKMAN, N., KIMMEL, C., JONES, M., and ADAMS, R. (2003). Shaping the zebrafish notochord. *Development* **130**, 873–887.

GOTO, T., and KELLER, R. (2002). The planar cell polarity gene strabismus regulates convergence and extension and neural fold closure in *Xenopus*. *Dev Biol* **247**, 165–181,

HABAS, R., DAWID, I. B., and HE, X. (2003). Coactivation of Rac and Rho by Wnt/Frizzles signaling is required for vertebrate gastrulation. *Genes Dev* **17**, 295–309.

HABAS, R., KATO, Y., and HE, X. (2001). Wnt/Frizzled activation of rho regulates vertebrate gastrulation and requires a novel formin homology protein Daam1. *Cell* **107**, 843–854.

HARRIS, A., WILD, P., and STOPAK, D. (1980). Silicone rubber substrata: a new wrinkle in the study of cell locomotion. *Science* **208**, 177–179.

HEISENBERG, C.P., TADA, M., RAUCH, G.J., SAUDE, L., CONCHA, M.L., GEISLER, R., STEMPLE, D.L., SMITH, J.C., and WILSON, S.W. (2000). Silberblick/Wnt11 mediates convergent extension movements during zebrafish gastrulation. *Nature* **405**, 76–81.

JACOBSON, A.G., and GORDON, R. (1976). Changes in the shape of the developing vertebrate nervous system analyzed experimentally, mathematically and by computer simulation. *J Exp Zool* **197**, 191–246.

JACOBSON, A.G., and MOURY, J.D. (1995). Tissue boundaries and cell behavior during neurulation. *Dev Biol* **171**, 98–110.

JESSEN, J.R., TOPCZEWSKI, J., BINGHAM, S., SEPICH, D.S., MARLOW, F., CHANDRASEKHAR, A., and SOLNICA-KREZEL, L. (2002). Zebrafish trilobite reveals new roles for Strabismus in gastrulation and neuronal movements. *Nat. Cell Biol.*

KATSUMI, A., MILANINI, J., KIOSSES, W.B., DEL POZO, M.A., KAUNAS, R., CHIEN, S., HAHN, K.M., and SCHWARTZ, M.A. (2002). Effects of cell tension on the small GTPase Rac. *J Cell Biol* **158**, 153–64.

KELLER, R. (2002). Shaping the vertebrate body plan by polarized embryonic cell movements. *Science* **298**, 1950–1954.

KELLER, R., COOPER, M.S., DANILCHIK, M., TIBBETTS, P., and WILSON, P.A. (1989). Cell intercalation during notochord development in *Xenopus laevis. J Exp Zool* **251**, 134–54.

KELLER, R., and DAVIDSON, E. (2004). In "*Cell Motility: From Molecules to Organisms*" (A. RIDLEY, M. PECKHAM, P. CLARK, Eds.). New York, John Wiley & Sons Inc.

KELLER, R., DAVIDSON, L., EDLUND, A., ELUL, T., EZIN, M., SHOOK, D., and SKOGLUND, P. (2000). Mechanisms of convergence and extension by cell intercalation. *Philos Trans R Soc Lond B Biol Sci* **355**, 897–922.

KELLER, R., DAVIDSON, L.A., and SHOOK, D.R. (2003). How we are shaped: The biomechanics of gastrulation. *Differentiation* **71**, 171–205.

KELLER, R., SHIH, J., and DOMINGO, C. (1992). The patterning and functioning of protrusive activity during convergence and extension of the Xenopus organiser. *Dev Suppl*, 81–91.

KOLEGA, J. (1986). Effects of mechanical tension on protrusive activity and microfilament and intermediate filament organization in an epidermal epithelium moving in culture. *J Cell Biol* **102**, 1400–11.

LANE, M.C., and KELLER, R. (1997). Microtubule disruption reveals that Spemann's organizer is subdivided into two domains by the vegetal alignment zone. *Development* **124**, 895–906.

MARLOW, F., TOPCZEWSKI, J., SEPICH, D., and SOLNICA-KREZEL, L. (2002). Zebrafish Rho kinase 2 acts downstream of Wnt11 to mediate cell polarity and effective convergence and extension movements. *Curr Biol* **12**, 876–884.

MARSDEN, M., and DESIMONE, D.W. (2001). Regulation of cell polarity, radial intercalation and epiboly in Xenopus: novel roles for integrin and fibronectin. *Development* **128**, 3635–3647.

MEDINA, A., REINTSCH, W., and STEINBEISSER, H. (2000). Xenopus frizzled 7 can act in canonical and non-canonical Wnt signaling pathways: implications on early patterning and morphogenesis. *Mech Dev* **92**, 227–237.

MIYAMOTO, D. M., and CROWTHER, R. J. (1985). Formation of the notochord in living ascidian embryos. *J Embryol Exp. Morphol.* **86**, 1–17.

MLODZIK, M. (2002). Planar cell polarization: do the same mechanisms regulate Drosophila tissue polarity and vertebrate gastrulation? *Trends in Genetics* **18**, 564–571.

MUNRO, E. M., and ODELL, G. M. (2002). Polarized basolateral cell motility underlies invagination and convergent extension of the ascidian notochord. *Development* **129**, 13–24.

MYERS, D. C., SEPICH, D. S., and SOLNICA-KREZEL, L. (2002). BMP activity gradient regulates convergent extension during zebrafish gastrulation. *Dev Biol.* **243**, 81–98.

NINOMIYA, H., ELINSON, R., and WINKLBAUER, R. (2004). Antero-posterior tissue polarity links mesoderm convergent extension to axial patterning. *Nature* **430**, 364–367.

PARK, M., and MOON, R. T. (2002). The planar cell-polarity gene stbm regulates cell behaviour and cell fate in vertebrate embryos. *Nat Cell Biol* **4**, 20–25.

SEPICH, D. S., MYERS, D. C., SHORT, R., TOPCZEWSKI, J., MARLOW, F., and SOLNICA-KREZEL, L. (2000). Role of the zebrafish trilobite locus in gastrulation movements of convergence and extension. *Genesis* **27**, 159–173.

SHIH, J., and KELLER, R. (1992a). Cell motility driving mediolateral intercalation in explants of Xenopus laevis. *Development* **116**, 901–914.

SHIH, J., and KELLER, R. (1992b). Patterns of cell motility in the organizer and dorsal mesoderm of Xenopus laevis. *Development* **116**, 915–930.

SOKOL, S. (2000). A role for Wnts in morpho-genesis and tissue polarity. *Nat Cell Biol* **2**, E124–125.

TADA, M., and SMITH, J. C. (2000). Xwnt11 is a target of Xenopus Brachyury: regulation of gastrulation movements via Dishevelled, but not through the canonical Wnt pathway. *Development* **127**, 2227–238.

TAHINCI, E., and SYMES, K. (2003). Distinct functions of Rho and Rac are required for convergent extension during Xenopus gastrulation. *Dev Biol* **259**, 318–335.

WALLINGFORD, J. B., ROWNING, B. A., VOGELI, K. M., ROTHBACHER, U., FRASER, S. E., and HARLAND, R. M. (2000). Dishevelled controls cell polarity during Xenopus gastrulation. *Nature* **405**, 81–85.

WINKLBAUER, R., MEDINA, A., SWAIN, R. K., and STEINBEISSER, H. (2001). Frizzled-7 signalling controls tissue separation during Xenopus gastrulation. *Nature* **413**, 856–860.

ZALLEN, J., and WIESCHAUS, E. (2004). Patterned gene expression directs bipolar planar polarity in Drosophila. *Dev. Cell* **6**, 343–355.

7

On the Border of Understanding Cell Migration

Jocelyn A. McDonald and Denise J. Montell

7.1
Introduction

The *Drosophila* border cells have emerged in recent years as a simple model system for studying the mechanisms that control cell migration *in vivo*. Work in several laboratories has revealed multiple extracellular signals that are required for proper migration. These signals specify the migratory population, guide the cells to their proper destination, and facilitate migration at the proper stage of development. In addition, some progress has been made in understanding the changes in cell adhesion and organization of the actin cytoskeleton that are essential for border cell motility, though many interesting questions remain.

The border cells are a small group of epithelial follicle cells that migrate during ovarian development (Fig. 7.1). The *Drosophila* ovary is made up of multiple strings of egg chambers, each of which will produce an egg. In the center of each egg chamber are the germline cells consisting of a single oocyte at the very posterior and 15 nurse cells at the anterior, which nourish the oocyte (Fig. 7.1A). Early in oogenesis, somatic follicle cells form an epithelium surrounding the germline (Fig. 7.1A). A pair of specialized follicle cells at the anterior and posterior poles of the egg chamber, called the polar cells, are distinguishable at these stages by their rounded morphology and expression of specific markers. At mid-oogenesis, most of the follicle cells stack up to form a columnar epithelium in contact with the oocyte and will eventually secrete the eggshell. At the same time, the anterior polar cells recruit 4 to 8 cells to surround them, forming the border cell cluster (see below). The border cells round up, separate from the epithelium, and migrate as a cohesive cluster between the nurse cells, ending up at the anterior border of the oocyte (Fig. 7.1A–C). It is here that the border cells will contribute to the formation of a structure in the eggshell called the micropyle, which has a pore through which the sperm will enter and fertilize the egg.

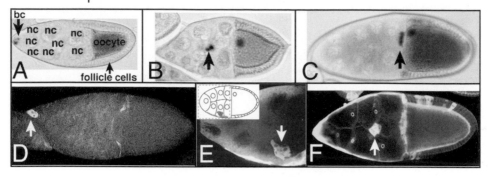

Fig. 7.1 Border cell migration in wild type and mutant egg chambers. (A–C) Nomarski optics of egg chambers stained for β-galactosidase activity from the *PZ6356* enhancer trap, which stains the border cells (arrows) and the oocyte nucleus (B–C). A) Stage 8 egg chamber in which the border cells have not migrated yet and are located at the anterior tip of the egg chamber. The border cells (bc), nurse cells (nc), oocyte, and follicle cells are labeled. B) Stage 9 egg chamber in which the border cells are in the process of migrating. C) Stage 10 egg chamber in which the border cells have completed their migration. D) Stage 10 egg chamber expressing DN-EcR in the border cells. The border cells (arrow) fail to migrate when any member of the ecdysone pathway is disrupted. Genotype shown is *slbo-GAL4*, UAS-mCD8-GFP/ UAS-DN-EcR (F645A) [18]. E) Example of a stage 10 egg chamber with misguided border cells in which follicle cells misexpress PVF1 ligand (green). The border cells (red, arrow) are located on the side of the egg chamber, adjacent to the cells misexpressing PVF1. Inset shows a schematic of the egg chamber. Genotype shown is *hs-flp*, *Pvf1^{EP1624}; AyGAL4, UAS-lacZ/UAS-Pvf1*. Reprinted from [19]. F) Stage 10 egg chamber in which the border cells (green) are mutant for *Pvr*. Loss of *Pvr*, using the *Pvr^{c2195}* allele, results in mild border cell (arrow) migration defects [4, 19]. Anterior is located to the left in all panels. (This figure also appears with the color plates.)

Over the past decade, many genes have been identified that play an important role in border cell migration, helping us to understand this process in some detail [1–8]. The first gene identified was *slow border cells* (*slbo*), a homolog of the C/EBP leucine zipper transcription factor. Border cell migration is severely disrupted in *slbo* mutant egg chambers [9]. The identification of *slbo* and additional genes from a variety of subsequent screens, as well as changes in the morphology of the border cells, has defined several distinct features of border cell migration. First, border cell identity is specified. Next, the border cells become migratory, extend protrusions between the nurse cells, and delaminate from the epithelium.

The border cells then move between the nurse cells and are actively guided toward the oocyte, during which time actin cytoskeletal rearrangement and dynamic cell adhesion play a big role. In this review, we will summarize what is known about each of these steps, emphasizing recent work and highlighting future directions. Finally, we will summarize new studies implicating border cell migration as a model for epithelial-based tumor metastasis.

7.2
The Who, What, When, and Where of Border Cell Migration

How do the border cells know *who* they are and *what* tells them to become migratory and invasive? Only 6–10 cells out of the approximately 900 follicle cells become border cells, leading to the question of why the border cells migrate whereas other follicle cells do not. The JAK/STAT signaling pathway contributes to specifying which cells become the border cells and promotes their migratory behavior [2, 7]. *Drosophila* STAT activates transcription downstream of a well-conserved pathway; upon binding of a cytokine (in *Drosophila* called Unpaired (UPD)) to its receptor (Domeless), constitutively bound Janus Kinase {JAK; *hopscotch* (*hop*)} becomes auto-phosphorylated and recruits and phosphorylates STAT, which then translocates to the nucleus where it activates transcription {reviewed in [10]}. Loss-of-function mutations in any member of this pathway, including *Stat92E*, *domeless*, *Jak/hop*, or *upd* cause severe border cell migration defects [2, 7, 11]. The activating ligand, UPD, is localized specifically to the polar cells [2, 7, 12], which are located at the center of the cluster. Polar cells were previously proposed to recruit the border cells because mutants in which extra polar cells form have extra border cells associated with them [13]. The STAT protein is concentrated in nuclei of the migratory cells that directly contact the polar cells (D. L. Silver and D. J. M., unpublished). Loss of UPD or other proteins in the JAK/STAT pathway causes a reduction in the number of border cells recruited to the cluster [2, 7], indicating that UPD secreted from the polar cells likely recruits cells to become border cells. Ectopic expression of either JAK or UPD in follicle cells that normally do not migrate, induces these cells to become migratory [7], confirming that the JAK/STAT pathway is both necessary and sufficient to recruit cells to become migratory. The JAK/STAT pathway also specifies border cell identity, because the expression of genes known to be required for border cell migration, such as *slbo* and *DE-cadherin*, are reduced in *Stat92E* mutant border cells [7]. It is unknown whether recruitment and specification are one step or whether these are separable processes.

Once border cell identity has been specified and these cells acquire the ability to migrate, they have to be told *when* to migrate. However, *upd* and *Stat92E* are expressed early in oogenesis [2, 7], long before border cells migrate, indicating that one or more additional signals prevent precocious migration or activate migration at the appropriate time. At least one of these signals may be the *Drosophila* steroid hormone ecdysone [1]. Ecdysone binds to a receptor heterodimer that consists of an RXR homolog Ultraspiracle (Usp) and a hormone-binding subunit (the ecdysone receptor; EcR) [14]. Ecdysone titers rise at mid-oogenesis [15] and loss of EcR in the germline results in degeneration of egg chambers specifically at these stages [16, 17], indicating that ecdysone promotes egg chamber development at this time. A screen for mutants that affect border cell migration identified a hormone receptor coactivator called *taiman* (*tai*) [1]. Loss of *tai* or *usp* [1], or expression of a dominant negative (DN) form of EcR [18], cause strong border cell migration defects (Fig. 7.1 D). In addition, TAI, USP, and EcR are all expressed at high levels in the border cells throughout migration and in other follicle cells. TAI is a true ecdysone receptor coactivator, since it binds EcR directly, in a hormone-dependent manner, and enhances ecdysone-dependent transcription in cultured cells [1]. Together these data indicate that the rise in ecdysone levels at mid-oogenesis might contribute to promoting border cell migration at the appropriate stage. Further, it is likely that other morphological changes that occur at mid-oogenesis are ecdysone-dependent; follicle cells fail to rearrange and surround the oocyte when they are mutant for *tai* or receive less synthesized ecdysone in *ecdysoneless* mutants (J.A.M. and D.J.M., unpublished). No transcriptional targets of the ecdysone pathway in the border cells have been identified as of yet, so future work will be needed to address this point. Recently, a screen was performed for mutations that exhibit dominant genetic interactions with *tai* [19]; cloning these genes may help identify members of this pathway. It appears that there must be at least one additional timing cue because expression of SLBO expression is not induced prior to stage 9, yet its expression is independent of ecdysone. Additional studies will be required to identify this timing mechanism.

The border cells never deviate from their central path between the nurse cells on their way to the oocyte, even though they pass several nurse cell–nurse cell junctions down which they could migrate, leading to the question of what tells the border cells *where* to go. It was predicted that the oocyte would be the source of an attractive signal, based on the observation that when the oocyte is mispositioned within the egg chamber in specific mutant backgrounds, the border cells continue to migrate to the oocyte no matter where it is located [20, 21]. The identity of the attractive cue(s) remained a mystery until relatively recently, when one gene was identified in two unrelated screens. A candidate guidance factor was identified in a screen for genes that disrupt border cell migration when misexpressed in the germline [3, 4]. This screen was designed with the idea that chemo-attractants likely need to be present in a gradient in

order to effectively guide migration. Global misexpression of a chemo-attractant should disrupt this gradient and thus disrupt migration. Global
misexpression of a *Drosophila* protein that is equally related to mammalian platelet-derived growth factor (PDGF) and vascular-endothelial
growth factor (VEGF), called PDGF/VEGF related factor 1 (PVF1), completely disrupts border cell migration [4]. *Pvf1* was also identified in a
screen for genes that exhibit dominant genetic interactions with *tai* [19].
PVF1 is expressed specifically in the germline and is expressed at highest levels in the oocyte [4], which is the target of migration. Its receptor,
PVR, is a receptor tyrosine kinase localized to the cell cortex of all follicle
cells including the border cells [4, 19].

Based on the localization of PVF1 in the oocyte and its receptor in the
border cells, it seemed likely that PVF1 was an attractive cue for the border cells to migrate to the oocyte. To test this directly, McDonald et al.
[19] misexpressed PVF1 in random groups of follicle cells at the anterior
end of the egg chamber and asked whether the border cells would redirect their migration to this new source of ligand. When PVF1 is expressed ectopically, either in the presence or absence of endogenous ligand, the border cells redirect their migration in 13–20% of egg chambers (Fig. 7.1 E) [19]. When the number of expressed transgenes is
doubled, presumably increasing the concentration of ectopic PVF1, almost half the border cells are misguided [19], indicating that PVF1 is a
concentration-dependent guidance cue. This effect is specific, because
misexpression of a related ligand, PVF2, does not have any effect on border cell guidance [19]. However, loss of *Pvf1* causes only a mild migration defect, with most border cells completing their migration [4, 19].
Therefore, PVF1 is an attractive cue for the border cells to migrate to the
oocyte, but additional cues must contribute to their guidance.

What other attractive cues participate in guiding the border cells to the
oocyte? Although there are two additional PVF homologs in *Drosophila*,
PVF2 and PVF3, they do not seem to contribute to border cell migration
since loss of the single receptor results in the same mild border cell migration defects as loss of *Pvf1* (Fig. 7.1 F) [4, 19]. Therefore, other cues
must participate in guiding the border cells. A clue that the epidermal
growth factor receptor (EGFR) pathway is involved came from experiments in which global misexpression of an EGFR ligand, Vein, in the
germline causes strong border cell migration defects [3, 4]. EGFR is expressed in all follicle cells, including the border cells [22], similar to
PVR. When EGFR and PVR are simultaneously disrupted using dominant–negative (DN) transgenes, border cell migration is disrupted to a
greater extent than when either EGFR or PVR are disrupted alone [4].
Therefore, the EGFR and PVR signaling pathways function redundantly
to guide the border cells to the oocyte.

There are four EGFR ligands in *Drosophila*, Vein, Gurken (GRK), Spitz
(SPI), and Keren (KRN) {reviewed in [23]}, but it is not known which

one contributes to guiding the border cells. Vein is not a good candidate for an attractive ligand, because it is expressed after border cell migration to the oocyte is complete in the follicle cells located over the dorsal–anterior corner of the oocyte [24]. GRK, which is localized to the dorsal–anterior corner of the oocyte, adjacent to the oocyte nucleus [25], is known to guide the dorsal migration of the border cells [3] and therefore was also proposed to guide the border cells in their posterior-directed migration to the oocyte [4, 23, 26]. However, GRK does not misdirect the border cells when ectopically expressed nor does GRK potentiate the effect of PVF1 [19]. Additional work will be needed to determine whether another EGFR ligand, such as SPI or KRN, acts to guide the border cells to the oocyte, either on its own or in conjunction with GRK.

7.3
Regulating the Cytoskeleton

How do the border cells integrate all of the information from the JAK/ STAT, ecdysone, and PVR/EGFR pathways and convert this information into actual physical movement? Although the answers to these questions are not known, it is clear that remodeling the actin cytoskeleton is critical to regulating cellular motility. One group of molecules generally implicated in regulating actin during migration is the Rho family of small GTPases. Rac, a member of this family, can induce reorganization of the actin cytoskeleton in cultured fibroblasts and stimulates the formation of lamellipodia in migrating cells (reviewed in [27, 28]). All Rho GTPases undergo a transition from an active state, in which GTP is bound, to an inactive state, in which GTP is hydrolyzed to GDP. Two groups of proteins regulate this transition; GTPase-activating proteins (GAPs) increase the rate of GTP hydrolysis, whereas guanine nucleotide exchange factors (GEFs) catalyze the exchange of GDP for GTP (reviewed in [27, 28]). Rac is required for many types of cell migrations and is known to regulate border cell migration, since expression of a dominant–negative form of Rac (Rac^{N17}; DN-Rac) specifically in the border cells causes complete failure of migration [29]. Loss of all three *Drosophila* Rac genes, using loss-of-function mutant alleles, causes the same phenotype as DN-Rac, confirming that Rac is required for border cell migration [30].

The exact mechanism by which Rac regulates actin in the migrating border cells is not known and F-actin accumulation appears to be grossly normal in border cells expressing DN-Rac [29, 30]. As discussed above, PVF1 through its receptor PVR, guides the border cells to the oocyte. To better understand the role of PVR in regulating border cell migration, a constitutively active form of PVR (λ-PVR) was expressed in follicle cells [4]. When PVR is activated to high levels, the levels of F-actin increase dramatically, similar to what occurs when a constitutive activated form of

Rac (Rac^{V12}) is expressed [4]. Duchek et al. [4] reasoned that activation of PVR could affect border cell migration by regulating actin rearrangement via Rac. To test this, they coexpressed λ-PVR and DN-Rac in follicle cells and found a decrease in F-actin accumulation compared to λ-PVR alone. These results indicate that Rac is required for PVR-induced actin polymerization. A putative Rac guanine nucleotide exchange factor (GEF), the *Drosophila* DOCK180 homolog *myoblast city* (*mbc*), is expressed in all follicle cells, including the border cells [4]. Loss of *mbc* causes a mild migration defect and, like DN-Rac, decreases the F-actin accumulation caused by λ-PVR, indicating that it could be acting as a Rac GEF in the border cells. However, MBC is probably not the only GEF for Rac, since loss of *mbc* does not cause as strong an effect on border cell migration as loss of Rac itself. There are multiple GEFs in *Drosophila* and it will be interesting in the future to discover which additional GEFs participate in Rac-mediated border cell migration.

To identify additional components of the Rac pathway, as well as to better understand how Rac interacts with the actin cytoskeleton, Geisbrecht and Montell [30] performed a screen for genes that, when overexpressed, suppress the loss of border cell migration caused by DN-Rac. One gene identified in this screen is *thread*, which encodes the *Drosophila* inhibitor of apoptosis protein 1 (DIAP1) [31]. All follicle cells, including the border cells, express DIAP1, and loss of *thread* causes border cell migration defects. The effect on cell migration is independent of its effects on apoptosis, since loss of *thread* or expression of DN-Rac does not lead to cell death in either the border cells or other follicle cells. Loss of *thread* causes a decrease in F-actin accumulation in follicle cells, indicating that DIAP1 may play a role in regulating actin. In support of this, *thread* mutations show a strong genetic interaction with *chickadee*, the gene that encodes the *Drosophila* homolog of profilin, a molecule that maintains the cellular pool of polymerization-competent, monomeric actin. Like DIAP1, overexpression of profilin suppressed the loss of border cell migration due to DN-Rac. Rac, DIAP1, and profilin were found in a biochemical complex in a *Drosophila* cell line, indicating that DIAP1 contributes to Rac-mediated border cell migration via the actin cytoskeleton.

IAP proteins are known to inhibit apoptosis by blocking the activity of caspases, which are cysteine-dependent aspartic proteases that are responsible for carrying out the cell death program (reviewed in [32]). This leads to the question of whether other caspase inhibitors play a similar role to DIAP1 in regulating cell migration. However, the p35 protein from baculovirus, another class of caspase inhibitor that blocks apoptosis, does not rescue the DN-Rac border cell migration defects [30], further supporting the conclusion that the function of DIAP1 in border cell migration is distinct from its ability to protect cells from apoptosis. One caspase enzyme, the *Drosophila* homolog of caspase 9, called Dronc, is inhibited by DIAP1 but not by p35 [33]. Geisbrecht and Montell [30] found that inhibition of Dronc activity rescued the DN-Rac border cell migration defects. Since caspases cleave a variety of cytoskeletal proteins including Rac and actin [34, 35], one possibility is that DIAP1-mediated inhibition of Dronc protects Rac and/or actin from cleavage and allows them to function in border cell migration. Future studies will be needed to determine the physiologically relevant Dronc target(s) and to elucidate how caspase activity in border cells lacking DIAP1 is restricted so that the cells survive.

Several other actin-binding proteins have recently been shown to contribute to border cell migration. One such actin-binding protein, filamin, is known to bind to signaling molecules such as the Rho GTPases, providing a possible link between the actin cytoskeleton and signaling pathways (reviewed in [36]). *Drosophila* filamin (*cheerio*) is highly expressed in the border cells, both in the cytoplasm and at the membrane, primarily between the border cells [8]. Loss of *cheerio* causes a delay in migration [8]. Another actin-binding protein, cortactin, appears to play a minor role in border cell migration [37]. Cortactin is a target of phosphorylation by the nonreceptor tyrosine kinase Src (reviewed in [38, 39]) and can bind to the Arp2/3 complex (reviewed in [39]). Although cortactin is enriched in migrating border cells and loss of *cortactin* causes a slight delay early in migration, most *cortactin*-deficient border cells complete their migration by the appropriate stage [37]. Cofilin/ADF, another actin-binding protein, is known to accelerate actin filament depolymerization and is thought to play a key role in maintaining adequate levels of monomeric actin in migrating cells. Interestingly, cofilin can also promote actin polymerization at the leading edge (reviewed in [40]). Loss of the *Drosophila* cofilin (*twinstar*), using a temperature sensitive combination of alleles, causes a very strong migration defect [41]. Together, these data indicate that, not surprisingly, many proteins that regulate the actin cytoskeleton participate in border cell migration. However, several questions remain, including whether other actin-binding proteins, such as the Arp2/3 complex, regulate border cell migration and how the activities of these actin-binding proteins are spatially regulated in the border cells.

7.4
Adhesive Forces Provide Traction for Border Cell Migration

Precise regulation of adhesion is crucial for cells to migrate at the proper speed. Border cells migrate upon the surfaces of other cells, the nurse cells, and must have strong enough adhesion to gain traction at the front while simultaneously detaching at the rear. Thus dynamic regulation of adhesion, between migrating cells and their substrate, is an important feature of cell migration. The homophilic cell–cell adhesion molecule DE-cadherin appears to act as a critical adhesion molecule for border cell migration because loss of DE-cadherin (encoded by the gene *shotgun*) from either the border cells or the nurse cells disrupts migration completely [42]. DE-cadherin protein is expressed at higher levels in the border cells than in other follicle cells, but is dynamically localized within the border cell cluster [1, 42]. Before migration, DE-cadherin is localized uniformly on the lateral surfaces of border cells and other follicle cells. However, when the border cells begin their migration, DE-cadherin levels at the interface between border cells and nurse cells drop dramatically while high levels are maintained between the border cells. Thus the border cells appear to be specialized for post-transcriptional down-regulation of E-cadherin specifically at this interface.

What regulates the precise levels of DE-cadherin at the interface between border cells and nurse cells to promote normal migration? Two proteins that bind to E-cadherin, p120-catenin and β-catenin, may play a role in regulating E-cadherin levels. p120-catenin binds to the juxtamembrane domain (JD) of E-cadherin, whereas β-catenin, which is called Armadillo (Arm) in *Drosophila*, binds to the C terminus and connects E-cadherin to the actin cytoskeleton via its binding partner α-catenin (reviewed in [43]. p120-catenin was previously proposed to either positively regulate adhesion or to downregulate adhesion (reviewed in [43]). Either of these mechanisms could contribute to regulating E-cadherin-dependent border cell migration. To test the role of p120-catenin *in vivo*, Pacquelet et al. [44] assessed whether two different mutant E-cadherin constructs could rescue E-cadherin loss-of-function phenotypes such as border cell migration defects. The first construct mutated several amino acids in the JD to disrupt p120-catenin binding (DE-cadherin-AAA) and the second construct deleted the Arm binding domain (DE-cadherin-δβ). DE-cadherin-AAA rescued the loss of border cell migration caused by loss of E-cadherin, whereas DE-cadherin-δβ did not. Normal border cell migration was also observed when endogenous p120-catenin levels were reduced by expressing an RNAi transgene [44]. Together these results indicate that p120-catenin binding is dispensable for normal E-cadherin function although Arm binding is essential. These results were confirmed by a recent study, using a loss-of-function allele of p120-catenin, which found that p120-catenin function is dispensable throughout *Drosophila* development [45]. Neither study rules out a subtle role for p120-catenin in regulating E-cadherin during border cell migration, for example if it acts redundantly with other factors. In support of this idea, loss of p120-catenin strongly enhanced the larval cuticle defects caused by loss of *shotgun*/DE-cadherin [45].

Several signals appear to destabilize E-cadherin at the border cell/nurse cell interface. For example, the insect steroid hormone ecdysone is required for the proper distribution of E-cadherin; loss of the ecdysone receptor co-activator *tai* results in a dramatic increase in E-cadherin levels at some border cell/nurse cell contacts [1]. In addition, loss or overexpression of PVF1, results in similar increases in E-cadherin at the border cell/nurse cell membrane [19]. It is unknown whether the increase in E-cadherin in either *tai* or *Pvf1* mutant border cells is directly responsible for the observed border cell migration phenotypes, nor is it known how either pathway regulates E-cadherin distribution. One or both pathways could regulate tyrosine phosphorylation of Arm, since phosphorylation of β-catenin has been correlated with destabilization of E-cadherin (reviewed in [46]). Phosphorylated β-catenin binds with a lower affinity to E-cadherin, which in turn uncouples E-cadherin from the cytoskeleton and reduces adhesion between cells (reviewed in [46]). It is interesting to note that Arm is highly tyrosine-phosphorylated in the *Drosophila* ovary ([47]; E. Pinheiro, J.A.M., and D.J.M., unpublished). Since PVF1 signals through a receptor tyrosine kinase, PVR, this pathway could directly phosphorylate Arm. The source of PVF is in the germline, providing a plausible explanation for how E-cadherin could be specifically destabilized at the border cell/germ cell interface. However, testing this hypothesis may be difficult, since assessing the phosphorylation state of β-catenin specifically at particular subcellular locations within the border cells is presently not feasible.

In order for the border cells to migrate, a dynamic balance between destabilization and stabilization of E-cadherin is needed. The unconventional myosin, myosin VI (MyoVI), plays a role in stabilizing the E-cadherin/Armadillo (Arm) complex at the border cell membranes [5]. MyoVI is found in both cytoplasmic and membrane-associated pools in the border cells. Geisbrecht and Montell [5] observed strong border cell migration defects in border cells depleted of MyoVI. In follicle cells depleted of MyoVI, the levels of DE-cadherin and Armadillo are reduced, whereas the expression of a number of other molecules was normal. In addition, MyoVI levels are reciprocally reduced in cells lacking ARM or DE-cadherin. MyoVI, ARM and DE-cadherin are found in a biochemical complex within the ovary, supporting the idea that these three proteins stabilize each other.

What is the mechanism by which MyoVI stabilizes E-cadherin/β-catenin to promote migration? MyoVI, unlike most myosins, moves toward the minus ends of actin filaments. Since the plus ends (fast-growing ends) of actin filaments are usually located at the periphery of the cell, untethered MyoVI molecules would be expected to move toward the cell interior. However, the possibility also exists that MyoVI molecules that are tethered to the membrane via their association with E-cadherin and ARM could propel actin filaments outward, generating a protrusive force. Alternatively, MyoVI may play a more static role in holding protein complexes together, stabilizing specific actin arrangements and/or anchoring cell membranes in particular locations. These possibilities have all been suggested based on the phenotypes associated with reduction in MyoVI function in a variety of cell types and organisms and based on the biophysical properties of this motor (reviewed in [48]).

MyoVI may provide a protrusive force or may stabilize actin based protrusions and by either mechanism could play a role in the generation of the long cellular extensions (LCEs) that have been observed in migrating border cells (Fig. 7.2) [49]. At the initiation of migration, a single border cell extends an LCE, defined as being ~ 20–$40\ \mu m$ in length, between the nurse cells towards the oocyte (Fig. 7.2 A). Later during migration, one or multiple border cells extend protrusions of varying, but shorter, lengths (10–15 μm in length) [5, 49]. Fulga and Rørth [49] found that directional guidance cues are required for the formation of LCEs. Downregulation of either of the guidance receptors, PVR or EGFR, or general upregulation of guidance ligands, such as PVF1, disrupts the formation of LCEs (Fig. 7.2 B). Thus LCEs may play a role in sensing gradient cues of growth factors from the oocyte. LCE formation and/or stabilization is also dependent on E-cadherin, since LCEs are not observed when DE-cadherin is lost from either the germline or the border cells (Fig. 7.2 B). To test whether actin–myosin forces contribute to LCE formation, Fulga and Rørth [49] expressed high levels of actin in the border cells. When actin is overexpressed in the border cells, LCEs are longer and less straight than normal and border cell migration is impaired (Fig. 7.2 C). In addition, loss of non-muscle myosin II regulatory light chain (encoded by *spaghetti squash*) from the border cells causes an increase in the formation of extremely long, curvy LCEs (Fig. 7.2 C) [49] and defects in border

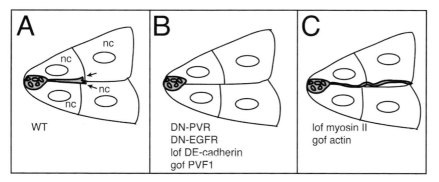

Fig. 7.2 Schematic of egg chambers showing the formation of long cellular extensions (LCEs) in migrating border cells [49]. A) In wild-type egg chambers a single LCE forms from the border cell cluster at the start of migration and extends ~ 20–$40\ \mu m$, or approximately the length of one nurse cell (nc). Short, actin-rich projections form at the end of LCEs (arrows). B) LCEs are lost in DN-PVR, DN-EGFR, loss-of-function (lof) *shotgun*/DE-cadherin egg chambers, and when PVF1 is overexpressed (gof). C) LCEs grow extremely long and are no longer straight in lof *spaghetti squash*/non-muscle myosin II egg chambers and when actin is overexpressed.

cell migration [49, 50]. Thus, actin and myosin regulate LCE length and rigidity. The border cells may use LCEs to pioneer a route to the oocyte, possibly by sensing attractive cues secreted from the oocyte, and LCEs may use the mechanical force provided by actin and myosin to pull the rest of the border cell cluster towards the oocyte.

7.5
Relationship of Border Cell Migration to Tumor Metastasis

Border cell migration resembles tumor invasion and metastasis in that the border cells detach from a polarized epithelium, invade other cells (the nurse cells) and migrate a long distance before attaching to other cells (the oocyte). It is interesting to note that many of the genes required for border cell migration also contribute to cancer. For example, *taiman* is a homolog of AIB1 [1], a gene that is amplified in some breast and ovarian cancers. Such observations raise the question whether studies of border cell migration can serve as a starting point to identify new molecules and/or new functions for known molecules in tumor invasion and metastasis. Two recent studies suggest that this is the case [51, 52].

Drosophila STAT promotes the motility of the border cells, and one of the mammalian STATs, STAT3 is known to be constitutively activated in a variety of cancers. Therefore, Silver et al. [51] tested the hypothesis that STAT3 contributes to the motility of ovarian carcinoma cells. An antibody to phosphorylated-STAT3 was used to examine the expression of activated STAT3 in a tissue array of ovarian carcinoma biopsies. Activated STAT3 is not expressed detectably in normal ovarian epithelium whereas it is expressed at highest levels in high-grade carcinomas. STAT3 is activated to high levels in several ovarian cancer cell lines, and decreasing activated STAT3, either using a pharmacological inhibitor of JAK2 or siRNA of STAT3, causes a reduction in the ability of these cells to migrate in a modified Boyden Chamber assay. This study suggests a role for activated STAT3 in promoting ovarian tumor cell motility.

A second study examined the function of MyoVI in ovarian cancer [52]. Human MyoVI expression was examined in tissue arrays of normal ovary and epithelial ovarian tumors. MyoVI is not detectable in normal ovarian tissue, whereas high levels of MyoVI are found in high-grade ovarian carcinomas. The function of MyoVI in ovarian carcimoma cell migration was examined by inhibiting MyoVI, using either antisense sequences or siRNA, in the ES-2 ovarian cancer cell line. MyoVI-depleted ES-2 cells exhibit reduced migration, both in a wound-healing "scratch" assay and in a Boyden chamber assay. To test whether inhibition of MyoVI could prevent the spread of ovarian carcinoma cells in mice, ES-2 cells stably transfected with antisense-MyoVI were injected into nude mice. Both control (MyoVI-positive) ES-2 cells and the antisense MyoVI

ES-2 cells cause tumor formation at the site of injection, but MyoVI-depleted cells do not spread to other tissues to the same extent as the MyoVI-positive ES-2 cells. Therefore, MyoVI promotes tumor dissemination just as *Drosophila* MyoVI promotes border cell migration. In the future, it will be interesting to test whether additional genes that regulate border cell migration have a role in epithelial-based tumor invasion and migration.

7.6
Acknowledgment

We thank M. Starz-Gaiano for providing the egg chamber in Fig. 1 D.

7.7
References

1 BAI, J., Y. UEHARA, and D. J. MONTELL, Regulation of invasive cell behavior by Taiman, a *Drosophila* Protein Related to AIB1, a steroid receptor coactivator amplified in breast cancer. *Cell*, 2000, **103**, 1047–1058.

2 BECCARI, S., L. TEIXEIRA, and P. RØRTH, The JAK/STAT pathway is required for border cell migration during *Drosophila* oogenesis. *Mech Dev*, 2002, **111**, 115–123.

3 DUCHEK, P. and P. RØRTH, Guidance of cell migration by EGF Receptor signaling during *Drosophila* oogenesis. *Science*, 2001, **291**, 131–133.

4 DUCHEK, P., et al., Guidance of cell migration by the *Drosophila* PDGF/VEGF receptor. *Cell*, 2001, **107**, 17–26.

5 GEISBRECHT, E. and D. J. MONTELL, Myosin VI is required for E-cadherin-mediated border cell migration. *Nat Cell Biol*, 2002, **4**, 616–620.

6 LIU, Y. and D. J. MONTELL, Jing, a downstream target of slbo required for developmental control of border cell migration. *Development*, 2001, **128**, 321–330.

7 SILVER, D. L. and D. J. MONTELL, Paracrine signaling through the JAK/STAT pathway activates invasive behavior of ovarian epithelial cells in *Drosophila*. *Cell*, 2001, **107**, 831–841.

8 SOKOL, N. S. and L. COOLEY, *Drosophila* filamin is required for follicle cell motility during oogenesis. *Dev Biol*, 2003, **260**(1), 260–272.

9 MONTELL, D. J., P. RØRTH, and A. C. SPRADLING, Slow border cells, a locus required for a developmentally regulated cell migration during oogenesis, encodes *Drosophila* C/EBP. *Cell*, 1992, **71**, 51–62.

10 HOU, S. X., et al., The Jak/STAT pathway in model organisms: emerging roles in cell movement. *Dev Cell*, 2002, **3**(6), 765–778.

11 GHIGLIONE, C., et al., The *Drosophila* cytokine receptor Domeless controls border cell migration and epithelial polarization during oogenesis. *Development*, 2002, **129**(23), 5437–5447.

12 McGregor, J.R., R. Xi, and D.A. Harrison, JAK signaling is somatically required for follicle cell differentiation in *Drosophila*. *Development*, 2002, **129**(3), 705–717.

13 Liu, Y. and D.J. Montell, Identification of mutations that cause cell migration defects in mosaic clones. *Development*, 1999, **126**(9), 1869–1878.

14 Yao, T.P., et al., Functional ecdysone receptor is the product of EcR and Ultraspiracle genes. *Nature*, 1993, **366**(6454), 476–479.

15 Riddiford, L.M., *Hormones and Drosophila development*, in *The Development of Drosophila melanogaster* (M. Bate and A. Martinez Arias, Eds) 1993, Cold Spring Harbor Laboratory Press: Cold Spring Harbor, pp. 899–940.

16 Buszczak, M., et al., Ecdysone response genes govern egg chamber development during mid-oogenesis in *Drosophila*. *Development*, 1999, **126**(20), 4581–4589.

17 Carney, G.E. and M. Bender, The *Drosophila* ecdysone receptor (EcR) gene is required maternally for normal oogenesis. *Genetics*, 2000, **154**(3), 1203–1211.

18 Cherbas, L., et al., EcR isoforms in *Drosophila*: testing tissue-specific requirements by targeted blockade and rescue. *Development*, 2003, **130**(2), 271–284.

19 McDonald, J.A., E.M. Pinheiro, and D.J. Montell, Pvf1, a PDGF/VEGF homolog, is sufficient to guide border cells and interacts genetically with Taiman. *Development*, 2003, **130**, 3469–3478.

20 Gonzalez-Reyes, A. and D. St Johnston, Role of oocyte position in establishment of anterior-posterior polarity in *Drosophila*. *Science*, 1994, **266**(5185), 639–642.

21 Lee, T., L. Feig, and D.J. Montell, Two distinct roles for Ras in a developmentally regulated cell migration. *Development*, 1996, **122**, 409–418.

22 Sapir, A., R. Schweitzer, and B.Z. Shilo, Sequential activation of the EGF receptor pathway during *Drosophila* oogenesis establishes the dorsoventral axis. *Development*, 1998, **125**(2), 191–200.

23. Shilo, B.Z., Signaling by the *Drosophila* epidermal growth factor receptor pathway during development. *Exp Cell Res*, 2003, **284**(1), 140–149.

24 Wasserman, J.D. and M. Freeman, An autoregulatory cascade of EGF receptor signaling patterns the *Drosophila* egg. *Cell*, 1998, **95**(3), 355–364.

25 Neuman-Silberberg, F.S. and T. Schupbach, The *Drosophila* TGF-alpha-like protein Gurken: expression and cellular localization during *Drosophila* oogenesis. *Mech Dev*, 1996, **59**(2), 105–113.

26 Rørth, P., Initiating and guiding migration: lessons from border cells. *Trends Cell Biol*, 2002, **12**(7), 325–331.

27 Burridge, K. and K. Wennerberg, Rho and Rac take center stage. *Cell*, 2004. **116**(2), 167–179.

28 Raftopoulou, M. and A. Hall, Cell migration: Rho GTPases lead the way. *Dev Biol*, 2004, **265**(1), 23–32.

29 Murphy, A.M. and D.J. Montell, Cell type-specific roles for Cdc42, Rac, and RhoL in *Drosophila* oogenesis. *J Cell Biol*, 1996, **133**, 617–630.

30 GEISBRECHT, E. and D.J. MONTELL, A role for *Drosophila* IAP1-mediated caspase inhibition in Rac-dependent cell migration. *Cell*, In press.

31 HAY, B.A., D.A. WASSARMAN, and G.M. RUBIN, *Drosophila* homologs of baculovirus inhibitor of apoptosis proteins function to block cell death. *Cell*, 1995, **83**(7), 1253–1262.

32 SALVESEN, G.S. and J.M. ABRAMS, Caspase activation – stepping on the gas or releasing the brakes? Lessons from humans and flies. *Oncogene*, 2004, **23**(16), 2774–2784.

33 MEIER, P., et al., The *Drosophila* caspase DRONC is regulated by DIAP1. *EMBO J*, 2000, **19**(4), 598–611.

34 MASHIMA, T., et al., Actin cleavage by CPP-32/apopain during the development of apoptosis. *Oncogene*, 1997, **14**, 1007–1012.

35 MARTIN, D.N. and E.H. BAEHRECKE, Caspases function in autophagic programmed cell death in *Drosophila*. *Development*, 2004, **131**(2), 275–284.

36 STOSSEL, T.P., et al., Filamins as integrators of cell mechanics and signalling. *Nat Rev Mol Cell Biol*, 2001, **2**(2), 138–145.

37 SOMOGYI, K. and P. RØRTH, Cortactin modulates cell migration and ring canal morphogenesis during *Drosophila* oogenesis. *Mech Dev*, 2004, **121**(1), 57–64.

38 WEED, S.A. and J.T. PARSONS, Cortactin: coupling membrane dynamics to cortical actin assembly. *Oncogene*, 2001, **20**(44), 6418–6434.

39 WEAVER, A.M., et al., Integration of signals to the Arp2/3 complex. *Curr Opin Cell Biol*, 2003, **15**(1), 23–30.

40 POLLARD, T.D. and G.G. BORISY, Cellular motility driven by assembly and disassembly of actin filaments. *Cell*, 2003, **112**(4), 453–465.

41 CHEN, J., et al., Cofilin/ADF is required for cell motility during *Drosophila* ovary development and oogenesis. *Nat Cell Biol*, 2001, **3**(2), 204–209.

42 NIEWIADOMSKA, P., D. GODT, and U. TEPASS, DE-Cadherin is required for intercellular motility during *Drosophila* Oogenesis. *J Cell Biol*, 1999, **144**(3), 533–547.

43 PEIFER, M. and A.S. YAP, Traffic control: p120-catenin acts as a gatekeeper to control the fate of classical cadherins in mammalian cells. *J Cell Biol*, 2003, **163**(3), 437–440.

44 PACQUELET, A., L. LIN, and P. RØRTH, Binding site for p120/delta-catenin is not required for *Drosophila* E-cadherin function in vivo. *J Cell Biol*, 2003, **160**(3), 313–319.

45 MYSTER, S.H., et al., *Drosophila* p120 catenin plays a supporting role in cell adhesion but is not an essential adherens junction component. *J Cell Biol*, 2003, **160**(3), 433–449.

46 LILIEN, J., et al., Turn-off, drop-out: Functional state switching of Cadherins. *Developmental Dynamics*, 2002, **224**, 18–29.

47 PEIFER, M., L.M. PAI, and M. CASEY, Phosphorylation of the *Drosophila* adherens junction protein Armadillo: roles for wingless signal and zeste-white 3 kinase. *Dev Biol*, 1994, **166**(2), 543–556.

48 FRANK, D.J., T. NOGUCHI, and K.G. MILLER, Myosin VI: a structural role in actin organization important for protein and organelle localization and trafficking. *Curr Opin Cell Biol*, 2004, **16** In press.

49 FULGA, T. A. and P. RØRTH, Invasive cell migration is initiated by guided growth of long cellular extensions. *Nat Cell Biol*, 2002, **4**(9), 715–719.

50 EDWARDS, K. A. and D. P. KIEHART, *Drosophila* nonmuscle myosin II has multiple roles in imaginal disc and egg chamber morphogenesis. *Development*, 1996, **122**, 1499–1511.

51 SILVER, D. L., et al., Activated signal transducer and activator of transcription (STAT) 3: Localization in focal adhesions and function in ovarian cancer cell motility. *Cancer Res*, 2004, **64**(10), 3550–3558.

52 YOSHIDA, H., et al., Lessons from border cell migration in the *Drosophila* ovary: A role for myosin VI in dissemination of human ovarian cancer. *Proc Natl Acad Sci USA*, 2004 In press.

8
Glia Cell Migration in *Drosophila*

Gundula Edenfeld and Christian Klämbt

8.1
Introduction

The appearance of a cell that was able to move towards a food source or away from unfavorable growth conditions may have been one of the major breakthroughs in evolution. Cell migration turned out to be one of the most important aspects of life; it is found in bacteria and unicellular organisms, as well as in complex multicellular organisms. Bacteria developed a cilium, which is able to propel them towards a signal; similarly cilia help sperm cells to find the oocyte. The transition of unicellular to multicellular life, however, required different means of migration, namely amoeboid movements. Here, receptors on one side of the cell are able to sense the external environment and in response form cell protrusions called lamellipodia that pull the cell towards a position. This archetypical mode of cell migration can be seen in *Dictyostelium* as well as in macrophages as they patrol through the body. It is also how growth cones, the moving tips of navigating neuronal axons, find their way.

Being multicellular required the development of selective adhesion mechanisms to keep cells at certain places. The establishment of a merely static system, however, results in a number of disadvantages.

In contrast to plants, which use motile cells only during sexual life, the development of an animal requires a large number of migratory processes: germ cells as well as their progenitor cells; cells of the immune system assigned to important defense tasks; mesodermal cells; tracheal cells, and of course the diverse cells of the nervous system. All these cells have to migrate to develop into a functional complex organism.

8.2
Migration Must be Controlled

Cell migration, especially within the body, must be tightly controlled and we can discriminate at least three different phases, including initiation,

8.5
Development of the Embryonic *Drosophila* Nervous System

The prime reason for the rapid gain in the understanding of glial cell specification and differentiation is based on the beautiful simplicity of the *Drosophila* nervous system. To date all central or peripheral nervous system (CNS or PNS) glial cells can unambigously be identified using a steadily growing set of different molecular markers (Freeman et al., 2003; Granderath and Klambt, 1999; Jacobs, 2000; Van De Bor and Giangrande, 2002). These tools allow the analysis of single cells in wild type and in different mutant backgrounds and have paved the way to set up genetic screens for genes required in gliogenesis.

The CNS of the fly embryo consists of two brain hemispheres located laterally in the head region and the ventral nerve cord which comprises three thoracic and eight abdominal neuromeres condensed to a single ventral nerve cord. At the end of the first third of embryogenesis a set of neuronal stem cells, the neuroblasts, has been allocated. Elegant studies in the laboratories of G. Technau and C. Doe now permit identification of the 30 different neuroblasts found in each hemineuromere and also the lineages of almost all of these stem cells (Bossing et al., 1996; Schmid et al., 1999; Schmidt et al., 1997). At the end of embryogenesis about 60 glial cells and 700 neurons are found per abdominal neuromere. The vast majority of these neurons project their axons on the contralateral side through one of the two commissures (Fig. 8.1 B). The CNS midline cells and in particular the midline glia serve as an intermediate target guiding commissural axons initially towards and then across to the other side. But how does a growth cone know via which of the two commissures it should cross the midline? How are the many fiber tracts organized after crossing the midline? Part of the answer to these questions are the midline glial cells.

8.6
The Midline Glia

All midline cells stem from 7–8 segmentally arranged stem cells located at the boundary between the neuroectodermal and the mesodermal anlage (Bossing and Technau, 1994; Klämbt et al., 1991). In contrast to neuroblasts, these progenitor cells do not fully delaminate from the neuroepithelium but stay anchored to it with thin cell processes. The nature of these endfeet is unclear but it is interesting to note that they are only retracted when the migration of the glial cells towards axonal membranes has started.

As soon as the midline glia has delaminated into the interior of the embryo, basal cell processes start to extend posterior to the site where

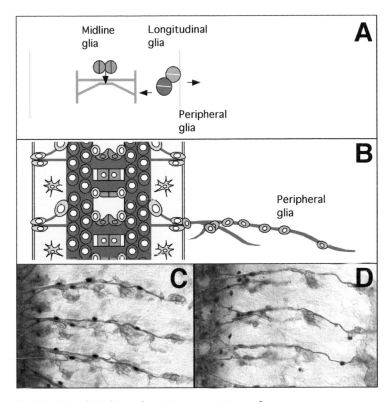

Fig. 8.1 A) and B) show schematic representations of ventral nerve cord development. A) In early development first axons form the segmental commissures (grey). The midline glia (red) start to migrate posteriorly whereas the progeny of the lateral glioblast migrate towards the forming commissures. The peripheral glial cells (green) migrate along the segmental nerves as soon as axons have left the CNS (B). B) In a mature embryonic nervous system the midline glia (red) has covered the commissures, the longitudinal glia covers the longitudinal connectives and the peripheral glial cells (green) are found in the PNS. Some additional glial cells are indicated in grey. C, D) *Drosophila* embryonic PNS axons labeled anti-Futsch and subsequent HRP immunohistochemistry; subset of peripheral glia nuclei visualized via an enhancer trap insertion into *gliotactin* (blue). C) Note the stereotyped positions of the glial nuclei along the peripheral nerves. D) In *pilage* mutant embryos, glial cells stall at the CNS/PNS transition zone and do not migrate into the periphery. (This figure also appears with the color plates.)

commissural axons will cross the midline. Initially, only axons within the posterior commissure will cross the midline; only when the glia has contacted the first commissural axons is the anterior commissure established. Initially, both commissure bundles are very closely associated, separated only by axons of the VUM midline neurons and the processes of the midline glia. These cellular extensions use the growing axons of the VUM midline neurons as a monorail that guides the migration of the midline glia in between the anterior and posterior commissures. Only this intercalating migration of the midline glia finally results in the separation of anterior and posterior commissures into two distinct axon bundles. If the midline glia is absent or not properly differentiated, a typical CNS phenotype develops where the two segmental commissures never separate and thus appear fused (Klämbt et al., 1991).

8.6.1
Determination and Early Differentiation of the Midline Glia

The decision to become a midline cell first requires the expression of the bHLH/PAS type transcription factors Single minded and Tango (Crews, 1998; Jacobs, 2000). The allocation of the glial cell fate within the midline domain is then brought about by a combination of *single minded* and different segmentation genes. Subsequently, *tramtrack* expression is activated that blocks neuronal development in glial cells and concomitantly glial differentiation is initiated by the *Drosophila* EGF-receptor homolog (DER) which, within the embryonic CNS, is expressed only in these cells (Scholz et al., 1997). In response to DER activation, two antagonizing ETS-domain transcription factors, PointedP2 and Yan become phosphorylated in the nucleus (Brunner et al., 1994; Rebay and Rubin, 1995). Upon phosphorylation PointedP2 directly acitvates glial differentiation, while Yan normally acts a transcriptional repressor that only upon phosphorylation translocates to the cytoplasm were it becomes degraded.

Finally, the correct number of midline glia cells needs to be specified. Of the six midline glial cells initially formed, only three or four survive to the end of embryogenesis. The regulation of the final glial cell number is controlled by the activity of the EGF-receptor and the small hormone ecdysone (Bergmann et al., 2002; Giesen et al., 2003).

8.6.2
Genetic Analysis of Midline Glia Migration

Migration of the midline glia is a prerequisite for the correct formation of the two segmental commissures found in the ventral nerve cord of the *Drosophila* embryo. As soon as migration of the midline glia cells is disrupted, a common phenotype of fused commissures develops. In order to further understand glial differentiation and in particular glial cell migration we conducted a large-scale genetic screen looking for mutations that lead to such axonal defects and identified more than 20 genes (Hummel et al., 1999 a, b).

8.6.3
Mutations in *kette* Affect the Differentiation of Neurites that Guide Glial Migration

To disrupt glial migration, the identified genes could either act autonomously in the glia or they could be required in the neuronal substrate used by the glial cell during their normal migration. Analysis of the *kette* mutant phenotype showed that all midline cells are specified normally and revealed defective neurite extension. In addition, genetic evidence suggested that *kette* is likely to act in CNS neurons (Hummel et al., 2000). We subsequently showed that *kette* mutants disrupt the function of the *Drosophila* homolog of the Hem-2/Nap1 protein. *kette* is initally expressed ubiquitously in the embryo but expression becomes restricted to CNS neurons by the end of embryogenesis. In the CNS midline, the VUM neurons show distinct projection defects, which correlate to the migratory defects of the midline glial cells. Within the mesoderm *kette* is required for muscle fusion and the subsequent migration of the muscles to their attachment sites in the ectoderm. Further analyses showed that *kette* performs its function by regulating the dynamics of the F-actin cytoskeleton and loss of *kette* function leads to an increase of F-actin within the cells.

The dynamic actin cytoskeleton plays a central role in cell morphology and motility (Borisy and Svitkina, 2000; Pantaloni et al., 2001). Rearrangement of F-actin is evoked by extracellular stimuli and sets of actin-associated proteins tightly control the ability to switch from a monomeric (G-actin) to a filamentous form (F-actin). Polymerization of F-actin starts with the de novo nucleation of an actin trimer, a process, which requires the action of the Arp2/3 complex (Higgs and Pollard, 2001). This complex in turn is regulated by a set of activators, such as the members of the WASP (Wiskott-Aldrich syndrome protein) and WAVE families (Suetsugu et al., 2002). WASP proteins are auto-inhibited, whereas WAVE is trans-inhibited through its association with a large protein complex that also comprises Kette (Eden et al., 2002). In agreement with a proposed role of Kette in trans-inhibiting WAVE we found most of the Kette protein in the cytosol, however, small amounts of Kette can be detected at the cell membrane where Kette resides in its active form. Genetic experiments revealed that Kette not only participates in the inhibition of WAVE, which normally induces the formation of lamellipodia, but is also able to activate WASP whose activity is required for filopodia formation (Bogdan and Klämbt, 2003).

Regulation of the dynamic F-actin cytoskeleton is essential during cell migration and requires a close link to the plasma membrane and transmembrane receptors involved in the perception of extracellular signals – secreted for example by the midline glia. In the case of receptor tyrosine kinases (RTKs), autophosphorylation of tyrosine residues leads to the recruitment of SH2 (Src homology 2) domain containing adapter proteins such as Nck to the membrane. Among the Nck-interacting proteins are WASP, the non-receptor tyrosine kinase Abl (Li et al., 2001) and Kette (also called Nck-associated protein 1) (Bogdan and Klämbt, 2003; Hummel et al., 2000; Li et al., 2001). However, the membrane receptor involved in the perception a glial-derived signal is unknown to date.

8.6.4
Lateral Glia Development and Migration

The lateral glial cells of the embryonic ventral nerve cord derive from the neural ectoderm where they originate from either pure glioblasts or neuro-glioblasts that generate mixed lineages (Bossing et al., 1996; Schmid et al., 1999; Schmidt et al., 1997). As described for the midline glia, many of the lateral glial cells migrate during their development. The lateral glioblast generates the longitudinal glial cells that cover the longitudinal connectives. This glioblast is born at the lateral edge of the nerve cord and its progeny migrate towards the forming connectives (Jacobs et al., 1989).

Not much is known about the mechanism that directs the specification of glial progenitor cells but it finally results in the activation of the gene *glial cells missing* (*gcm*), which acts as a master regulator of lateral glial cell development (Hosoya et al., 1995; Jones et al., 1995; Vincent et al., 1996). *gcm* function may not even require a neural ground state and can induce at least some markers of glial differentiation in the mesoderm when expressed very early during development (Akiyama-Oda et al., 1998; Bernardoni et al., 1998). *gcm* encodes a transcription factor that directly activates its own expression as well as the expression of *pointedP1* and *tramtrack* (Giesen et al., 1997; Granderath et al., 2000; Miller et al., 1998). As mentioned earlier, *tramtrack* activity is needed to ensure that the neuronal differentiation program is not activated in the glia (Giesen et al., 1997). PointedP1 and Repo, another transcription factor, in turn direct the expression of glial differentiation markers and thus route the committed precursor cells into terminal glial differentiation (Klaes et al., 1994; Yuasa et al., 2003).

Migration of the longitudinal glial cells towards the midline appears to be directed by the same signals that guide commissural growth cones across the midline. During migration the longitudinal glia transiently express the Roundabout receptor, which receives its repulsive ligand Slit from the CNS midline cells. Activation of the receptor halts their ventral migration short of the midline where glial cells start their differentiation (Kinrade et al., 2001). Again, the correct migration of longitudinal glial cells is required for the normal formation of longitudinal axon tracts. Mutations that interfere with longitudinal glia development also lead to severe axonal phenotypes (Jacobs, 1993) likewise does a targeted ablation of all or subsets of glial cells result in an abnormal axonal tract formation (Hidalgo and Booth, 2000; Hidalgo et al., 1995). Concomitantly, glial cells are dependent on the presence of axons (Booth et al., 2000; Hidalgo et al., 2001). In summary a number of clear mutant phenotypes will allow selection of mutations that affect glial migration and will further advance our understanding of glial cell migration in the future.

8.7
PNS Glia Embryo and Larvae

Glial migration in the CNS usually does not extend over large distances. In contrast to this are glial cells in the PNS, which often cover considerable distances. In early embryonic development, peripheral glial cells are born in the CNS and migrate to the lateral edge of the ventral nerve cord (Ito et al., 1995; Klämbt and Goodman, 1991). Here they seem to be intermediate targets for motoaxons that are navigating into the periphery. Once the motoaxons have extended their processes, the peripheral glial cells follow the routes of these pioneer neurons to finally occupy stereotyped positions along the segmental and intersegmental nerve (Sepp et al., 2000, Fig. 1 B, C). As described for other glial cells a reciprocal interaction exists between peripheral glial and PNS neurons (Sepp and Auld, 2003a; Sepp et al., 2001).

To date not much is known about how peripheral glial cell migration is controlled. The pivotal role of a dynamic F-Actin cytoskeleton as an integral part of cell motility has recently been shown using an Actin-GFP marker expressed in glial cells (Sepp and Auld, 2003b). The peripheral glial cells appear to migrate as a continuous chain of cells, with the leading glial cells exploring the extracellular surroundings with filopodia-like Actin containing projections. Using transgenic expression models the small GTPases RhoA and Rac1 were found to have distinct roles in peripheral glial cell migration and nerve ensheathment whereas Cdc42 does not appear to have a prominent role in peripheral glial development. The open question remains how activity of Rac1 mediated F-actin dynamics is controlled in glial cells.

Not only within the embryonic PNS, but also in the larval PNS, glial cells migrate over considerable distances. In the larval PNS, the developing photoreceptor cells send their axons through the optic stalk to the brain. Here, glial cells are born which migrate outwards to the receptor cells along the optic stalk (Choi and Benzer, 1994). Recent analysis has shown that these glial cells are guided by a signal released from the developing photoreceptor cells in the eye disc and invade the eye discs even in the absence of photoreceptor axons. Since the glial cells are able to migrate to isolated patches of differentiating photoreceptors without axons providing a continuous physical substratum, it has been suggested that glial cells are attracted into the eye disc by a chemotaxic mechanism (Rangarajan et al., 1999). Motility of the glial cells appears in part to be controlled by *hedgehog*, however, there is still some debate on whether *hedgehog* itself is a guidance molecule or whether it sets the stage by regulating other genes (Hummel et al., 2002; Rangarajan et al., 2001).

8.7.1
Genetic Analysis of PNS Glia Migration

To further understand the molecular mechanisms that orchestrate glia migration during development, we have initiated another genetic screen looking for mutations affecting the migration of the peripheral glial cells. To specifically assess the migration of these cells we used an enhancer trap marker that allows the detection of only four glial cell nuclei along the peripheral nerves. The position of the cell nucleus is an excellent first criterion for the position of the respective glial cell. Following a conventional EMS mutagenesis protocol we have analyzed more than 2400 strains that carried at least one lethal hit on the X-chromosome. In addition, we have screened a collection of third chromosomal mutations generated in the laboratory (Schimmelpfeng, Hummel CK unpublished). All mutations were balanced over a dominantly marked chromosome that directly allowed to score the mutant embryos. In total we have selected 28 mutant lines that show clear defects in number and position of the different peripheral glial cells (Tab. 8.1).

In most of the mutants that lead to an excess of glial cells, the additional cells are intermingled in-between the wild type positions, indicating that the position of a given glial cell is not fixed. In contrast, two mutations lead to an apparent duplication of one particular peripheral glial cell. This finding suggests that cell fate can be specified down to each individual cell. In one allele we noted a reduced number of glial cells. Here we may have hit a gene, which like *gcm* is required for the correct development of the glial lineage.

Tab. 8.1 Screen summary of X-chromosomal mutants

Number of mutants displaying defects in		
Differentiation	more glial cells	8
	less glial cells	1
Migration	excess migration	4
	less migration	10
	irregular migration	14

In 19 alleles we noted no defects in glial cell number but observed an abnormal migration pattern. This rather low and therefore unexpected number reflects the astonishing accuracy in normal glial patterning. In addition it indicates that the mutants identified affect important aspects of peripheral glial cell migration (Edenfeld and Klämbt in preparation).

8.8
Conclusion

We have now focused on those mutations that display excess or generally reduced glial migration. To further determine a more general relevance of the identified genes during glial migration we have analyzed the migration of all other glial cells in the embryonic nervous system, too. We also generated mutant cell clones during larval development in order to determine whether a given mutation is required during larval stages and if so whether it is acting cell-autonomously or not during the migration of glial cells. Analysis of the different genes identified so far is expected to advance our understanding of glial migration possibly not only in the *Drosophila* system.

In summary, a number of examples of cell migration have been described in the *Drosophila* system. Genetic analysis of the genes underlying the different migratory phenomena has only just begun. In principle we may find genes that interfere with extrinsic signals guiding directed migration or encode receptors required to read the respective signals. We are expecting genes that code further components of the cytoskeleton linking adhesive contacts to the cellular cytoskeleton.

8.9
References

AKIYAMA-ODA, Y., T. HOSOYA, and Y. HOTTA (**1998**) Alteration of cell fate by ectopic expression of Drosophila glial cells missing in non neural cells. *Dev Genes Evol. 208*, 578–585.

BERGMANN, A., M. TUGENTMAN, B.Z. SHILO, and H. STELLER (**2002**) Regulation of cell number by MAPK-dependent control of apoptosis: a mechanism for trophic survival signaling. *Dev Cell. 2*, 159–170.

BERNARDONI, R., A.A. MILLER, and A. GIANGRANDE (**1998**) Glial differentiation does not require a neural ground state. *Development 125*, 3189–3200.

BOGDAN, S., and C. KLÄMBT (**2003**) Kette regulates actin dynamics and genetically interacts with Wave and Wasp. *Development 130*, 4427–4437.

BOOTH, G.E., E.F. KINRADE, and A. HIDALGO (**2000**) Glia maintain follower neuron survival during Drosophila CNS development. *Development 127*, 237–244.

BORISY, G.G., and T.M. SVITKINA (**2000**) Actin machinery: pushing the envelope. *Curr Opin Cell Biol. 12,* 104–112.

BOSSING, T., and G.M. TECHNAU (**1994**) The fate of the CNS midline progenitors in Drosophila as revealed by a new method for single cell labeling. *Development 120,* 1895–1906.

BOSSING, T., G. UDOLPH, C.Q. DOE, and G.M. TECHNAU (**1996**) The embryonic central nervous system lineages of *Drosophila melanogaster.* I. Neuroblast lineages derived from the ventral half of the neuroectoderm. *Dev Biol. 179,* 41–64.

BROIHIER, H.T., L.A. MOORE, M. VAN DOREN, S. NEWMAN, and R. LEHMANN (**1998**) zfh 1 is required for germ cell migration and gonadal mesoderm development in *Drosophila. Development 125,* 655–666.

BRUNNER, D., K. DUCKER, N. OELLERS, E. HAFEN, H. SCHOLZ, and C. KLÄMBT (**1994**) The ETS domain protein pointed P2 is a target of MAP kinase in the sevenless signal transduction pathway. *Nature 370,* 386–389.

CANTERA, R. (**1993**) Glial cells in adult and developing prothoracic ganglion of the hawk moth Manduca sexta. *Cell Tissue Res. 272,* 93–108.

CHO, N.K., L. KEYES, E. JOHNSON, J. HELLER, L. RYNER, F. KARIM, and M.A. KRASNOW (**2002**) Developmental control of blood cell migration by the Drosophila VEGF pathway. *Cell 108,* 865–876.

CHOI, K.W., and S. BENZER (**1994**) Migration of glia along photoreceptor axons in the developing *Drosophila* eye. *Neuron 12,* 423–431.

CREWS, S.T (**1998**) Control of cell lineage specific development and transcription by bHLH PAS proteins. *Genes Dev. 12,* 607–620.

DICKSON, B.J (**2002**) Molecular mechanisms of axon guidance. *Science 298,* 1959–1964.

EDEN, S., R. ROHATGI, A.V. PODTELEJNIKOV, M. MANN, and M.W. Kirschner (**2002**) Mechanism of regulation of WAVE1-induced actin nucleation by Rac1 and Nck. *Nature 418,* 790–793.

EDWARDS, J.S., L.S. SWALES, and M. BATE (**1993**) The differentiation between neuroglia and connective tissue sheath in insect ganglia revisited: the neural lamella and perineurial sheath cells are absent in a mesodermless mutant of Drosophila. *J Comp Neurol. 333,* 301–308.

FORBES, A., and R. LEHMANN (**1999**) Cell migration in *Drosophila. Curr Opin Genet Dev. 9,* 473–478.

FREEMAN, M.R., J. DELROW, J. KIM, E. JOHNSON, and C.Q. DOE (**2003**) Unwrapping glial biology: Gcm target genes regulating glial development, diversification, and function. *Neuron 38,* 567–580.

GIESEN, K., T. HUMMEL, A. STOLLEWERK, S. HARRISON, A. TRAVERS, and C. KLÄMBT (**1997**) Glial development in the Drosophila CNS requires concomitant activation of glial and repression of neuronal differentiation genes. *Development 124,* 2307–2316.

GIESEN, K., U. LAMMEL, D. LANGEHANS, K. KRUKKERT, I. BUNSE, and C. KLÄMBT (**2003**) Regulation of glial cell number and differentiation by ecdysone and Fos signaling. *Mech Dev. 120,* 401–413.

GRANDERATH, S., I. BUNSE, and C. KLÄMBT (**2000**) gcm and pointed synergistically control glial transcription of the drosophila gene loco [In Process Citation]. *Mech Dev. 91,* 197–208.

GRANDERATH, S. and C. KLÄMBT (**1999**) Glia development in the embryonic CNS of Drosophila. *Curr Opin Neurobiol. 9,* 531–536.

HIDALGO, A., and G. E. BOOTH (2000) Glia dictate pioneer axon trajectories in the Drosophila embryonic CNS. *Development, 127,* 393–402.

HIDALGO, A., E. F. KINRADE, and M. GEORGIOU (2001) The Drosophila neuregulin vein maintains glial survival during axon guidance in the CNS. *Dev Cell. 1,* 679–690.

HIDALGO, A., J. URBAN, and A. H. BRAND (1995) Targeted ablation of glia disrupts axon tract formation in the *Drosophila* CNS. *Development 121,* 3702–3712.

HIGGS, H. N. and T. D. POLLARD (2001) Regulation of actin filament network formation through ARP2/3 complex: activation by a diverse array of proteins. *Annu Rev Biochem. 70,* 649–676.

HOSOYA, T., K. TAKIZAWA, K. NITTA, and Y. HOTTA (1995) Glial cells missing: a binary switch between neuronal and glial determination in *Drosophila. Cell 82,* 1025–1036.

HOWARD, K., M. JAGLARZ, N. ZHANG, J. SHAH, and R. WARRIOR (1993) Migration of Drosophila germ cells: analysis using enhancer trap lines. *Dev Suppl.* 213–218.

HOYLE, G. (1986) Glial cells of an insect ganglion. *J Comp Neurol. 246,* 85–103.

HUMMEL, T., S. ATTIX, D. GUNNING, and S. L. ZIPURSKY (2002) Temporal control of glial cell migration in the Drosophila eye requires gilgamesh, hedgehog, and eye specification genes. *Neuron 33,* 193–203.

HUMMEL, T., K. LEIFKER, and C. KLÄMBT (2000) The Drosophila HEM-2/NAP1 homolog KETTE controls axonal pathfinding and cytoskeletal organization. *Genes Dev. 14,* 863–873.

HUMMEL, T., K. SCHIMMELPFENG, and C. KLÄMBT (1999a) Commissure formation in the embryonic CNS of *Drosophila*: I Identification of the required gene functions. *Dev. Biol. 209,* 381–398.

HUMMEL, T., K. SCHIMMELPFENG, and C. KLÄMBT (1999b) Commissure formation in the embryonic CNS of *Drosophila* : II Function of the different midline cells. *Development 126,* 771–779.

ITO, K., J. URBAN, and G. M. TECHNAU (1995) Distribution, classification and development of Drosophila glial cells during late embryogenesis. *Roux's Arch Dev Biol. 204,* 284–307.

JACOBS, J. R. (1993) Perturbed glial scaffold formation precedes axon tract malformation in Drosophila mutants. *J Neurobiol. 24,* 611–626.

JACOBS, J. R. (2000) The Midline Glia of Drosophila: a molecular genetic model for the developmental functions of Glia. *Prog Neurobiol. 62,* 475–508.

JACOBS, J. R., Y. HIROMI, N. H. PATEL, and C. S. GOODMAN (1989) Lineage, migration, and morphogenesis of longitudinal glia in the Drosophila CNS as revealed by a molecular lineage marker. *Neuron 2,* 1625–1631.

JONES, B. W., R. D. FETTER, G. TEAR, and C. S. GOODMAN (1995) Glial cells missing: a genetic switch that controls glial versus neuronal fate. *Cell 82,* 1013–1023.

KINRADE, E. F., T. BRATES, G. TEAR, and A. HIDALGO (2001) Roundabout signalling, cell contact and trophic support confine longitudinal glia and axons in the *Drosophila* CNS. *Development 128,* 207–216.

KLAES, A., T. MENNE, A. STOLLEWERK, H. SCHOLZ, and C. KLÄMBT (1994) The Ets transcription factors encoded by the Drosophila gene pointed direct glial cell differentiation in the embryonic CNS. *Cell 78*, 149–160.

KLÄMBT, C. and C. S. GOODMAN (1991) The diversity and pattern of glia during axon pathway formation in the Drosophila embryo. *Glia 4*, 205–213.

KLÄMBT, C., J. R. JACOBS, and C. S. GOODMAN (1991) The midline of the *Drosophila* central nervous system: a model for the genetic analysis of cell fate, cell migration, and growth cone guidance. *Cell 64*, 801–815.

LEMKE, G. (2001) Glial control of neuronal development. *Annu Rev Neurosci. 24*, 87–105.

LI, W., J. FAN, and D. T. WOODLEY (2001) Nck/Dock: an adapter between cell surface receptors and the actin cytoskeleton. *Oncogene 20*, 6403–6417.

MILLER, A. A., R. BERNARDONI, and A. GIANGRANDE (1998) Positive autoregulation of the glial promoting factor glide/gcm. *EMBO J. 17*, 6316–6326.

MONTELL, D. J. (2003) Border-cell migration: the race is on. *Nat Rev Mol Cell Biol. 4*, 13–24.

MOORE, L. A., H. T. BROIHIER, M. VAN DOREN, L. B. LUNSFORD, and R. LEHMANN (1998) Identification of genes controlling germ cell migration and embryonic gonad formation in Drosophila. *Development 125*, 667–678.

PANTALONI, D., C. LE CLAINCHE, and M. F. CARLIER (2001) Mechanism of actin-based motility. *Science 292*, 1502–1506.

RAMON Y CAJAL, S. and D. SÁNCHEZ Y SÁNCHEZ (1915) Contribution al conocrimiento de los centros nerviosos de los insectos. Parte I. Retina y centros opitcos. *Trab. Lab. Invertebr. Biol. Univ. Madrid. 13*, 1–168.

RAMON-CUETO, A. and J. AVILA (1997) Differential expression of microtubule associated protein 1B phosphorylated isoforms in the adult rat nervous system. *Neuroscience 77*, 485–501.

RANGARAJAN, R., H. COURVOISIER, and U. GAUL (2001) Dpp and Hedgehog mediate neuron-glia interactions in Drosophila eye development by promoting the proliferation and motility of subretinal glia. *Mech Dev. 108*, 93–103.

RANGARAJAN, R., Q. GONG, and U. GAUL (1999) Migration and function of glia in the developing Drosophila eye. *Development 126*, 3285–3292.

RAZ, E. (2003) Primordial germ-cell development: the zebrafish perspective. *Nat Rev Genet. 4*, 690–700.

REBAY, I. and G. M. RUBIN (1995) Yan functions as a general inhibitor of differentiation and is negatively regulated by activation of the Ras1/MAPK pathway. *Cell 81*, 857–866.

RIBEIRO, C., V. PETIT, and M. AFFOLTER (2003) Signaling systems, guided cell migration, and organogenesis: insights from genetic studies in Drosophila. *Dev Biol. 260*, 1–8.

RØRTH, P. (2002) Initiating and guiding migration: lessons from border cells. *Trends Cell Biol. 12*, 325–331.

SCHMID, A., A. CHIBA, and C. Q. DOE (1999) Clonal analysis of Drosophila embryonic neuroblasts: neural cell types, axon projections and muscle targets. *Development 126*, 4653–4689.

SCHMIDT, H., C. RICKERT, T. BOSSING, O. VEF, J. URBAN, and G. M. TECHNAU (1997) The embryonic central nervous system lineages of *Drosophila melanogaster*. II. Neuroblast lineages derived from the dorsal part of the neuroectoderm. *Dev Biol. 189*, 186–204.

SCHOLZ, H., E. SADLOWSKI, A. KLAES, and C. KLÄMBT (1997) Control of midline glia development in the embryonic *Drosophila* CNS. *Mech Dev. 64*, 137–151.

SEPP, K. J. and V. J. AULD (2003a) Reciprocal interactions between neurons and glia are required for *Drosophila* peripheral nervous system development. *J Neurosci. 23*, 8221–8230.

SEPP, K. J. and V. J. AULD (2003b) RhoA and Rac1 GTPases mediate the dynamic rearrangement of actin in peripheral glia. *Development 130*, 1825–1835.

SEPP, K. J., J. SCHULTE, and V. J. AULD (2000) Developmental dynamics of peripheral glia in Drosophila melanogaster. *Glia 30*, 122–133.

SEPP, K. J., J. SCHULTE, and V. J. AULD (2001) Peripheral glia direct axon guidance across the CNS/PNS transition zone. *Dev Biol. 238*, 47 63.

STARZ-GAIANO, M., N. K. CHO, A. FORBES, and R. LEHMANN (2001) Spatially restricted activity of a Drosophila lipid phosphatase guides migrating germ cells. *Development 128*, 983–991.

STRAUSFELD, N. J. (1976) *Atlas of an insect brain*. Springer, Berlin Heidelberg New York.

SUETSUGU, S., H. MIKI, and T. TAKENAWA (2002) Spatial and temporal regulation of actin polymerization for cytoskeleton formation through Arp2/3 complex and WASP/WAVE proteins. *Cell Motil Cytoskeleton 51*, 113–122.

UV, A., R. CANTERA, and C. SAMAKOVLIS (2003) Drosophila tracheal morphogenesis: intricate cellular solutions to basic plumbing problems. *Trends Cell Biol. 13*, 301–309.

VAN DE BOR, V. and A. GIANGRANDE (2002) glide/gcm: at the crossroads between neurons and glia. *Curr Opin Genet Dev. 12*, 465–472.

VAN DOREN, M., H. T. BROIHIER, L. A. MOORE, and R. LEHMANN (1998) HMG CoA reductase guides migrating primordial germ cells. *Nature 396*, 466–469.

VINCENT, S., J.-L. VONESCH, and A. GIANGRANDE (1996) *glide* directs glial fate commitment and cell fate switch between neurones and glia. *Development 122*, 131–139.

YUASA, Y., M. OKABE, S. YOSHIKAWA, K. TABUCHI, W. C. XIONG, Y. HIROMI, and H. OKANO (2003) Drosophila homeodomain protein REPO controls glial differentiation by cooperating with ETS and BTB transcription factors. *Development 130*, 2419–2428.

9

The Neural Crest: Migrating from the Border

Marianne Bronner-Fraser

9.1
Introduction

In the early embryo, the nervous system arises as the neural plate forms a neural tube that then transforms into the brain, spinal cord and peripheral ganglia. The neural plate arises in the midline of the ectoderm and folds upon itself to form the neural tube, a cylindrical structure that spans the rostrocaudal length of the embryo. At the edges or border region of the neural plate lie presumptive neural crest cells. These are a transient population of cells, so named because they arise on the "crest" of the closing neural tube.

Formation of neural crest cells at the border between the neural plate and non-neural ectoderm occurs by an inductive interaction between these two tissues (Moury and Jacobson, 1990; Selleck and Bronner-Fraser, 1995). Recently, the molecular mechanisms underlying this induction process have been described. In birds, Wnt appears to be both necessary and sufficient to cause neural crest formation from competent neural plate tissue (Garcia-Castro et al., 2002). Because Wnt6 is expressed appropriately in the non-neural ectoderm, this is likely to represent an endogenous inducer of cells with the ability to form neural crest cells. In Xenopus, Wnt family members also are strong inducers of neural crest markers when injected in neuralized animal caps (Saint-Jeannet et al., 1997; LaBonne and Bronner-Fraser, 1998).

Although precursors with the potential to form neural crest initially are contained within the dorsal portion of the neural tube, neural crest cells subsequently delaminate from the neuroepithelium by undergoing an epithelial to mesenchymal conversion. Around the time of neural tube closure, these cells lose their epithelial connections to other neural tube cells and become a migratory mesenchymal population that moves quickly away from the central nervous system. The transcription factor, Slug, is expressed in premigratory and early migrating cells and represents an early neural crest marker (Fig. 9.1). Slug has been associated with epithelial–mesenchymal conversions in a number of cell types and

Cell Migration. Edited by Doris Wedlich
Copyright © 2005 WILEY-VCH Verlag GmbH & Co. KGaA, Weinheim
ISBN: 3-527-30587-4

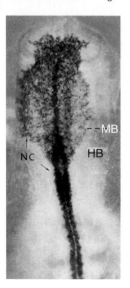

Fig. 9.1 Slug is a marker for early migrating and premigratory neural crest cells. Whole mount *in situ* hybridization of a 9 somite stage embryo showing the distribution pattern of the transcription factor Slug, which is an early neural crest marker. At the level of the midbrain (MB), neural crest cells (NC, arrow) have already begun to migrate and express the Slug, whereas in the caudal hindbrain (HB) and trunk, Slug stains premigratory neural crest cells. (Adapted from Sechrist et al. 1995.)

its function appears to be necessary for the emigration of neural crest cells (Nieto et al., 1994; delBarrio and Nieto, 2002). Initiation of neural crest cell migration proceeds in a rostral to caudal progression along most of the neural axis, following upon the heals of the head-to-tailward closure of the neural tube. After emigration, these cells then migrate extensively and undergo significant rearrangements. They move in a highly patterned fashion through neighboring tissues to localize in specific and diverse sites. The molecular nature of the cues guiding cell movement are beginning to be understood and appear to differ between different regions of the nervous system.

Upon completion of migration, neural crest cells differentiate into an astonishing array of derivatives. All of the dorsal root, sympathetic, parasympathetic and enteric ganglia are derived from neural crest cells. Furthermore, most cranial sensory ganglia receive a contribution from the neural crest, with the remaining cells derived from the ectodermal placodes. In addition to forming neurons and glia of the peripheral nervous system, neural crest cells form melanocytes, cranial cartilage and adrenal chromaffin cells. This wide variety of cell types arises from multipotent cells that have stem cell-like properties (Bronner-Fraser and Fraser, 1988; Stemple and Anderson, 1992; Lee et al., 2004).

9.2
Different Populations of Neural Crest Cells along the Rostrocaudal Axis

Neural crest cells emerge from the dorsal neural tube and migrate along pathways that are characteristic of their axial level of origin (Bronner-Fraser et al., 1991; Le Douarin, 1986; Le Douarin et al., 1992). The different populations of neural crest cells arising along the neural axis have been designated as cranial, vagal, trunk and lumbosacral (Fig. 9.2). Distinct cell types differentiate from these different populations. For example, vagal neural crest cells emerge from the caudal hindbrain and migrate into and along the rostrocaudal extent of the gut where they form the enteric ganglia. In contrast, trunk neural crest cells never enter the gut even if transplanted to vagal levels (Erickson and Goins, 2000; Le Douarin and Teillet, 1973, 1974).

At cranial levels, some neural crest cells contribute to the cranial sensory ganglia and the parasympathetic ciliary ganglion of the eye whereas others migrate ventrally to form many of the cartilaginous elements of the facial skeleton. Precise quail/chick grafting experiments have determined the regions of neural tube from which neural crest cells arise to contribute to cartilaginous elements (Le Douarin and Teillet, 1973, 1974). Cranial neural crest cells contribute to the quadrate, Meckel's cartilage and surrounding membrane bones, cartilage of the tongue, and to membrane bones of the upper jaw and skull (Noden, 1978; Couly et al., 1992; 1993). Cranial neural crest cells can be subdivided further into regions designated as caudal forebrain, midbrain, rostral hindbrain and caudal hindbrain (which overlaps with rostral vagal) neural crest cells; each group has a somewhat different pattern of migration and prospective derivatives. For example, neural crest cells originating in the midbrain migrate primarily as a broad, unsegmented sheet under the ectoderm; they contribute to derivatives ranging from ciliary ganglion and trigeminal

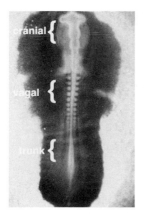

Fig. 9.2 Neural crest cells are regionalized along the neural axis, with different populations giving rise to different derivatives. In this whole mount view, the cranial, vagal and trunk neural crest regions are indicated. Each gives rise to a population of neural crest cells that migrates along distinct pathways and forms different types of derivatives. More caudally, lumbosacral neural crest will form (not shown).

ganglia to the periocular skeleton, connective tissue and membranous bones of the face (LeDouarin, 1982). Neural crest cells arising in the hindbrain migrate ventrally and enter the branchial arches to form the bones of the jaw.

Vagal neural crest cells migrate long distances to form the enteric nervous system, which also receives a contribution from the lumbosacral neural crest. Within the gut, the earliest-generated neural crest cells move as a wave from anterior to posterior to populate the bowel. The failure of neural crest migration in the aganglionic bowel of lethal spotted mutant mice relates to a defect in mesenchymal components and not the neural crest cells themselves. In mice and humans, aganglionic bowel (Hirschprung's disease) arises from mutations in endothelins and endothelin receptors (Hosoda et al., 1994).

Trunk neural crest cells migrate along two main pathways (Weston, 1963): dorsolaterally between the ectoderm and somite; and ventrally through the rostral half of the somitic sclerotome (Rickmann et al., 1985). Cells following the dorsolateral pathway form melanocytes whereas those following the ventral pathway give rise the peripheral nervous system of the trunk, including the chain of sympathetic ganglia and the dorsal root ganglia, as well as the chromaffin cells of the adrenal medulla (Fig. 9.3). In addition to neurons, these cells generate glia of the peripheral ganglia and Schwann cells that ensheathe and myelinate peripheral axons.

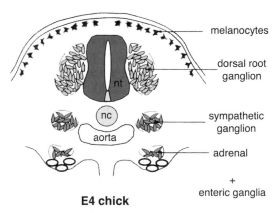

E4 chick

Fig. 9.3 Schematic diagram showing derivatives of trunk neural crest cells. Trunk neural crest cells migrate along two pathways: ventrally to contribute to dorsal root ganglia, sympathetic ganglia and adrenomedullary cells and dorsally to form melanocytes.

9.3

Both Intrinsic and Extrinsic Cues Influence Neural Crest Migration

Because neural crest cells that arise from different axial levels assume different fates, it is possible that they are differentially specified prior to migration. Alternatively, their migratory pathways may contribute to their fate decisions. These two possibilities are not mutually exclusive.

One strategy used to test the role of migratory pathways in neural crest cell fate decisions is to graft neural tubes from one axial level to a new, "heterotopic," location. For example, vagal neural crest cells normally contribute to the enteric ganglia of the gut whereas trunk neural crest cells fail to invade the gut even though it is immediately below their migratory pathway. Vagal neural crest cells grafted in place of the trunk neural tube form normal trunk derivatives. In addition, the grafted vagal cells invade the gut and form enteric ganglia (LeDouarin, 1982). Therefore, vagal neural crest cells can respond to normal trunk neural crest migratory cues, but also do something unique to the vagal neural crest by migrating directionally to the gut. This indicates some intrinsic differences between the two populations. Similarly, when cranial neural tubes are grafted in place of the trunk neural tube, donor cranial neural crest cells form some normal derivatives in the trunk like dorsal root and sympathetic ganglia while others fail to migrate and differentiate into ectopic cartilage masses. In the reverse experiment, grafts of trunk neural tube to the head region result in some contribution to the cranial ganglia, though there appear to be little or no neuronal differentiation from the donor cells. Similarly, trunk neural crest cells fail to form cartilage, even when grafted directly into the branchial arches (Lwigale et al., 2004).

The results of these experiments suggest that environmental factors, in fact, can influence neural crest cells derived from different axial levels to express a broader range of fates than they would normally express if left in place. However, they also reveal intrinsic differences between populations. Neural crest cell migration and fate decisions are likely to be governed by a combination of intrinsic and extrinsic information. Factors encountered along the migratory pathways may play an important regulatory role in cell fate specification. Indeed, there are different distributions of important inducing factors, including Wnts, BMPs, neuregulins and glucocorticoids along neural crest migratory pathways and these have been shown to influence neural crest cell fate decisions (Anderson, 1992; Lee et al., 2004). When added to multipotent neural crest stem cells in culture, these factors can drive these cells into sensory, sympathetic, glial, or chromaffin lineages, respectively. According, BMP-7 is present in the dorsal aorta adjacent to the location where trunk neural crest cells differentiate into sympathetic neurons (Shah et al., 1996) and Wnts are expressed in the dorsal neural tube adjacent to which dorsal root ganglia form (Garcia-Castro et al., 2002). Similarly, glucorticoids are produced by the adrenal cortex which surrounds the adrenal medullary cells.

The time at which neural crest cells emigrate from the neural tube also may play an important role in neural crest cell fate decisions. Whereas the first cells to migrate from the neural tube tend to move most ventrally, later emigrating neural crest cells contribute to progressively more dorsal derivatives (Weston, 1963; Serbedzija et al., 1989). Although early and late migrating cranial neural crest cells appear to have similar developmental potential (Baker et al., 1997), the last-emigrating trunk neural crest give rise to pigment cells but appear to have limited capacity to form sympathetic neurons (Artinger et al., 1992). Well past the normal time of neural crest cell emigration, however, neural tube cells with potential to form multiple neural crest derivatives persist and appear to migrate out of the neural tube very late through the dorsal roots (Sharma et al., 1995). In the embryo, these late-emigrating cells appear to contribute to a subpopulation of neural crest derived cells in the dorsal root ganglia (Korade and Frank, 1996).

9.4
Trunk and Hindbrain Neural Crest Migrate in a Segmental Fashion

After neural crest cells leave the neural tube, they migrate through various tissues in the periphery and then form multiple derivatives. In the trunk, they condense to form segmentally-arranged sensory and sympathetic ganglia of the peripheral nervous system. For each of the somites lying adjacent to the neural tube, one sensory and one sympathetic ganglion forms. Careful examination revealed that neural crest cells selectively migrate through the rostral but not caudal half of each somitic sclerotome (Rickman et al., 1986; Fig. 9.4A). Experimental manipulation of the somites suggests that the somites themselves are responsible for the segmental organization of the ganglia. For example, removal of the somites results in failure of segmentation such that one large and unsegmented ganglion-like structure forms. On the other hand, inversion of the segmental plate in the rostrocaudal dimension reverses the pattern of neural crest migration (Bronner-Fraser and Stern, 1991), such that neural crest cells migrate through the caudal (original rostral) halves of the rotated somites. This suggests that the information necessary to guide neural crest cells and motor axons is intrinsic to the somites.

These experiments not only suggest that there is intrinsic patterning information within the somites but also that there are different domains within individual somites. In fact, the caudal-half of each somite appears to be inhibitory for both neural crest migration and motor axon guidance, whereas the rostral half is permissive. This was demonstrated via experiments in which somites constructed by grafting either "all rostral"

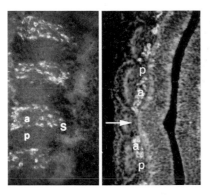

Fig. 9.4 Trunk neural crest cells migrate through the anterior (rostral) but not posterior (caudal) half of each somite. A) In a longitudinal section stained with the neural crest marker, HNK-1, neural crest cells can be observed migrating through the anterior (a) but not posterior (p) half of each somite (s). B) If the somites are inverted by surgically rotating the segmental plate prior to somitogenesis, the pattern of migration is inverted such that neural crest cells migrate through the part of the somite that was originally anterior, but is inverted after the manipulation. Arrow indicates a small somite at the border of the surgical rotation (adapted from Bronner-Fraser, 1986, and Stern and Bronner-Fraser, 1991, respectively).

or "all caudal" halves together. The former had a complete absence of segmentation and the latter had a complete absence of migration (Stern and Keynes, 1987). In addition, rostrocaudal rotation of the segmental plate prior to somite formation results in somites with reversed rostrocaudal polarity; during neural crest migration, the cells migrate selectively through the somites that were originally rostral but were now located caudally following rotation (Bronner-Fraser and Stern, 1989; Fig. 9.4 B). These findings suggest the existence of cues that are attractive cues for neural crest cells in the rostral-half sclerotome, inhibitory cues in the caudal-half sclerotome and/or a combination thereof. These experiments also show that the rostrocaudal polarity of the somites is already established at the segmental plate stage. It is interesting to note that the dorsoventral polarity of the somites is subject to interactions with adjacent tissues such as the ectoderm and ventral neural tube/notochord up until much later stages (Fan and Tessier-Lavigne, 1994; Munsterberg et al., 1995; Munsterberg and Lassar, 1995).

In addition to the trunk regions, other populations of neural crest cells also exhibit segmental manner patterns of migration. For example, those cells emigrating from the hindbrain migrate in a segmental comprised of three broad streams adjacent to rhombomere 2 (r2), r4 and r6 (Lumsden et al., 1991). In contrast, there is no apparent neural crest migration adjacent to r3 and r5. Although it has been proposed that neural crest cells arising from r3 and r5 may be eliminated by selective cell death (Graham et al., 1993), focal injections of DiI at the levels of r3 and r5 reveal that rhombomeres generate neural crest cells, but these cell migrate either rostrally or caudally to join the neural crest streams emerging from even numbered rhombomeres (Sechrist et al., 1993).

9.5
Attraction and Repulsion in Neural Crest Cell Guidance

9.5.1
Extracellular Matrix and Adhesion Molecules

Changes in expression of adhesion molecules presage emigration of neural crest cells from the neural tube. For example, neural tube cells express high levels of N-cadherin and cadherin-6b (cad-6b; Nakagawa et al., 1995) prior to neural crest migration. As the cells become premigratory, N-cadherin and cad-6b levels are down-regulated in the dorsal neural tube and cadherin-7 is up-regulated on early migrating neural crest cells. This suggests that a shift in cell surface and adhesive properties may accompany the onset of migratory behavior. RhoB is another molecule expressed by premigratory and early migrating neural crest cells (Liu and Jessell, 1998). It is expressed early in response to a BMP signal and may be involved in the epithelial–mesenchymal transition occurring during neural crest emigration.

After leaving the neural tube, neural crest cells enter migratory pathways that are lined with extracellular matrix molecules including fibronectin, laminin, collagens and proteoglycans. As many of these molecules interact with integrin receptors that are abundant on the neural crest cell surface, these are likely to serve as a suitable migratory substrate. Evidence that cell–matrix interactions are important for neural crest development comes from experiments in which antibodies to integrins were microinjected into embryo at the time of cranial neural crest migration (Bronner-Fraser, 1985, 1986). The subsequent defects in craniofacial development suggest that integrins are important for normal development of this region. Furthermore, antibodies to the α_4 integrin subunit caused defects in neural crest cell movement in hindbrain explants (Kil et al., 1998). Interestingly, the pattern of migration remains segmental despite disruption of integrin–matrix interactions in these explants. These results suggest that perturbing integrin function alters the properties of migratory cells, but not the overall migratory pattern. This suggests that cell–matrix interactions play a permissive, rather than instructive, role in segmental migration of neural crest cells.

9.5.2
Eph Receptor/Ephrin Interactions

In the developing nervous system, Eph receptor tyrosine kinases and their ligands have been shown to be expressed in intriguing patterns that are often complementary. Eph receptors are functionally divided into two subclasses, Eph A receptors, which interact primarily with a GPI-linked subclass of ligands, and Eph B receptors, which interact primarily with a

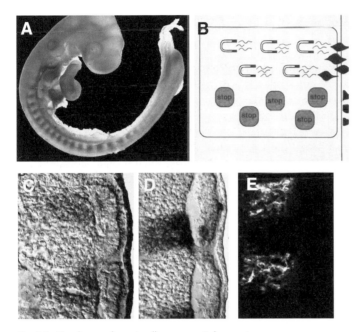

Fig. 9.5 Trunk neural crest cells express Eph receptors and ephrins in the posterior (caudal) half o each somite inhibit neural crest entry. A) Whole mount in situ hybridization of an E2.5 chick embryo showing a metameric pattern of EphB3 in the trunk somites. B) Schematic diagram showing that neural crest cells appear to be inhibited from migrating into the posterior (caudal) half of each somites by inhibitory cues like ephrins. However, there may also be attractive cues in the anterior half somite. C) Longitudinal section through the trunk somites of an E2.5 chick embryo showing EphB3 staining in the anterior half of each somite. D) Longitunial section showing that ephrin B1 is expressed in a reciprocal pattern in the posterior (caudal) half of each somite. E) HNK-1 staining showing that neural crest cells are migrating in the domain that expresses EphB3 but excluded from the domain expressing ephrin B1 (adapted from Krull et al., 1997).

transmembrane subclass of ligands (Gale et al., 1996). These receptor/ligand interactions play important roles in axon guidance as well as in cell migration, including that of neural crest cells.

A complementary pattern of Eph receptors and ephrin ligands is evident in the sclerotomal portion of the somites during the time of neural crest migration. An Eph receptor, EphB3 in chick and EphB1 in rat, is expressed on neural crest cells as they migrate through the somites (Wang and Anderson, 1997; Krull et al., 1997). The cognate ephrin B1 ligand is expressed in a reciprocal pattern in the caudal portion of the sclerotome through which neural crest cells fail to migrate (Fig. 9.5). Addition of exogenous ligand as a competitive inhibitor disrupts the segmental pattern of neural crest migration, demonstrating that the ligand is sufficient to inhibit neural crest migration. This suggests that interactions between Eph receptors and ephrin ligands restrict neural crest migration; this in turn leads to the subsequent metameric migration of neural crest cells and the subsequent organization of neural crest-derived sensory and sympathetic ganglia.

9.5.3
Chondroitin Sulfate and other Molecules may Inhibit Neural Crest Cells from the Perinotochordal Region

Other molecules implicated in neural crest cell guidance include chondroitin sulfate proteoglycans (Oakley et al., 1994) and Semaphorins (Eickholt et al., 1999). Neural crest cells fail to migrate in the region surrounding the notochord. The notochord secretes a thick perinotochordal matrix which contains chondroitin sulfate proteoglycan and other molecules that may act to inhibit crest cell migration (Pettway et al., 1992). Similarly, Semaphorin 3A appears to inhibit hindbrain neural crest cell migration in vitro (Eickholt et al., 1999).

9.5.4
Erb4/Neuregulin

Interactions between neuregulin, with its ligand Erb4 may be involved in keeping neural crest cells derived from r4 out of the mesodermal territory adjacent to r3. Thus, this represents another example of a signal that restricts migrating neural crest cells. In ErbB4 mutant mice, neural crest cells derived from r4 invade the region adjacent to r3. Interestingly, this mutation is non-neural crest cell autonomous. While ErbB4 is expressed in r3, neuregulin is confined to r4 (Golding et al., 2000).

9.5.5
Slit/Robo Interactions

Slit proteins have been shown to play important functions in axon guidance in both vertebrates and invertebrates (Brose et al., 1999; Kidd et al., 1999; Li et al., 1999). Slits are glycoproteins that are potent chemorepellants for midline axons in Drosophila as well as forebrain, dentate gyrus and olfactory axons in mammals (Bagri et al., 2002; Brose et al., 1999; Kidd et al., 1999; Li et al., 1999; Nguyen Ba-Charvet et al., 1999). In addition to functioning as repulsive factors during neuronal and glia cell migration in mammals (Hu, 1999; Kinrade et al., 2001; Wu et al., 1999; Zhu et al., 1999), they also can regulate axon elongation and branching (Wang et al., 1999).

Slits are expressed in places where they could influence migrating neural crest cells: in the dorsal neural tube from which neural crest cells emigrate and near the entrance to the gut, which is selectively invaded by vagal but not trunk neural crest cells. Consistent with this, only trunk neural crest cells possess Robo receptors for Slit (deBellard et al., 2003). When trunk neural crest cells either in vitro or in vivo are confronted by membrane-bound Slit2, their movement is inhibited (Fig. 9.6). These results account, at least in part, for the differential ability of vagal but not trunk neural crest cells to invade and innervate the gut. The presence of Slit family members at the entrance of the gut mesenchyme coupled with the presence of Robo receptors on trunk and their absence from vagal neural crest provides the first molecular explanation for the differences in migratory behavior of these two subpopulations via Slit chemorepulsion.

In addition to being a chemorepellent, soluble Slit2 selectively enhances the distance migrated by trunk neural crest. This suggests a dual role for Slit2 in a migratory cell type, both inhibiting movement when in direct confrontation and enhancing motility when in solution. Such dual function for Slit2 concurs with observations on axon guidance where Slit family members have been shown to play multiple roles, causing both faster axon growth and induction of branching (Ozdinler and Erzurumlu, 2002; Wang et al., 1999).

The dual functions of Slit2 in both repulsion and stimulation of migration in the same type of cell are not necessarily contradictory; an optimal chemorepellent might logically be expected to stimulate rapid movement away from its source. The results suggest that neural crest cells may not only avoid but also rapidly move away from sources of chemorepellants like Slit.

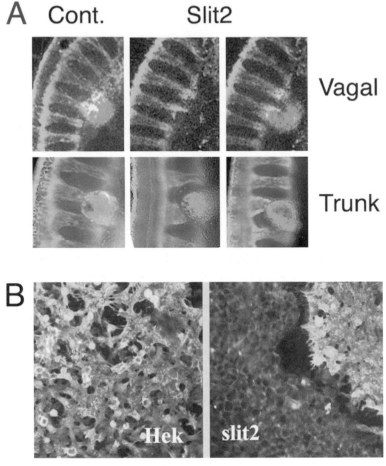

Fig. 9.6 Slit inhibits trunk neural crest migration.
A) Injection of cells producing Slit2 (red) at trunk levels
results in formation of a border between neural crest
cells, recognized with the HNK-1 antibody (green), and
Slit2 cells (red); in contrast, similar injections at vagal
levels lead to intermixing of Slit2 (red) and HNK-1 positive cells (green). B) HNK-positive neural crest cells
(green) intermix with control HEK cells, identified by
DAPI staining (blue). In contrast, neural crest cells
avoid and form a border with Slit-2 transfected cells
(blue = DAPI staining) cells (adapted from deBellard
et al., 2003). (This figure also appears with the color
plates).

9.6
Conclusions

Neural crest cells are a highly migratory cell type that emerges from the neural tube via an epithelial/mesenchymal transition and then embarks upon extensive migrations throughout the body of the embryo. These cells interact with a number of different extracellular and cell surface bound molecules as well as tissues along their pathways. For example, as neural crest cells move through the somites, ephrin ligands are selectively expressed in the caudal half of each somitic sclerotome whereas their cognate Eph receptors are expressed on neural crest cells migrating through the rostral half. Eph receptor/ephrin ligand interactions appear to restrict neural crest cells such that they migrate in a segmental fashion, leading to the subsequent metameric distribution of neural crest-derived sensory and sympathetic ganglia. This highlights the importance of inhibitory cues in controlling the pattern of migration. A similar inhibitory mechanism may restrict trunk neural crest cells from invading the gut. Slits are inhibitory proteins expressed at the entrance to the dorsal mesentery of the gut, among other locations. Trunk neural crest cells express Robo receptors that appear to lead to their exclusion from the gut. In contrast, vagal neural crest cells, which invade and eventually differentiate into enteric neurons that innervate the gut, lack Robo receptors and therefore are insensitive to the expression of Slit in this site. Taken together, these experiments suggest that neural crest cells are highly migratory and invasive. A predominant means of controlling their migratory pattern may occur via inhibitory signals that restrict them from particular sites, thus leading to the exquisite patterning of their migration.

9.7
References

ANDERSON, D.J. (**1993**) Molecular control of cell fate in the neural crest: the sympathoadrenal lineage. *Annu. Rev. Neurosci. 16*, 129–158.

ARTINGER, K.B. and BRONNER-FRASER, M.E. (**1992**) Partial restriction in the developmental potential of late emigrating avian neural crest cells. *Dev. Biol. 149*, 149–157.

BAGRI, A., MARIN, O., PLUMP, A.S., MAK, J., PLEASURE, S.J., RUBENSTEIN, J.L., and TESSIER-LAVIGNE, M. (**2002**) Slit proteins prevent midline crossing and determine the dorsoventral position of major axonal pathways in the mammalian forebrain. *Neuron. 33*, 233–248.

BRONNER-FRASER, M.E. (**1985**). Alterations in neural crest migration by an antibody that affects cell adhesion. *J. Cell Biol. 101*, 610–617.

BRONNER-FRASER, M. (**1986**). An antibody to a receptor for fibronectin and laminin perturbs cranial neural crest development. *Dev. Biol. 117*, 528–536.

Bronner-Fraser, M. and Stern, C. (**1991**) Effects of mesodermal tissues on avian neural crest cell migration. *Dev. Biol. 143*, 213–217.

Brose, K., Bland, K. S., Wang, K. H., Arnott, D., Henzel, W., Goodman, C. S., Tessier-Lavigne, M., and Kidd, T. (**1999**) Slit proteins bind Robo receptors and have an evolutionarily conserved role in repulsive axon guidance. *Cell 96*, 795–806.

Couly, G. F., Coltey, P. M., and Le Douarin, N. M. (**1992**) The developmental fate of the cephalic mesoderm in quail-chick chimeras. *Development 114*, 1–15.

Couly, G. F., Coltey, P. M., and Le Douarin, N. M. (**1993**) The triple origin of the skull in higher vertebrates: a study in quail-chick chimeras. *Development 117*, 409–429.

DeBellard, M., Rao, Y., and Bronner-Fraser, M. (**2003**) Dual function of Slit2 in repulsion and enhanced migration of trunk neural crest cells. *J. Cell Biol. 162*, 269–280.

del Barrio M. G. and Nieto M. A. (**2002**) Overexpression of Snail family members highlights their ability to promote chick neural crest formation. *Development 129*, 1583–1593.

Eickholt, B. J., Mackenzie, S. L., Graham, A., Walsh, F. S., and Doherty, P. (**1999**) Evidence for collapsin-1 functioning in the control of neural crest migration in both trunk and hindbrain regions. *Development 126*, 2181–2189.

Erickson, C. A. and Goins, T. L. (**2000**) Sacral neural crest cell migration to the gut is dependent upon the migratory environment and not cell-autonomous migratory properties. *Dev. Biol. 219*, 79–97.

Fan C. M. and Tessier-Lavigne M. (**1994**) Patterning of mammalian somites by surface ectoderm and notochord evidence for sclerotome induction by a hedgehog homolog. *Cell 79*, 1175–1186.

Gale, N. W., Holland, S., Valenzuela., D., Flenniken, A., Pan, L., Henkemeyer, M., Strebhardt, K., Hirai, H., Wilkinson, D. G., Pawson, T., Davis, S., and Yancopoulos, G.D. (**1996a**) Eph receptors and ligands comprise two major specificity subclasses, and are reciprocally compartmentalized during embryogenesis. *Neuron.* 9–19.

García-Castro, M., Marcelle, C., and Bronner-Fraser, M. (**2002**) Ectodermal Wnt function as a neural crest inducer. *Science 297*, 848–851.

Golding, J., Trainor, P., Krumlauf, R., and Gassman, M. (**2000**) Defects in pathfinding by cranial neural crest cells in mice lacking the Neuregulin receptor ErbB4. *Nat. Cell. Biol. 2*, 103–109.

Graham, A., Heyman, I., and Lumsden, A. (**1993**) Even-numbered rhombomeres control the apoptotic elimination of neural crest cells from odd-numbered rhombomeres in the chick hindbrain. *Development 119*, 233–245.

Hosoda K., Hammer R. E., Richardson J. A., Baynash A. G., Cheung J. C., Giaid A., and Yanagisawa M. (**1994**) Targeted and natural (piebald-lethal) mutations of endothelin-B receptor gene produce megacolon associated with spotted coat color in mice. *Cell 79*, 1267–1276.

Hu, H. (**1999**) Chemorepulsion of neuronal migration by Slit2 in the developing mammalian forebrain. *Neuron. 23*, 703–711.

Kidd, T., Bland, K. S., and Goodman, C. S. (**1999**) Slit is the midline repellent for the robo receptor in *Drosophila*. *Cell 96*, 785–794.

KIL, S. H., KRULL, C. E., CANN, G., CLEGG, D., and BRONNER-FRASER, M. (**1998**) The α_4 subunit of integrin is essential for neural crest migration *Dev. Biol. 202*, 29–42.

KINRADE, E. F., BRATES, T, TEAR, G., and HIDALGO, A. (**2001**) Roundabout signalling, cell contact and trophic support confine longitudinal glia and axons in the Drosophila CNS. *Development 128*, 207–216.

KORADE, Z. and Frank, E. (**1996**) Restriction in cell fates of developing spinal cord cells transplanted to neural crest pathways. *J. Neurosci. 16*, 7638–7648.

KRULL, C. E., LANSFORD, R., GALE, N. W., COLLAZO, A., MARCELLE, C., YANCOPOULOS, G. D., FRASER, S. E., and BRONNER-FRASER, M. (**1997**) Interactions of Eph-related receptors and ligands confer rostrocaudal pattern to trunk neural crest migration. *Curr. Biol. 7*, 571–580.

LABONNE, C. and BRONNER-FRASER, M. (**1998**) Neural crest cell induction in Xenopus: Evidence for a two-signal model. *Development 125*, 2403–2414.

LEDOUARIN, N. M. (**1982**) *The neural crest*. Cambridge Univ. Press, New York.

LE DOUARIN, N. M. and TEILLET, M. A. (**1973**). The migration of neural crest cells to the wall of the digestive tract in avian embryo. *J. Embryol. Exp. Morphol. 30*, 31–48.

LE DOUARIN, N. M. and TEILLET, M. A. (**1974**) Experimental analysis of the migration and differentiation of neuroblasts of the autonomic nervous system and of neuroectodermal mesenchymal derivatives, using a biological cell marking technique. *Dev. Biol. 41*, 162–184.

LE DOUARIN, N. M. (**1986**) Cell line segregation during peripheral nervous system ontogeny. *Science 231*, 1515–1522.

LEE, H. Y., KLEBER, M., HARI, L., BRAULT ,V., SUTER, U., TAKETO, M. M., KEMLER, R., and SOMMER, L.. (**2004**) Instructive role of Wnt/beta-catenin in sensory fate specification in neural crest stem cells. *Science 303*, 1020–1023.

LI, H. S., CHEN, J. H., WU, W., FAGALY, T., ZHOU, L., YUAN, W., DUPUIS, S., JIANG, Z. H., NASH, W., GICK, C., ORNITZ, D. M., WU, J. Y., and RAO, Y. (**1999**) Vertebrate slit, a secreted ligand for the transmembrane protein roundabout, is a repellent for olfactory bulb axons. *Cell. 96*, 807–818.

LIU, J. P. and JESSELL (**1998**) A role for rhoB in the delamination of neural crest cells from the dorsal neural tube. *Development 125*, 5055–5067.

LWIGALE, P. Y., CONRAD, G., and BRONNER-FRASER, M. (**2004**) Graded potential of neural crest to form cornea, sensory neurons and cartilage along the rostrocaudal axis. *Development* (in press).

LUMSDEN, A., SPRAWSON, N., and GRAHAM A. (**1991**) Segmental origin and migration of neural crest cells in the hindbrain region of the chick embryo. *Development 113*, 1281–1291.

MOURY, J. D. and JACOBSON, A. G. (**1990**) The origins of neural crest cells in the axolotl. *Dev. Biol. 141*, 243–253.

MUNSTERBERG, A., KITAJEWSKI, J., BUMCROT, D., MCMAHON, and A., LASSAR, A.(**1995**) Combinatorial signaling by Sonic hedgehog and Wnt family members induces myogenic bHLH gene-expression in the somite. *Genes Dev. 9*, 2911–2922.

MUNSTERBERG, A. and LASSAR, A. (1995) Combinatorial signals from the neural-tube, floor plate and notochord induce myogenic bHLH gene-expression in the somite. *Development 121*, 651–660.

NAKAGAWA, S. and TAKEICHI, M. (1998) Neural crest emigration from the neural tube depends on regulated cadherin expression. *Development 125*, 2963–2971.

NIETO, A.M., SARGENT, M.G., WILKINSON, D.G., and COOKE, J. (1994) Control of cell behavior during vertebrate development by Slug, a zinc finger gene. *Science 264*, 835–839.

NODEN, D.M. (1978) The control of avian cephalic neural crest cyto-differentiation. I. Skeletal and connective tissues. *Dev. Biol. 67*, 296–312.

NGUYEN BA-CHARVET, K.T., BROSE, K., MARILLAT, V., KIDD, T., GOODMAN, C.S., TESSIER-LAVIGNE, M., SOTELO, C., and CHEDOTAL, A. (1999) Slit2-Mediated chemorepulsion and collapse of developing forebrain axons. *Neuron. 22*, 463–473.

OAKLEY, R.A. and TOSNEY, K.W. (1991) Peanut agglutinin and chondroitin-6-sulfate are molecular markers for tissues that act as barriers to axon advance in the avian embryo. *Dev. Biol. 147*, 187–206.

OZDINLER, P.H. and ERZURUMLU, R.S. (2002) Slit2, a branching-arborization factor for sensory axons in the Mammalian CNS. *J. Neurosci. 22*, 4540–4549.

PETTWAY, Z., DOMOWICZ, M., SCHWARTZ, N.B., and BRONNER-FRASER, M. (1996) Age-dependent inhibition of neural crest migration by the notochord correlates with alterations in the S103L chondroitin sulfate proteoglycan. *Exp. Cell. Res. 225*, 195–206.

RICKMANN, M., FAWCETT, J.W., and KEYNES, R.J. (1985) The migration of neural crest cells and growth cones of motor axons through the rostral half of the chick somite. *J. Embryol. Exp. Morphol. 90*, 437–455.

SAINT-JEANNET, J.P., HE, X, VARMUS, H.E., and DAWID, I.B. (1997) Regulation of dorsal fate in the neuraxis by Wnt-1 and Wnt-3a. *Proc. Natl. Acad. Sci. USA 94*, 13713–13718.

SECHRIST, J., SERBEDZIJA, G., SCHERSON, T., FRASER, S., and BRONNER-FRASER, M. (1993) Segmental migration of the hindbrain neural crest does not arise from segmental generation. *Development 118*, 691–703.

SECHRIST, J., NIETO, A., ZAMANIAN, R., and BRONNER-FRASER, M. (1995) Regulative response of the cranial neural tube after neural fold ablation: spatiotemporal nature of neural crest generation and up-regulation of *Slug*. *Development 121*, 4103–4135.

SELLECK, M.A.J. and BRONNER-FRASER, M. (1995) Origins of the avian neural crest: the role of neural plate-epidermal interactions. *Development 121*, 526–538.

SERBEDZIJA, G., BRONNER-FRASER, M., and FRASER, S.E. (1989) Vital dye analysis of the timing and pathways of avian trunk neural crest cell migration. *Development 106*, 806–816.

SHAH, N.M., GROVES, A.K., and ANDERSON, D.J. (1996) Alternative neural crest cell fates are instructively promoted by TGFbeta super-family members. *Cell 85*, 331–343.

SHARMA, K., KORADE, Z., and FRANK, E. (1995) Late-migrating neuroepithelial cells from the spinal cord differentiate into sensory ganglion cells and melanocytes. *Neuron. 14*, 143–152.

STEMPLE, D.L. and ANDERSON, D.J. (1993) Lineage diversification of the neural crest: in vitro investigations. *Dev. Biol. 159*, 12–23.

STERN, C. D. and KEYNES, R. J. (1987) Interactions between somite cells: the formation and maintenance of segment boundaries in the chick embryo. *Development* 99, 261–272.

WANG, H. U. and ANDERSON, D. J. (1997) Roles of Eph family transmembrane ligands in repulsive guidance of trunk neural crest migration and motor axon outgrowth. *Neuron.* 18, 383–396.

WANG, K. H., BROSE, K., ARNOTT, D., KIDD, T., GOODMAN, C. S., HENZEL, W., and TESSIER-LAVIGNE, M. (1999) Biochemical purification of a mammalian slit protein as a positive regulator of sensory axon elongation and branching. *Cell.* 96, 771–784.

WESTON, J. A. (1963) A radiographic analysis of the migration and localization of trunk neural crest cells in the chick. *Dev. Biol.* 6, 279–310.

WU, J. Y., FENG, L., PARK, H. T., HAVLIOGLU, N., WEN, L., TANG, H., BACON, K. B., JIANG, Z., ZHANG, X., and RAO, Y. (2001) The neuronal repellent Slit inhibits leukocyte chemotaxis induced by chemotactic factors. *Nature* 410, 948–952.

WU, W., WONG, K., CHEN, J., JIANG, Z., DUPUIS, S., WU, J. Y., and RAO, Y. (1999) Directional guidance of neuronal migration in the olfactory system by the protein Slit. *Nature* 400, 331–336.

YUAN, W., ZHOU, L., CHEN, J. H., WU, J. Y., RAO, Y., and ORNITZ, D. M. (1999) The mouse SLIT family: secreted ligands for ROBO expressed in patterns that suggest a role in morphogenesis and axon guidance. *Dev. Biol.* 212, 290–306.

ZHU, Y., YU, T., ZHANG, X. C., NAGASAWA, T., WU, J. Y., and RAO, Y. (2002) Role of the chemokine SDF-1 as the meningeal attractant for embryonic cerebellar neurons. *Nat. Neurosci.* 5, 719–720.

10
Primordial Germ Cell Migration in Zebrafish

Erez Raz

10.1
Introduction

The gonad is composed of two major cell populations: somatic cells and germ cells that differentiate into gametes. In many organisms the germ cells are specified early in development and in a position that is distinct from that where the gonad develops. Therefore, following their specification, the germ cells, termed primordial germ cells (PGCs) at this stage, have to migrate through the developing embryo towards their target. The understanding of the cellular and molecular basis for this process is therefore a central topic in developmental biology that also serves as a general model for long-range cell migration. Here, PGC migration in zebrafish will be reviewed with reference to relevant studies in other organisms.

10.2
PGC Migration in Zebrafish: Hints from Embryological Studies

Specification of zebrafish PGCs occurs very early in development as cells that inherit specific maternally provided RNAs and proteins collectively termed 'germ plasm' become germline cells [1, 2]. Interestingly, germ plasm material is found in four different locations whose position, relative to the early dorsoventral embryonic axis, is random. Therefore, zebrafish PGCs migrate towards their target from the different points at which they were specified. Despite this complication, the concurrent tissue movements and patterning in the early embryo, virtually all the PGCs manage to reach one of two bilateral PGC clusters by the end of the first day of development [1, 3]. This robustness of PGC migration in zebrafish has been most convincingly demonstrated by transplanting PGCs to positions where they are not normally found and showing that the cells could nevertheless reach their target [4].

As a first step in defining the mechanisms governing PGC migration in zebrafish, the process has been described in fixed embryos using mo-

Cell Migration. Edited by Doris Wedlich
Copyright © 2005 WILEY-VCH Verlag GmbH & Co. KGaA, Weinheim
ISBN: 3-527-30587-4

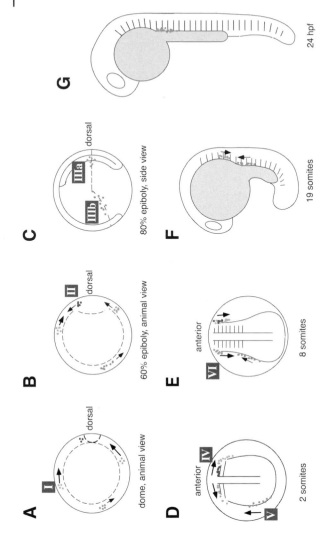

Fig. 10.1 The six steps of early PGC migration in zebrafish. Schematic drawings of embryos from dome stage (4.5 h post fertilization (hpf)) to 24 hpf showing the positions and movements of the four PGC clusters. At dome stage, four clusters of PGCs are found close to the blastoderm margin in a symmetrical 'square' shape. All possible orientations of the square relative to the dorsal side of the embryo can be observed. At 24 hpf, the PGC clusters are located at the anterior end of the yolk extension, which corresponds to the 8th to10th somite level. In most embryos, all PGCs have reached this region by the end of the first day of development. (This figure also appears with the color plates.)

lecular markers to visualize the cells at different steps of their journey to the gonad (Fig. 10.1) [1, 3]. This analysis of the migration route taken by the PGCs revealed some important principles. Firstly, the apparent random arrangement of the PGCs was transformed early in development into a configuration that was not random, as a result of evacuation of certain tissues and embryonic structures. It therefore became apparent that the position of the cells at different stages of their migration reflected a response to cues provided by somatic tissues. Secondly, analysis of PGC migration in mutants in which specific embryonic structures did not develop was consistent with the notion that attractive cues direct the migration [3, 5]. This idea stems from the finding that defects in differentiation or alterations in the location of the target tissues, rather than defects in structures the PGCs evacuate, affected the migration path of the cells (Fig. 10.2). Another important observation was that the cells do not migrate directly towards their final target, but are first found in tissues that they later vacate [1, 3, 5, 6]. This idea of attraction by intermediate targets and dynamic changes in the distribution of the targets within the embryo provided a conceptual framework for understanding the robustness and flexibility of zebrafish PGC migration. Specifically, irrespective of their birth position, all the PGCs are found in proximity to a large area that constitutes an intermediate target which collects the cells, bringing them to a common cellular environment (the borders of the trunk mesoderm). Alterations in the shape and position of intermediate targets and corresponding cell migration in response to these alterations culminates in the formation of two PGC clusters on either side of the body axis (Fig. 10.3).

Consistent with this model, analysis of PGC migration in live embryos revealed that throughout their migration, PGCs exhibit dynamic alterations in cellular morphology and changes in their position relative to so-

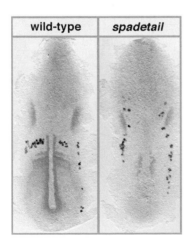

| wild-type | spadetail |

Fig. 10.2 At the 3-somite stage (11 h post fertilization), alignment of PGCs at their intermediate target at level of the 1st somite as seen in wild-type embryos (the two groups of cell perpendicular to the body axis) but is lost in *spadetail* mutants in which the differentiation of somatic cells in the target region is defective. Germ cells in blue (*vasa*) and somatic expression of *myoD*, *papc* and *pax8* in red. (This figure also appears with the color plates.)

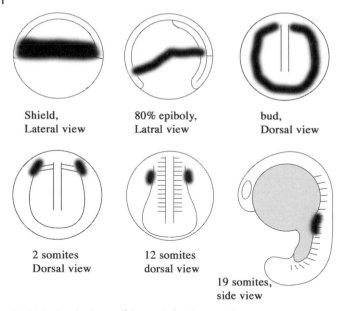

Fig. 10.3 On the basis of the analysis of germ-cell migration in wild-type and mutant embryos, domains that are more attractive for the germ cells are labeled in black.

matic cells. This behavior is characteristic of cells performing active migration in response to directional cues. The idea that attractive signals guide active PGC migration is not unique to zebrafish. Earlier studies have convincingly demonstrated that such cues direct PGCs in mouse and *Drosophila* [7, 8]. In addition to the concept of attractive signals, *Drosophila* PGCs depend on repulsive cues that lead to concentration of the migrating cells in proximity to their final target [9, 10].

The analysis of zebrafish PGC migration in wild-type and mutant embryos suggested that attractants are expressed in certain domains in the embryos and that they guide the cells towards their intermediate and final targets. Nevertheless, the class of mutations used in this analysis exhibits gross defects in early development of certain somatic tissues. Thus, although the mutations provided important clues regarding the nature and position of the signals guiding PGCs, they affect this process only secondarily. Identification of genes involved specifically in the production of directional cues for zebrafish PGCs and of genes that function within the PGCs themselves to receive and interpret the signals required genetic screens designed particularly for this purpose.

10.3
Guidance Cues for Zebrafish PGCs

Molecules whose activity is specifically required for PGC migration were first identified in genetic screens in *Drosophila* [8, 9, 11, 12]. The identification of these molecules, whose function will be discussed in more detail below, underscored the central role of somatic tissues in guiding PGCs. Specifically, genes whose function is to repel PGCs (the *wunen* genes) are expressed in domains that PGCs leave, whereas genes important for the generation of attractive cues (the *columbus* gene) are expressed in PGC targets. The success of the *Drosophila* screens prompted genetic screens in zebrafish aimed at identification of the corresponding molecules that guide PGCs in this vertebrate model organism.

Two genetic screens aimed at identification of the molecular cues controlling PGC migration were conducted in zebrafish (Fig. 10.4). The first approach used genomic information that included translation start sites of different genes. This information was used to design antisense oligonucleotides that were injected into embryos, thereby inhibiting the translation of specific genes. The injected embryos were subsequently analyzed for the resulting PGC-migration phenotype [13]. The second approach followed the "F$_2$ genetic screen" scheme in which the position and number of the PGCs in progeny of mutagenized fish were monitored followed by cloning of a mutated locus (Fig. 10.4) [14]. The gene identified in both approaches encodes the CXCR4b, a 7-transmembrane G-protein coupled chemokine receptor that is expressed in zebrafish PGCs throughout their migration. Significantly, CXCR4 and its ligand SDF-1/CXCL12 have previously been shown to be essential for stem cell homing and mobilization [15, 16], leukocyte trafficking [17, 18], neuronal cell migration [19], nerve growth cone guidance [20] and to determine the metastatic destination of tumor cells [21]. In these processes, it was suggested that SDF-1 directs cells towards their targets supporting correct formation and function of organs and tissues. Thus, PGC migration in zebrafish provides an excellent *in vivo* vertebrate model for these CXCR4-dependent processes.

Two *sdf*-1 genes have been identified in zebrafish, *sdf*-1a [13] and *sdf*-1b [14] and both have been suggested to play a role in guiding the PGCs. The discussion here will focus on SDF-1a for which higher phenotypic penetration upon injection of the corresponding antisense oligonucleotides has been reported. The expression pattern of the RNA of *sdf*-1a corresponds strikingly to the proposed distribution of intermediate and final targets for the migrating PGCs [13]. A clear demonstration of this point is the alterations in the spatial distribution of *sdf*-1a transcripts during somitogenesis stages that are mirrored by posterior migration of the PGC clusters (Fig. 10.5) [13]. Indeed, similar to the phenotype of the CXCR4b knock-down [13] or mutation [14], knocking down SDF-1a activity using

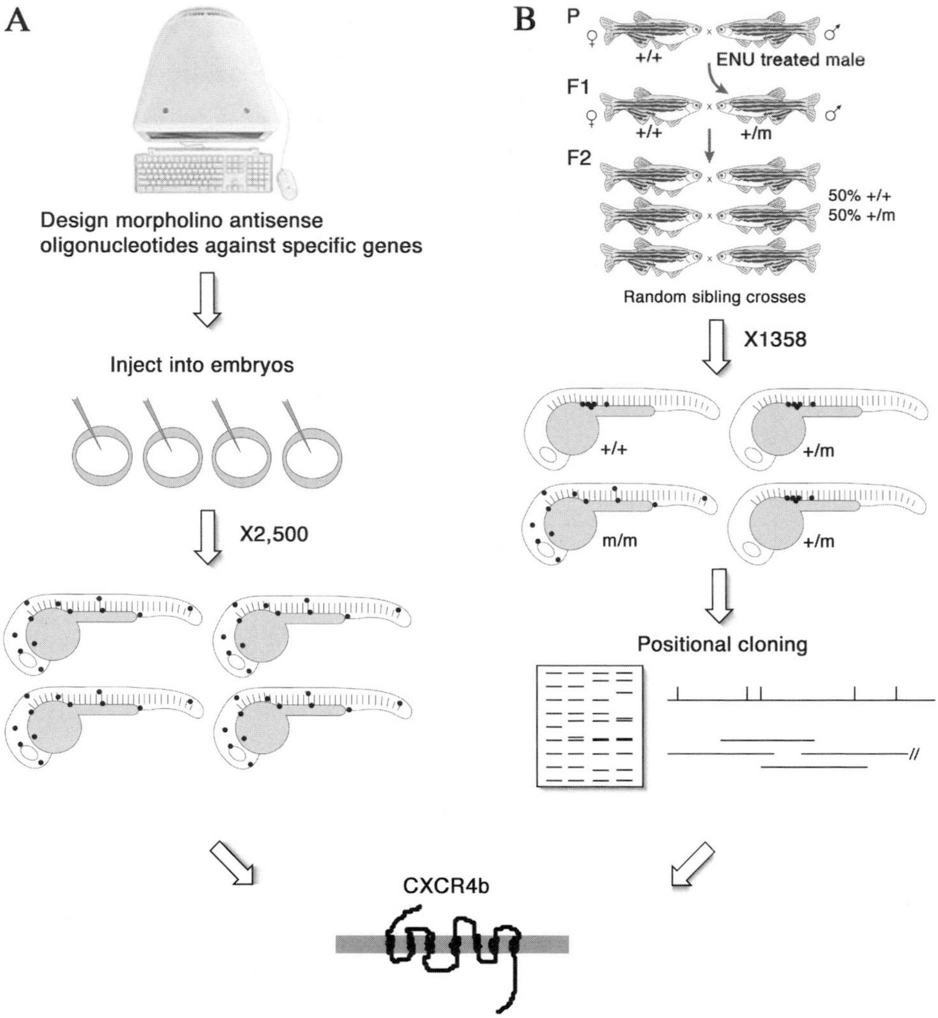

A

Design morpholino antisense
oligonucleotides against specific genes

⇩

Inject into embryos

X2,500

B   P   ♀ +/+   ×   ♂ ENU treated male

F1 ♀ +/+   ×   +/m ♂

F2   50% +/+
     50% +/m

Random sibling crosses

⇩ X1358

+/+   +/m

m/m   +/m

⇩

Positional cloning

CXCR4b

Fig. 10.4 Identification of the chemokine receptor Cxcr4b
by genetic screens for genes that affect germ-cell migration. A) A modified antisense oligonucleotide-based
screen where knock-down of CXCR4b resulted in a PGC
migration phenotype. B) In a large-scale chemical mutagenesis screen, a mutation in which the germ cells failed
to cluster at the region of the gonad was identified. Positional cloning of the affected gene identified *cxcr4b* as the
mutated gene.

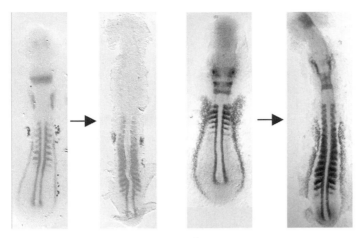

Fig. 10.5 During somitogenesis stages the PGC migrate as cell clusters from the level of the first 3 somites to the level of the 8th somite (left two panels, purple labeled cells). At the same time, *sdf*-1a expression pattern changes in a similar pattern (right two panels, blue labeled are on either side of the embryo). Reproduced with permission from [13] and [3]. (This figure also appears with the color plates.)

antisense oligonucleotides results in a strong migration phenotype [13]. Without the SDF-1a cues PGCs fail to reach their targets, but migrate relative to their somatic neighbors. Thus, the role of this chemokine is to provide the migratory PGCs with directional cues, but it is not required for motility *per se* (Fig. 10.6). The function of SDF-1a is thus similar to that of cAMP in *Dictyostelium* migration that appears to be required for providing directional cues to otherwise motile cells and differs from neutrophils where the ligand provides a signal for direction as well as for motility [22]. Finally, the dominant role SDF-1a plays in guiding zebrafish PGCs has been demonstrated by expressing the chemokine in ectopic positions within embryos in which the endogenous activity of the ligand was knocked down. The result of this experiment clearly demonstrated that SDF-1a is sufficient to attract the migrating cells to domains where they are normally not found.

The role of CXCR4 in directing zebrafish PGC migration represents a conserved function of this chemokine receptor. A role for CXCR4 and its ligand has been demonstrated in guiding PGC migration in mouse [23, 24]. Nevertheless, in contrast with the findings in zebrafish where SDF-1/CXCR4 signaling appears to be required for all stages of migration, some steps in PGC migration in the mouse embryo are independent of CXCR4 signaling, pointing to the involvement of other signaling pathways. An additional difference between the role of CXCR4 in zebrafish

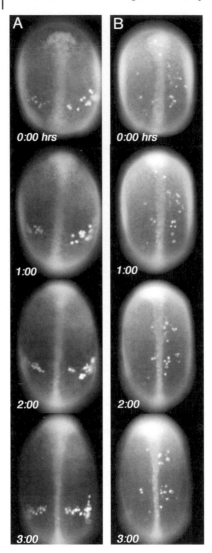

Fig. 10.6 PGC motility is independent of SDF-1a signaling. Snapshots from a 3-hour-long time-lapse movie showing that in the embryo injected with *sdf*-1a antisense oligonucleotides (B), the ectopic PGCs migrate relative to somatic tissues similar to the PGCs in the control embryo (A). Reproduced with permission from [13].

and mouse is that in contrast with zebrafish, in mice knocked out for CXCR4 or SDF-1 the number of PGCs is reduced, implicating this pathway in supporting cell survival in this organism. Although chemokines have not been identified in the genome of *Drosophila*, a 7-transmembrane receptor encoded by the locus *trapped in endoderm* (*tre*1) is required for the migration of the PGCs across the midgut epithelium suggesting that the conservation of the mechanisms of PGC migration can be extended to *Drosophila* as well.

As mentioned above, several genes that specifically function in guiding PGC migration have been identified in genetic screens in *Drosophila*. Of these, the *columbus* locus that encodes a 3-hydroxy-3-methylglutaryl coenzyme A reductase (HMGCoAr) is required to attract PGCs towards their target. *Columbus* is expressed in target tissues and is sufficient for generating attractive cellular environment for *Drosophila* PGCs as demonstrated by ectopic expression of the gene [8]. Genetic analysis of enzymes that function downstream of HMGCoAr in *Drosophila* revealed that the pathway controls the production of isoprenoids. These lipids could be used to modify the putative proteins serving as the actual attractants [25]. Interestingly, pharmacological inhibition of HMGCoAr as well as inhibition of geranyl-geranyl transferase 1 (GGT1) leads to PGC migration defects in zebrafish [26]. It is therefore tempting to speculate that the control of PGC migration in zebrafish and *Drosophila* shares another common molecular pathway where lipid modification of proteins plays a role. Nevertheless, as it is not yet clear if the HMGCoAr pathway indeed functions in the target tissue in zebrafish as it does in *Drosophila*, this proposition should be taken with caution.

10.4
PGC Migration: Cellular Behavior Aspects

A major advantage of studying cell migration in zebrafish is the optical clarity of the embryos that facilitates high quality *in vivo* data acquisition using relatively simple microscopy. Labeling zebrafish PGCs with GFP in transgenic animals or by injection of specific RNAs provided the first glance into the behavior of these cells in the intact organism [5, 6]. In these early studies it was demonstrated that throughout their migration the PGCs show formation of dynamic cellular extensions that are morphological features characteristic of motile cells. This description has recently been extended by the detailed analysis of three specific stages of PGC migration [27].

10.4.1
Migration of PGCs as Individual Cells

During the first phase of their migration zebrafish PGCs move as individual cells towards their intermediate targets [3]. At a later stage the cells form two cell clusters and remain in this configuration for an extended period of time before resuming their migration as a cluster towards their final target [5, 6].

The migration of PGCs as single cells can serve as a model for a variety of other cell migration process such as neural-crest cell migration, migration of metastatic cells as well as for pathfinding of pioneer axons.

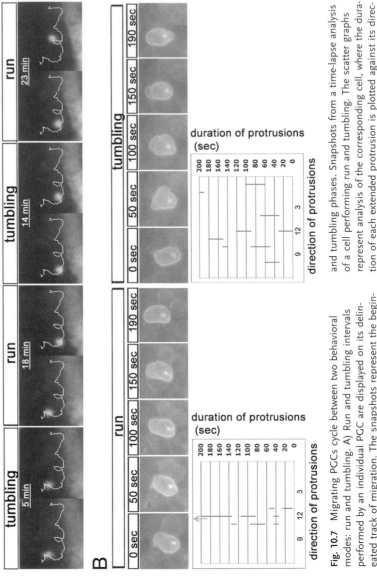

Fig. 10.7 Migrating PGCs cycle between two behavioral modes: run and tumbling. A) Run and tumbling intervals performed by an individual PGC are displayed on its delineated track of migration. The snapshots represent the beginning and end of two consecutive tumbling and run phases within the track. B) Cellular morphology of PGCs during run and tumbling phases. Snapshots from a time-lapse analysis of a cell performing run and tumbling. The scatter graphs represent analysis of the corresponding cell, where the duration of each extended protrusion is plotted against its direction. Reproduced with permission from [27].

Individually migrating PGCs were found to alternate between two modes of behavior [27]. In the first, the cells move forward (a behavior termed 'run') and in the second they remain in one position (a behavior termed 'tumbling') (Fig. 10.7). During 'run' phases the PGCs assume a polar elongated morphology while they extend cellular protrusions primarily in the direction of the migration. In contrast, 'tumbling' phases are characterized by loss of cellular polarity as judged by the uniform distribution of pseudopodia around the cell perimeter. This behavior of the PGCs is an autonomous cell property inherent to their migratory nature as it is executed also in cells in which SDF-1a/CXCR4b signaling is inhibited.

The functional importance of the 'run' phase is that it is the phase when PGCs actively migrate relative to somatic tissues on their way to the target. The role of 'tumbling' in PGC migration was revealed when a correlation between this phase and alterations in migration direction was demonstrated. Specifically, the vast majority of turns in migration direction occur following tumbling phases. Therefore, the tumbling phases could facilitate an alteration in direction by erasing prior cell polarity allowing the cells to sample the environment and correct its migration course. Consistently, in wild-type embryos, the post-tumbling migration direction is strongly biased towards the general migration direction preceding tumbling, presumably reflecting the distribution of the ligand at that place and time. Indeed, PGCs in embryos in which the SDF-1a signaling is inhibited also use tumbling as a point at which the migration direction is altered. In this case however, the direction the cells follow after they exit the tumbling is random with respect to their prior migration direction.

10.4.2
PGC Behavior in the Clustering Positions

Another type of behavior exhibited by PGCs is in positions where they stop migrating transiently (in the case of intermediate targets) or permanently (in the case of final targets). This behavior is characteristic of cells that migrate and stop in locations where they participate in building organs and tissues, and is relevant for diseases such as cancer where cells colonize distant locations and form secondary tumors.

Close examination of cell behavior at positions where the cells halt showed that in the cells extend protrusions in all directions and remain on the spot. As the positions in which the PGCs stop migrating are locations where high level of SDF-1a RNA is observed, the peak levels of SDF-1a at these positions presumably lead to uniform stimulation of the receptor and retention of the cells there.

10.4.3
PGC Behavior when Migrating as a Cluster

Migration as a cell cluster is a common theme in development. For example, tracheal cells in *Drosophila* remain firmly connected with each other throughout the migration process that is directed by specific cells within the cluster [28, 29]. Similarly, the migration of border cells during *Drosophila* oogenesis occurs in a cell cluster that is guided by one of the cells in the cluster [30]. In contrast with these examples, migration of cells in cluster could reflect the sum of single cell behaviors each of which responds individually to alterations in the spatial distribution of SDF-1a.

Migration of PGCs as a cluster occurs in mid-somitogenesis stages when corresponding to an alteration in the expression profile of SDF-1a the cluster of PGCs migrates posteriorly (Fig. 10.5) [5, 6, 13]. Observing PGCs at the relevant stages, it was found that the anterior cells in the cluster, the cells overlapping with the moving front of the *sdf*-1a expression domain are those that move posteriorly. The cells that then become positioned at the location of the receding SDF-1a are the cells that exhibit the next wave of posterior migration. These findings underscore the dominant role of the response of individual cells to local changes in SDF-1a distribution and argue against coordinated migration of cells in the cluster.

10.4.4
PGC Behavior compared with that of other Migrating Cell Types

As mentioned above, unlike cells that exhibit stable connections among members of the migrating cells, PGCs migrate independently of one another. This type of migration may be more prone to mistakes as sometimes seen in wild-type and mutant embryos where individual cells arrive at regions other than the gonad, regions that invariably express SDF-1a [3, 13]. Nevertheless, this type of guided migration provides a simple and flexible mode of transporting group of cells within the embryo; important parameters such as the cluster shape and velocity are controlled by a single parameter, the spatial distribution of the chemokine SDF-1a.

Zebrafish PGC polarity and motility are gained in the absence of any apparent extrinsic asymmetric signal as these behaviors are also observed in the absence of SDF-1a signaling. This cellular behavior is strikingly different from that described for neutrophils in which the motility and cellular polarity depend on the chemotactic signal, but is similar to that described in *Dictyostelium discoideum* [22].

The migratory behavior of zebrafish PGCs that is characterized by alternations between forward movement and phases when cellular polarity is lost and a change in direction follows is reminiscent of the description of bacterial chemotaxis [31]. Yet, important differences between the two systems should be pointed out. First, in contrast with bacteria in which the direction of movement following the tumbling phase is not biased by the attractant, following tumbling, PGCs repolarize and migrate in a direction dictated by the distribution of SDF-1a. Second, in contrast with bacteria that effectively adapt to highly uniform concentrations of the attractant and maintain their motility, PGCs that arrive at their target stop migrating. This feature of PGC migration is of general importance in development allowing stable and precise organization of cells in their target tissues, a pivotal requirement for organogenesis.

10.5
Signaling Pathways Controlling Primordial Germ Cell Migration in Zebrafish

Zebrafish PGCs respond to the SDF-1a signal by alterations in their morphology and direction of migration. Hints regarding the identity of the cellular components important for translating the extracellular signal into these behaviors are provided by studies in other cell types. A role for Gi proteins in chemokine directed chemotaxis has been demonstrated *in vitro* [32, 33] and recent work showed that these proteins are important for zebrafish PGC migration as well [34]. The role of Gi proteins in PGC migration has been shown by expressing pertussis toxin (PTX), a potent inhibitor of Gi proteins in the cells and analyzing their behavior [34]. Indeed, zebrafish PGCs expressing PTX were randomly distributed throughout the embryo with virtually no cells found at the proper position. Analysis of the basis for this phenotype revealed that the manipulated cells were motile, but consistent with a specific effect on the ability of the cells to respond to the directional cues, the cells migrated in random directions ignoring locations that normally serve as intermediate and final targets.

Downstream to G proteins and central to directional cell migration in many migrating cells are proteins of the posphoinositide 3-kinase (PI3K) family. PI3K has been shown to be important for directed migration of neutrophils and *Dictyostelium* cells. In these cells, activation of PI3Ks results in the production of $3'$-phosphorylated phosphoinositides, PtdIns $(3,4)P_2$ [PI$(3,4)P_2$] and PtdIns$(3,4,5)P_3$ [PIP$_3$], at the leading edge of migrating cells in response to receptor activation and G-protein signaling [22, 35–38]. The resulting asymmetric localization of PI$(3,4)P_2$ and PIP$_3$ serves to recruit proteins that contain the Pleckstrin Homology domain (PH-domain) to the leading edge and these which are thought to activate downstream responses such as actin polymerization and ultimately, the establishment of cellular polarity [38–40].

When the the distribution of PI$(3,4)P_2$ and PIP$_3$ in zebrafish PGCs was analyzed, it was first reported that these compounds are asymmetrically distributed such that a higher level is detected at positions were cellular extensions were formed [14]. Careful subsequent analysis revealed that

$PI(3,4)P_2$ and PIP_3 are uniformly distributed around the membrane and are not enriched in positions where pseudopodia are formed [34]. Consistent with this observation, PGCs in which CXCR4b, the receptor that transmits the polarized guidance cues, was knocked down exhibit normal uniform distribution of $PI(3,4)P_2$ and PIP_3 around their membrane [34].

These findings argue against a role for polarized activation of PI3K in directing zebrafish PGC migration. Nevertheless, inhibition of PI3K activity by expressing a dominant negative form of the protein in the PGCs revealed that this pathway does play a role in the migration of these cells [34]. Specifically, PGCs depleted for $PI(3,4)P_2$ and PIP_3 exhibit slower migration speed. Cells expressing the dominant negative form of PI3K exhibit two major abnormalities. First, the overall shape of the cells was more round and less polarized compared with wild-type cells and second, in manipulated cells filopodia protrusions were shorter in size and persistence. Therefore, the cellular basis for the effect on migration speed may be the requirement for PI3K activity for the ability of the cells to attain the correct cellular morphology.

In summary, the PI3K pathway is important for attaining normal PGC morphology and filopodia stability, which is likely to be essential for achieving optimal migration speed. At the level of the developing embryo, these defects are translated into a comparably mild migration phenotype. The slower migration speed coupled with the dynamic alterations in the distribution of SDF-1a lead to a situation where some of the PGCs lose their track and end up in ectopic positions.

10.6
Conclusions and Future Directions

In recent years we have witnessed significant advances in the understanding of the molecular mechanisms controlling the directed migration and motility of zebrafish PGCs. The definition of the tissues that support this process was followed by the identification of the molecule that constitutes the guidance cue (SDF-1a) and that responsible for receiving the signal (CXCR4b). Together with the detailed description of the behavior of the cells during different phases of their migration we can now provide a conceptual framework accounting for the process that efficiently brings the PGCs to their target from different starting points.

In contrast with the detailed understanding of the process at the "developmental biology" level, relatively little is currently known regarding the mechanisms that promote the motility and directional migration of the cells at the "cell biology" level. Future studies aimed at understanding the dynamic alterations in the architecture of the cytoskeleton during different stages of zebrafish PGC migration constitute an attractive avenue for bridging the gap between these two disciplines.

10.7
References

1 YOON, C., KAWAKAMI, K., and HOPKINS, N., Zebrafish vasa homologue RNA is localized to the cleavage planes of 2- and 4-cell-stage embryos and is expressed in the primordial germ cells. *Development*, **1997**, *124*, 3157–3165.

2 KNAUT, H., et al., Zebrafish vasa RNA but not its protein is a component of the germ plasm and segregates asymmetrically before germline specification. *J Cell Biol*, **2000**, *149*, 875–888.

3 WEIDINGER, G., et al., Identification of tissues and patterning events required for distinct steps in early migration of zebrafish primordial germ cells. *Development*, **1999**, *126*, 5295–5307.

4 CIRUNA, B., et al., Production of maternal-zygotic mutant zebrafish by germ-line replacement. *Proc Natl Acad Sci USA*, **2002**, *99*, 14919–14924.

5 WEIDINGER, G., et al., Regulation of zebrafish primordial germ cell migration by attraction towards an intermediate target. *Development*, **2002**, *129*, 25–36.

6 KNAUT, H., et al., An Evolutionary Conserved Region in the vasa 3′UTR Targets RNA Translation to the Germ Cells in the Zebrafish. *Curr Biol*, **2002**, *12*, 454–466.

7 GODIN, I., WYLIE, C., and HEASMAN, J., Genital ridges exert long-range effects on mouse primordial germ cell numbers and direction of migration in culture. *Development*, **1990**, *108*, 357–363.

8 VAN DOREN, M., et al., HMG-CoA reductase guides migrating primordial germ cells. *Nature*, **1998**, *396*, 466–469.

9 ZHANG, N., et al., The *Drosophila* protein *Wunen* repels migrating germ cells. *Nature*, **1997**, *385*, 64–67.

10 STARZ-GAIANO, M., et al., Spatially restricted activity of a *Drosophila* lipid phosphatase guides migrating germ cells. *Development*, **2001**, *128*, 983–991.

11 MOORE, L.A., et al., Identification of genes controlling germ cell migration and embryonic gonad formation in Drosophila. *Development*, **1998**, *125*, 667–678.

12 ZHANG, N., et al., Identification and genetic analysis of *wunen*, a gene guiding *Drosophila melanogaster* germ cell migration. *Genetics*, **1996**, *143*, 1231–1241.

13 DOITSIDOU, M., et al., Guidance of primordial germ cell migration by the chemokine SDF-1. *Cell*, **2002**, *111*, 647–659.

14 KNAUT, H., et al., A zebrafish homologue of the chemokine receptor Cxcr4 is a germ-cell guidance receptor. *Nature*, **2003**, *421*, 279–282.

15 PELED, A., et al., Dependence of human stem cell engraftment and repopulation of NOD/SCID mice on CXCR4. *Science*, **1999**, *283*, 845–848.

16 PETIT, I., et al., G-CSF induces stem cell mobilization by decreasing bone marrow SDF-1 and up-regulating CXCR4. *Nat Immunol*, **2002**, *17*, 17.

17 AIUTI, A., et al., The chemokine SDF-1 is a chemoattractant for human CD34+ hematopoietic progenitor cells and provides a

new mechanism to explain the mobilization of CD34+ progenitors to peripheral blood. *J Exp Med*, **1997**, *185*, 111–120.

18 BLEUL, C., et al., A highly efficacious lymphocyte chemoattractant, stromal cell-derived factor 1 (SDF-1). *J Exp Med*, **1996**, *184*, 1101–1109.

19 ZOU, Y. R., et al., Function of the chemokine receptor CXCR4 in haematopoiesis and in cerebellar development. *Nature*, **1998**, *393*, 595–599.

20 XIANG, Y., et al., Nerve growth cone guidance mediated by G protein coupled receptors. *Nat Neurosci*, **2002**, *5*, 843–848.

21 MULLER, A., et al., Involvement of chemokine receptors in breast cancer metastasis. *Nature*, **2001**, *410*, 50–56.

22 DEVREOTES, and JANETOPOULOS, C., Eukaryotic chemotaxis: Distinctions between directional sensing and polarization. *J Biol Chem*, **2003**, *278*, 20445–20448.

23 ARA, T., et al., Impaired colonization of the gonads by primordial germ cells in mice lacking a chemokine, stromal cell-derived factor-1 (SDF-1). *Proc Natl Acad Sci USA*, **2003**, *100*, 5319–5323.

24 MOLYNEAUX, K., et al., The chemokine SDF1/CXCL12 and its receptor CXCR4 regulate mouse germ cell migration and survival. *Development*, **2003**, *130*, 4279–4286.

25 SANTOS, A. C. and LEHMANN, R., Isoprenoids control germ cell migration downstream of HMGCoA reductase. *Dev Cell*, **2004**, *6*, 283–293.

26 THORPE, J. L., et al., Germ cell migration in zebrafish is dependent on HMGCoA reductase activity and prenylation. *Dev Cell*, **2004**, *6*, 295–302.

27 REICHMAN-FRIED, M., MININA, S., and RAZ, E., Autonomous modes of behavior in primordial germ cell migration. *Dev Cell*, **2004**, *6*, 589–596.

28 RIBEIRO, C., EBNER, A., and AFFOLTER, A., In vivo imaging reveals different cellular functions for FGF and Dpp signaling in tracheal branching morphogenesis. *Dev Cell*, **2002**, *2*, 677–683.

29 SATO, M. and KORNBERG, T. B., FGF is an essential mitogen and chemoattractant for the air sacs of the drosophila tracheal system. *Dev Cell*, **2002**, *3*, 195–207.

30 FULGA, T. A. and RORTH, Invasive cell migration is initiated by guided growth of long cellular extensions. *Nat Cell Biol*, **2002**, *4*, 715–719.

31 BERG, H. C. and BROWN, T. A., Chemotaxis in *Escherichia coli* analysed by three-dimensional tracking. *Nature*, **1972**, *239*, 500–504.

32 LUTHER, S. A. and CYSTER, J. G., Chemokines as regulators of T cell differentiation. *Nat Immunol*, **2001**, *2*, 102–107.

33 THELEN, M., Dancing to the tune of chemokines. *Nat Immunol*, **2001**, *2*, 129–134.

34 DUMSTREI, K., MENNECKE, R., and RAZ, E., Signaling pathways controlling primordial germ cell migration in zebrafish. *J Cell Sci*, **2004**, *117*, 4787–4795.

35 CHUNG, C. Y., FUNAMOTO, S., and FIRTEL, R. A., Signaling pathways controlling cell polarity and chemotaxis. *Trends Biochem Sci*, **2001**, *26*, 557–566.

36 IIJIMA, M., HUANG, Y. E., and DEVREOTES, Temporal and spatial regulation of chemotaxis. *Dev Cell*, **2002**, *3*, 469–478.

37 MERLOT, S. and FIRTEL, R., Leading the way: Directional sensing through phosphatidylinositol 3-kinase and other signaling pathways. *J Cell Sci*, **2003**, *116*, 3471–3478.

38 PARENT, C., Making all the right moves: chemotaxis in neutrophils and Dictyostelium. *Curr Opin Cell Biol*, **2004**, *16*, 4–13.

39 CHEN, L., et al., Two phases of actin polymerization display different dependencies on PI(3,4,5)P3 accumulation and have unique roles during chemotaxis. *Mol Biol Cell*, **2003**, *14*, 5028–5037.

40 LEMMON, M.A., FERGUSON, K.M., and ABRAMS, C.S., Pleckstrin homology domains and the cytoskeleton. *FEBS Lett*, **2002**, *513*, 71–76.

11

HGF/SF c-Met Signaling in the Epithelial–Mesenchymal Transition and Migration of Muscle Progenitor Cells

Ute Schaeper and Walter Birchmeier

Abstract

Hepatocyte growth factor/scatter factor (HGF/SF) has long been known to induce cell dissociation and motility in epithelial cell lines. It acts by binding and activating the receptor tyrosine kinase c-Met. Gene ablation experiments in the mouse have established that HGF/SF and c-Met play also a fundamental role in migration of embryonic cells, in particular long range migration of muscle progenitor cells. This process is essential for the formation of certain skeletal muscle groups, like muscles of the limbs and diaphragm. Here, we discuss the signaling pathways activated by HGF/SF and c-Met and their implications in cell migration *in vivo*.

11.1
Introduction: Hepatocyte Growth Factor/Scatter Factor (HGF/SF) is a Pleiotropic Growth Factor

HGF/SF can function as a classical growth factor, by promoting the proliferation and survival of epithelial cells. Indeed, HGF/SF was discovered as a serum-derived factor which induces DNA synthesis of hepatocytes [1, 2]. Independently, Stoker *et al.* reported the identification of a "scatter factor" produced from embryonic fibroblasts, which induces cell dissociation and motility, so called "scattering" of epithelial cells [3]. Under normal serum conditions, Madin-Darby canine kidney (MDCK) cells grow as polarized cells in tightly packed colonies (Fig. 11.1). When scatter factor is applied to these cells, cell colonies spread out. This is followed by loss of cell–cell contact and a change in morphology from a polarized epithelial cell type to a motile, more fibroblast-like cell type (Fig. 11.1). This process, the change from epithelial cell type to mesenchymal fibroblast-like cell type is also referred to as epithelial–mesenchymal transition and can be observed in embryonic development and tumorigenesis [4]. Purification and molecular cloning of "scatter factor" and "hepatocyte growth factor" from platelets and fibroblast-conditioned medium, respectively, re-

Cell Migration. Edited by Doris Wedlich
Copyright © 2005 WILEY-VCH Verlag GmbH & Co. KGaA, Weinheim
ISBN: 3-527-30587-4

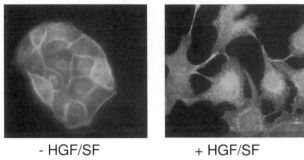

- HGF/SF   + HGF/SF

Fig. 11.1 HGF/SF induces scattering of epithelial cells.
Phalloidine staining of Madin Darby canine kidney
(MDCK) epithelial cells grown in the absence (left) or
presence (right) of HGF/SF.

vealed that these factors are identical and are thereafter referred to as
HGF/SF [5–7]. Independently, HGF/SF has also been identified in other
biological assays: as a morphogen, which induces "branching morphogenesis", the formation of branched tubular structures of epithelial cells
grown in three-dimensional matrixes [8–10]. In addition, HGF/SF promotes angiogenesis, the formation of new blood vessels and aids in tissue regeneration [11, 12]. Recently, it was shown that HGF/SF can also
act as an infection susceptibility-inducing factor for plasmodium para-
sites, the causative agent for malaria. When HGF/SF is applied to hepatocytes, or other epithelial cells, it makes them susceptible to infection by
the sporozoite stage of plasmodium. Infection of hepatocytes with sporo-
zoites is a prerequisite for establishment of malaria infection [13].

Sequence analyses and comparisons of HGF/SF cDNA with gene bank
files revealed that HGF/SF is a peptide growth factor most closely related
to serine proteases of the plasminogen subfamily [5, 7, 14, 15]. HGF/SF
contains an N-terminal domain, four kringle domains and a serine protease homology domain (Fig. 11.2). HGF/SF is synthesized and secreted
as an inactive precursor peptide chain. Activation occurs through cleavage at specific sites by serine proteases, like plasminogen activators, generating α- and β-chains linked by disulfide bonds. Despite its homology
to plasminogen, HGF/SF lacks proteolytic activity. However, HGF/SF and
plasminogen have in common, that they are both synthesized as inactive
precursor peptides and activated by proteolytic cleavage [16].

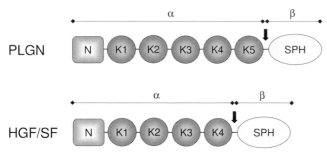

Fig. 11.2 HGF/SF is related to the serine protease plasminogen. HGF/SF has a domain structure that is similar to the serine protease plasminogen (PLGN). It is composed of an N-terminal domain (N), four kringle domains (K1–K4), and a catalytic inactive serine-protease homology domain (SPH). HGF/SF is secreted as an inactive precursor-peptide. Activation occurs by cleavage through proteases like plasminogen activator (indicated by arrow). Biologically active HGF/SF is composed of α- and β-chains linked by disulfide bonds.

11.2
HGF/SF is the Ligand for the Proto-Oncogene c-Met

c-Met was originally cloned as the cellular counterpart of the oncogenic variant Tpr-Met [17]. The Tpr-Met oncogene can transform immortalized fibroblasts and induce anchorage-independent growth and invasion. It was identified in a genetic screen for transforming genes that utilized a cDNA library, derived from a chemically mutagenized human osteosarcoma cell line [18]. Later it was shown that Tpr-Met is a gene product derived from two different genes generated by chromosomal translocation [17]. It contains the tyrosine kinase domain and substrate binding site of the receptor tyrosine kinase c-Met, but it lacks the extracellular and transmembrane region of the cell surface receptor [19]. Instead, Tpr-Met possesses a dimerization domain of another gene (Tpr, translocated promoter region) (Fig. 11.3). Due to this dimerization domain, Tpr-Met is constitutively active, independent of ligand binding.

Cellular Met (c-Met) in contrast, is a classical receptor tyrosine kinase activated by binding of extracellular ligand (Fig. 11.3). Ligand binding of receptor tyrosine kinases induces oligomerization, activation of intrinsic tyrosine kinase activity and autophosphorylation on tyrosine residues [20]. Using I^{125}-labeled HGF/SF , Bottaro *et al.* could demonstrate that HGF/SF binds the c-Met receptor expressed on hepatocytes, and that HGF/SF induces tyrosine phosphorylation of c-Met, establishing that HGF/SF is a ligand of the RTK c-Met [21]. c-Met is composed of two disulfide-linked subunits [22]. They are generated by cleavage of a single

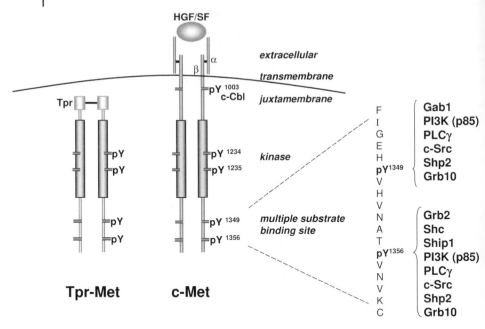

Tpr-Met   **c-Met**

Fig. 11.3 Tpr-Met is an oncogenic variant of c-Met. Cellular Met is a receptor tyrosine kinase activated by binding of HGF/SF. The ligand induces dimerization and transphosphorylation of Y1234 and Y1235 located in the activation loop of c-Met kinase. Phosphorylation at these sites is essential for Met-kinase activation. Y1349 and Y1356 are phosphorylation sites located in the multisubstrate binding site of c-Met. When phosphorylated by the c-Met kinase, they function as binding sites for several signaling molecules. Another substrate binding site, Y1003, is located in the juxtamembrane region of c-Met and involved in negative regulation of c-Met activity. It recruits the c-Cbl adaptor protein. c-Cbl promotes ubiquitination of c-Met and thereby enhances degradation of activated c-Met. Tpr-Met oncoprotein contains only the C-terminal part of c-Met including kinase domain and multiple substrate docking site. It lacks extracellular, transmembrane and the c-Cbl binding site (Y1003). In addition, it possesses a dimerization motiv of a different gene (translocated promoter region, tpr) at its N terminus. Due to this dimerization domain, Tpr-Met is constitutively active, independent of ligand binding. This figure was modified from Ref. [15]. (Reproduced with permission from Nature Reviews Molecular Cell Biology, Macmillan Magazines Ltd.)

c-Met precursor protein. The longer β-chain contains an extracellular domain, a transmembrane domain, a tyrosine kinase domain and a substrate docking site (Fig. 11.3). The shorter α-chain is exclusively extracellular and is required for HGF/SF binding.

c-Met receptor tyrosine kinase can indeed function as a transducer for HGF/SF-induced scatter activity. This was shown using chimeric c-Met receptor by which the extracellular domains of c-Met were replaced with the ligand-binding domain of the NGF-receptor, TrkA [23]. When expressed in MDCK cells and activated by NGF, the hybrid receptor induces cell scattering or branching morphogenesis similar to HGF/SF which binds endogenous c-Met. This supported the role of c-Met as a transducer of HGF/SF-mediated signals. Functional analyses of c-Met mutants or Tpr-Met oncoprotein mutants revealed that the tyrosine kinase activity as well as Y1349/Y1359 located close to the C terminus of c-Met are essential for all the biological activities of Met, including HGF/SF-induced motility of epithelial cells [24–28]. Y1349/Y1356 are located in the multiple substrate binding site of c-Met and constitute the major autophosphorylation sites of c-Met. Phosphorylation at these residues creates binding sites for several signaling molecules, like phosphoinositol-3-kinase (PI(3)K), phospholipase C-gamma (PLC-γ), the tyrosine kinase c-Src, and the adaptor proteins Shc and Grb2. In addition, the multiple docking site of c-Met recruits the Grb2 associated binder 1 (Gab1) [29]. Gab1 is unique, because it binds specifically to activated c-Met, but not to other RTKs. Gab1 contains a PH domain and protein interaction domains, like Grb2- and a c-Met binding site (MBS, Fig. 11.4). The c-Met binding site, together with the Grb2-binding site of Gab1, is required for c-Met association [30]. Phosphorylation of Gab1 by c-Met creates additional binding sites for signaling proteins, like PI(3)K, Crk and Shp2. It is thought that docking proteins enhance the signaling capacity of tyrosine kinases by creating a larger signaling platform.

Another phosphotyrosine docking site located in the juxtamembrane region of c-Met, tyrosine 1003 (Fig. 11.3), is involved in negative regulation of HGF/c-Met signaling. Mutation of this tyrosine residue to phenylalanine leads to constitutive scattering of MDCK cells, independent of HGF/SF binding [26]. Later it was shown that phosphophorylated Y1003 binds the c-Cbl proto-oncogene. c-Cbl has an important function in downregulation of c-Met signaling activity. By binding to activated c-Met, c-Cbl promotes c-Met ubiquitination, internalization into clathrin-coated pits and degradation by the proteasome pathway. Mutation of the c-Cbl-binding site enhances c-Met protein stability and c-Met signaling. The c-Cbl-binding site is not present in Tpr-Met, and lack of this negative feedback loop may thus also contribute to the oncogenic potential of the Tpr-Met oncoprotein [31].

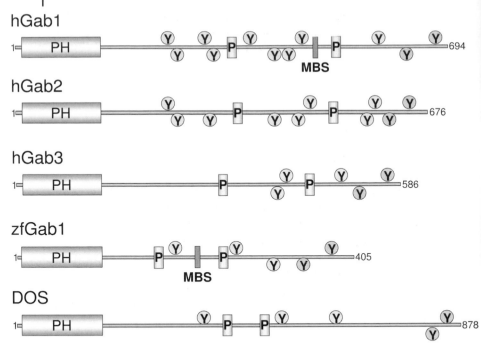

Fig. 11.4 Domain structure of Gab1-related docking proteins. Grb2 associated binder (Gab1) family members are substrates of tyrosine kinases. They are harbor an N-terminal pleckstrin homology domain (PH), and they contain multiple tyrosine phosphorylation sites that can function as docking sites (Y) for signaling proteins, for instance PI(3)K, Crk/CrkL and Shp2. Consensus binding sites for Shp2 are marked in green, PI(3)K binding sites in orange, and Crk/CrkL consensus binding sites in blue. Proline-rich regions (P) of Gab1 mediate interaction with Grb2. A c-Met binding site (MBS) is present in human Gab1 and in zebrafish Gab1 (zfGab1), but not found in Gab2 and Gab3 or *Drosophila* Dos. (This figure also appears with the color plates.)

11.3
HGF/SF acts as a Paracrine Growth Factor in Normal Tissue

HGF/SF and c-Met are rarely expressed by the same cell [13, 32]. c-Met is expressed primarily on epithelial cells, for instance in the embryonic kidney, liver, lung, pancreas, stomach, intestine and muscle cells. HGF/SF on the other hand is expressed in surrounding mesenchymal tissue in a timely and spatially restricted pattern, consistent with a function of HGF/SF as a paracrine growth factor. HGF/SF expression is upregulated in several organs after injury, like in acute liver injuries, or kidney disease. Exogenous application of HGF/SF accelerates organ regeneration after partial hepatectomy or kidney damage. Thus, HGF/SF can function in wound healing of liver and kidney [10, 33].

11.4
Functional Role of HGF/SF and c-Met in the Development of Skeletal Muscles

The function of HGF/SF and c-Met in embryonic development was analyzed by a gene knock-out approach in the mouse. Both c-met and hgf/sf null-mutations are embryonic lethal, causing defects in placental development [34, 35]. In particular the labyrinthine layer of the placenta, responsible for exchange of oxygen and nutrients between maternal and fetal blood, is impaired. In addition, the liver is also affected and reduced in size. Strikingly, c-met −/− as well as hgf/sf −/− mice completely lack muscles of the limbs, diaphragm and the tip of the tongue. In vertebrates, skeletal muscles of the trunk are derived from paraxial mesoderm, which segments into somites on either side of neurotube and notochord [36, 37]. The ventral part of the somites gives rise to the sclerotome, which will later form the ribs and axial skeleton, while the dorsal part gives rise to the dermomyotome, which will then form the dermis of the back and skeletal muscles of the trunk (Fig. 11.5). The dermomyotome is further divided into the epaxial region (dorso–medial lip), which will form the myotome and dorsal body wall muscle and the hypaxial part (dorso–lateral lip) which will generate ventral body wall muscles, but also the muscles of the limb, diaphragm and tongue [56]. In contrast to the body wall muscles, which are formed by extension of epithelial sheets of cells, limb muscles are generated from migratory progenitor cells [38]. At limb level, muscle progenitor cells of the ventro–lateral edge of the dermomyotome undergo epithelial–mesenchymal transition. They delaminate and migrate as "scattered" cells into the limb bud mesenchyme, where they further proliferate and differentiate to form the skeletal muscles of the limb.

In c-met −/− mice, muscle progenitor cells do not delaminate from the ventro–lateral lip of the dermomyotome and do not take up long range migration. As a consequence, these mice lack all muscles of the limb, diaphragm and tip of tongue [34, 39]. The phenotype of c-met −/− mice is identical to hgf/sf −/− mice, suggesting that HGF/SF is the single ligand required for c-Met activity [34, 35]. While c-Met is expressed throughout the dermomyotome in cells of the limb and interlimb level, HGF/SF is expressed in a specific pattern close to the migratory route of muscle progenitor cells [39]. Ectopic expression of HGF/SF at interlimb levels induces emigration of muscle progenitor cells, which would normally not migrate [40]. This indicates that HGF/SF acts in a paracrine fashion in directing delamination and migration of c-Met expressing epithelial cells.

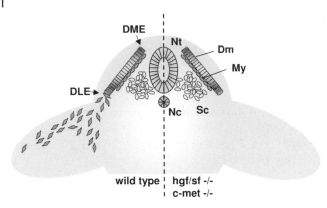

Fig. 11.5 Origin of limb muscle progenitor cells. Sclerotome (Sc) and dermomyotome (Dm) are derivatives of the somites. The Sclerotome gives rise to the skeleton and the dermomyotome (Dm) will form the dermis and musculature. Cells of the dorsal medial edge (DME) of the dermomyotome generate the myotome, which will form epaxial muscles, the musculature of the back. Cells from the dorsal lateral edge (DLE) of the dermomyotome form the hypaxial musculature, i.e., ventral body wall muscles and muscles of the limbs. At limb level, muscle progenitor cells delaminate from the DLE and migrate towards the limb anlage. In HGF/SF or c-Met –/– mice, muscle progenitor cells remain in the dermomyotome and skeletal muscles of the limbs do not form. Nt neurotube, nc notochord. This figure was modified from Ref. [15, 56]. (Reproduced with permission from Nature Reviews Molecular Cell Biology, Macmillan Magazines Ltd. and Curr Opin Cell Biol., Elsevier.) (This figure also appears with the color plates.)

Genetically, c-Met has been placed downstream of the Pax 3 homeobox transcription factor. Lack of all muscles derived from migratory precursor cells has also been observed in the spontaneous mouse mutant *Splotch* [41, 42]. *Splotch* mice carry a mutation in the pax 3 homeobox transcription factor (pax3^{Sp}). Homozygous pax3^{Sp} mice are embryonic lethal. They display multiple defects, like malformation of neural tissues, neural crest derived tissues, dermomyotome and myotome, leading to skeletal defects and malformation of the entire trunk musculature. As a consequence, epaxial muscles of the deep back and hypaxial nonmigratory muscles, like intercostal and abdominal muscles, are reduced. Migratory hypaxial muscles, like limbs and diaphragm, are completely absent [42]. c-Met may thus be a direct target gene of Pax3, since c-Met transcripts were not detected in the dermomyotome of *Splotch* mice [43]. Furthermore, Pax 3 has been shown to transactivate the c-Met reporter gene in vitro [44]. An activated form of Pax-3, Pax3-FKHR, has been identified in alveolar rhabdomyosarcomas (ARMS), a highly malignant class of pediatric soft tissue tumors with skeletal muscle characteristics [45]. Pax3-FKHR is the product of a chromosomal translocation, t(2;13), which leads to fusion of the C-terminal transactivation domain of FKHR to the N-terminal region of Pax3 involved in DNA binding. As a result of this gene-fusion, Pax3-FKHR is a stronger transcriptional activator than Pax3 [46]. When Pax3-FKHR is expressed in the Pax3-locus of the mouse, it leads to enhanced c-Met expression in muscle precursor cells and ectopic delamination of muscle progenitor cells, [47]. Possibly the higher c-Met expression levels induced by Pax3-FKHR mice leads to tyrosine kinase activation independent of HGF/SF binding. This phenotype is alleviated when one c-Met allele is removed.

11.5
c-Met Substrates are Required for Migration of Muscle Progenitor Cells

Biochemical analyses of several labs have shown, that downstream substrates of the c-Met tyrosine kinase are recruited by binding to Y1349 and Y1356, located in the multiple substrate docking site in the carboxy-terminal region of c-Met [25, 29]. A hypomorphic c-Met allele in which Y1349 and Y1356 are mutated to phenylalanine, causes multiple defects in embryonic development similar to c-Met −/− including lack of limb muscles [48]. This indicates that recruitment and activation of intracellular signaling molecules is essential for c-Met activity. Gab1 was recently identified as a docking protein that binds specifically to the multisubstrate binding site of c-Met and not other receptor tyrosine kinases [29, 30]. Gene ablation of Gab1 in the mouse causes significant reduction in the migration of muscle progenitor cells into the limb anlage [49]. Interestingly, muscle formation of the forelimbs and hind limbs is affected differently. In the forelimbs of Gab1 −/− mice, the extensor muscles are almost completely missing, while flexor muscles exist, but reach less far. In contrast, in hind limbs, flexor and extensor muscles appear equally affected reach less far and appear disorganized. Altogether, the muscle phenotype of gab1 −/− mice suggests that Gab1 functions in c-Met-induced migration of muscle progenitor cells, but additional signaling molecules may act in parallel. pY1349 and pY1356 of c-Met associate with many cytoplasmic signaling molecules, like phosphoinositide (3)kinase, phospholipase C-γ, c-Src [25, 27, 50]. pY1356 also mediates binding of the adaptor proteins Grb2 and Shc [25, 28]. Replacing the multiple docking site of c-Met with binding sites specific for PI(3)K or c-Src showed that neither activation of PI(3)K nor c-Src is sufficient for full migration of muscle progenitors. In contrast, substituting the multiple docking site of c-Met with two Grb2 binding sites results in full rescue of limb muscle development [48]. The Grb2-binding site of c-Met couples c-Met to the Ras pathway and is also required to enhance the association of c-Met with Gab1 [30]. Possibly, a combination of Grb2/Ras and Gab1-mediated signals is required for long range migration of muscle progenitor cells (Fig. 11.6).

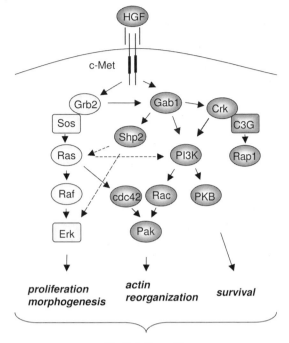

Cell Migration

Fig. 11.6 Signaling pathways downstream of Gab1 and Grb2 are implicated in cell migration. Both Grb2 and Gab1 bind directly to the multisubstrate binding site of c-Met. Grb2 recruits the GTP/GDP exchange factor Sos which leads to activation of Ras. Ras activation is required for HGF/SF dependent migration of MDCK cells, since microinjection of dominant–negative Ras blocks cell spreading and scattering of MDCK cells. Activated Ras by itself can induce cell spreading [53]. Ras activates several downstream signaling pathways, for instance the Erk/MAPK pathway via Raf 1, the PI(3)K pathway and the small GTP-binding proteins cdc42, Rac and Rho [53, 54]. Cdc42 and Rac1 are involved in filopodia and lamellipodia formation induced by HGF and also required for HGF/SF-induced cell migration. Activation of Rac 1 is in part be induced by PI(3)K. PI(3)K can also activate PKB/Akt which controls cell survival. Gab1 has been shown to activate several signaling molecules, like PI(3)K, Shp2 and CrkL. Activation of Shp2 is important for sustained activation of MAPK (Erk) by growth factors and is also required for HGF/SF dependent morphogenesis [30]. Crk/CrkL has many effectors, for instance Rap1 and PI(3)K and Rac1. Crk/CrkL can also promode lamellipodia formation and cell spreading of epithelial cells and may thus also be involved in HGF/SF-induced cell migration [55]. (This figure also appears with the color plates.)

11.6
Evolution of Migratory Muscle Progenitor Cells

The formation of appendicular muscles from migratory progenitors is a process which arose only recently in vertebrate evolution [51]. It is observed in tetrapods like birds and mammals, and in bony fish like zebrafish, but not in chondrichthyan species like the dogfish shark [52]. Here, the muscles of the fins are formed by continuous epithelial–myotomal extension. In contrast, fin muscles of the more "modern" teleost fish, i.e. zebrafish, arise from migratory progenitor cells that delaminate from the dermomyotome by epithelial–mesenchymal transition similar to tetrapods [52]. Interestingly, the similarity of limb muscle development in amniotes and the fin muscle development of modern fish can also be seen on the molecular level. In the mouse, the homeobox containing genes lbx and mox2 are expressed within the lateral hypaxial domain of mouse limb mesoderm. Gene ablation of lbx or mox2 results in loss of distinct muscle population of the limbs. In zebrafish, lbx and mox2 homologs exist and are also expressed in myogenic precursor cells at fin level. However lbx or mox 2 expression is not detected in the muscle progenitor cells of the evolutionary older dogfish shark [52]. Homologs of pax3, the tyrosine kinase c-met and gab1 and have also been identified in the zebrafish genome [47] (Fig. 11.4). The Gab1 homolog of zebrafish has a domain structure similar to mammalian Gab1. It contains a PH-domain, Grb2 binding sites and even a c-Met binding site. The c-Met binding site is not present in Gab1 homologs from Drosophila or *C. elegans*, which also lack c-Met (Fig. 11.4). This suggests that c-Met and the MBS of Gab1 have co-evolved.

11.7
Outlook

Gene knock-out experiments in the mouse have revealed a pivotal role of the HGF/SF-c-Met and Gab1 signaling cascade for limb muscle development, in particular, the emigration of muscle progenitor cells from the somites to the limb bud mesenchyme. Biochemical analyses, primarily from cultured cells, show that Gab1 is a substrate of c-Met. It is phosphorylated in response to HGF/SF stimulation and recruits additional signaling molecules, like PI(3)K, c-Src, Crk/CrkL and Shp2 (Fig. 11.6). This leads to activation and/or enhancement of multiple downstream signaling pathways, like the Ras/MAPK or PI(3)K/PKB/Akt pathway. Currently it is not known which signaling pathways downstream of c-Met and Gab1 are required for HGF/SF-induced cell migration in vivo. Additional genetic experiments will be necessary to address this question, for instance the generation of Gab1 "knock-in" mice, which express Gab1 point mutants deficient in binding and activating some of the downstream signaling molecules. Comparisons of the phenotypes of these

Gab1 hypomorphs will be helpful in dissecting the signaling pathway required for c-Met dependent cell migration *in vivo*.

11.8
Acknowledgments

We like to thank Katja Grossmann for illustrations.

11.9
References

1 MICHALOPOULOS, G., et al., Control of hepatocyte replication by two serum factors. *Cancer Res*, **1984**, *44(10)*, 4414–4419.

2 NAKAMURA, T., H. TERAMOTO, and A. ICHIHARA, Purification and characterization of a growth factor from rat platelets for mature parenchymal hepatocytes in primary cultures. *Proc Natl Acad Sci USA*, **1986**, *83(17)*, 6489–6493.

3 STOKER, M. and M. PERRYMAN, An epithelial scatter factor released by embryo fibroblasts. *J Cell Sci*, **1985**, *77*, 209–223.

4 BIRCHMEIER, W. and C. BIRCHMEIER, Mesenchymal–epithelial transitions. *Bioessays*, **1994**, *16(5)*, 305–307.

5 MIYAZAWA, K., et al., Molecular cloning and sequence analysis of cDNA for human hepatocyte growth factor. *Biochem Biophys Res Commun*, **1989**, *163(2)*, 967–973.

6 WEIDNER, K.M., et al., Evidence for the identity of human scatter factor and human hepatocyte growth factor. *Proc Natl Acad Sci USA*, **1991**, *88*(16), 7001–7005.

7 NAKAMURA, T., et al., Molecular cloning and expression of human hepatocyte growth factor. *Nature*, **1989**, *342(6248)*, 440–443.

8 MONTESANO, R., G. SCHALLER, and L. ORCI, Induction of epithelial tubular morphogenesis in vitro by fibroblast-derived soluble factors. *Cell*, **1991**, *66(4)*, 697–711.

9 MONTESANO, R., et al., Identification of a fibroblast-derived epithelial morphogen as hepatocyte growth factor. *Cell*, **1991**, *67(5)*, 901–908.

10 ROSARIO, M. and W. BIRCHMEIER, How to make tubes: signaling by the Met receptor tyrosine kinase. *Trends Cell Biol*, **2003**, *13(6)*, 328–335.

11 BUSSOLINO, F., et al., Hepatocyte growth factor is a potent angiogenic factor which stimulates endothelial cell motility and growth. *J Cell Biol*, **1992**, *119(3)*, 629–641.

12 GRANT, D.S., et al., Scatter factor induces blood vessel formation in vivo. *Proc Natl Acad Sci USA*, **1993**, *90(5)*, 1937–1941.

13 CARROLO, M., et al., Hepatocyte growth factor and its receptor are required for malaria infection. *Nat Med*, **2003**, *9(11)*, 1363–1369.

14 DONATE, L.E., et al., Molecular evolution and domain structure of plasminogen-related growth factors (HGF/SF and HGF1/MSP). *Protein Sci*, **1994**, *3(12)*, 2378–2394.

15 BIRCHMEIER, C., et al., Met, metastasis, motility and more. *Nat Rev Mol Cell Biol*, **2003**, *4(12)*, 915–925.

16 HARTMANN, G., et al., A functional domain in the heavy chain of scatter factor/hepatocyte growth factor binds the c-Met receptor and induces cell dissociation but not mitogenesis. *Proc Natl Acad Sci USA*, **1992**, *89(23)*, 11574–11578.

17 PARK, M., et al., Mechanism of met oncogene activation. *Cell*, **1986**, *45(6)*, 895–904.

18 COOPER, C. S., et al., Molecular cloning of a new transforming gene from a chemically transformed human cell line. *Nature*, **1984**, *311(5981)*, 29–33.

19 PARK, M., et al., Sequence of MET protooncogene cDNA has features characteristic of the tyrosine kinase family of growth factor receptors. *Proc Natl Acad Sci USA*, **1987**, *84(18)*, 6379–8633.

20 SCHLESSINGER, J., Cell signaling by receptor tyrosine kinases. *Cell*, **2000**, *103(2)*, 211–225.

21 BOTTARO, D. P., et al., Identification of the hepatocyte growth factor receptor as the c-met proto-oncogene product. *Science*, **1991**, *251(4995)*, 802–804.

22 GIORDANO, S., et al., Biosynthesis of the protein encoded by the c-met proto-oncogene. *Oncogene*, **1989**, *4(11)*, 1383–1388.

23 WEIDNER, K. M., M. SACHS, and W. BIRCHMEIER, The Met receptor tyrosine kinase transduces motility, proliferation, and morphogenic signals of scatter factor/hepatocyte growth factor in epithelial cells. *J Cell Biol*, **1993**, *121(1)*, 145–154.

24 ZHU, H., et al., Tyrosine 1356 in the carboxyl-terminal tail of the HGF/SF receptor is essential for the transduction of signals for cell motility and morphogenesis. *J Biol Chem*, **1994**, *269(47)*, 29943–29948.

25 PONZETTO, C., et al., A multifunctional docking site mediates signaling and transformation by the hepatocyte growth factor/scatter factor receptor family. *Cell*, **1994**, *77(2)*, 261–271.

26 WEIDNER, K. M., et al., Mutation of juxtamembrane tyrosine residue 1001 suppresses loss-of-function mutations of the met receptor in epithelial cells. *Proc Natl Acad Sci USA*, **1995**, *92(7)*, 2597–2601.

27 FIXMAN, E. D., et al., Efficient cell transformation by the Tpr-Met oncoprotein is dependent upon tyrosine 489 in the carboxy-terminus. *Oncogene*, **1995**, *10(2)*, 237–249.

28 FIXMAN, E. D., et al., Pathways downstream of Shc and Grb2 are required for cell transformation by the tpr-Met oncoprotein. *J Biol Chem*, **1996**, *271(22)*, 13116–13122.

29 WEIDNER, K. M., et al., Interaction between Gab1 and the c-Met receptor tyrosine kinase is responsible for epithelial morphogenesis. *Nature*, **1996**, *384(6605)*, 173–176.

30 SCHAEPER, U., et al., Coupling of Gab1 to c-Met, Grb2, and Shp2 mediates biological responses. *J Cell Biol*, **2000**, *149(7)*, 1419–1432.

31 PESCHARD, P., et al., Mutation of the c-Cbl TKB domain binding site on the Met receptor tyrosine kinase converts it into a transforming protein. *Mol Cell*, **2001**, *8(5)*, 995–1004.

32 SONNENBERG, E., K. M. WEIDNER, and C. BIRCHMEIER, Expression of the met b role of SF/HGF and c-Met in the development of skeletal muscle. *Development*, **1999**, *126(8)*, 1621–1629.

33 LAPING, N. J., Hepatocyte growth factor in renal disease: cause or cure? *Cell Mol Life Sci*, **1999**, *56(5/6)*, 371–377.

34 BLADT, F., et al., Essential role for the c-met receptor in the migration of myogenic precursor cells into the limb bud. *Nature*, **1995**, *376(6543)*, 768–771.

35 SCHMIDT, C., et al., Scatter factor/hepatocyte growth factor is essential for liver development. *Nature*, **1995**, *373(6516)*, 699–702.

36 BUCKINGHAM, M., et al., The formation of skeletal muscle: from somite to limb. *J Anat*, **2003**, *202(1)*, 59–68.

37 FRANCIS-WEST, P. H., L. ANTONI, and K. ANAKWE, Regulation for myogenic differentiation in the developing limb bud. *J Anat*, **2003**, *202(1)*, 69–81.

38 CHRIST, B., et al., On the origin, distribution and determination of avian limb mesenchymal cells. *Prog Clin Biol Res*, **1982**, *110 Pt B*, 281–291.

39 DIETRICH, S., et al., The role of SF/HGF and c-Met in the development of skeletal muscle. *Development*, **1999**, *126(8)*, 1621–1629.

40 BRAND-SABERI, B., et al., Scatter factor/hepatocyte growth factor (SF/HGF) induces emigration of myogenic cells at interlimb level in vivo. *Dev Biol*, **1996**, *179(1)*, 303–308.

41 DASTON, G., et al., Pax-3 is necessary for migration but not differentiation of limb muscle precursors in the mouse. *Development*, **1996**, *122(3)*, 1017–1027.

42 TREMBLAY, P., et al., A crucial role for Pax-3 in the development of the hypaxial musculature and the long-range migration of muscle precursors. *Dev Biol*, **1998**, *203(1)*, 49–61.

43 YANG, X.M., et al., Expression of the met receptor tyrosine kinase in muscle progenitor cells in somites and limbs is absent in Splotch mice. *Development*, **1996**, *122(7)*, 2163–2171.

44 EPSTEIN, J.A., et al., Pax3 modulates expression of the c-Met receptor during limb muscle development. *Proc Natl Acad Sci USA*, **1996**, *93(9)*, 4213–4218.

45 BARR, F.G., Gene fusions involving PAX and FOX family members in alveolar rhabdomyosarcoma. *Oncogene*, **2001**, *20(40)*, 5736–5746.

46 BENNICELLI, J.L., R.H. EDWARDS, and F.G. BARR, Mechanism for transcriptional gain of function resulting from chromosomal translocation in alveolar rhabdomyosarcoma. *Proc Natl Acad Sci USA*, **1996**, *93(11)*, 5455–5459.

47 RELAIX, F., et al., The transcriptional activator PAX3-FKHR rescues the defects of Pax3 mutant mice but induces a myogenic gain-of-function phenotype with ligand-independent activation of Met signaling in vivo. *Genes Dev*, **2003**, *17(23)*, 2950–2965.

48 MAINA, F., et al., Coupling Met to specific pathways results in distinct developmental outcomes. *Mol Cell*, **2001**, *7(6)*, 1293–1306.

49 SACHS, M., et al., Essential role of Gab1 for signaling by the c-Met receptor in vivo. *J Cell Biol*, **2000**, *150(6)*, 1375–1384.

50 PELICCI, G., et al., The motogenic and mitogenic responses to HGF are amplified by the Shc adaptor protein. *Oncogene*, **1995**, *10(8)*, 1631–1638.

51 HOLLWAY, G.E. and P.D. CURRIE, Myotome meanderings. Cellular morphogenesis and the making of muscle. *EMBO Rep*, **2003**, *4(9)*, 855–860.

52 NEYT, C., et al., Evolutionary origins of vertebrate appendicular muscle. *Nature*, **2000**, *408(6808)*, 82–86.

53 RIDLEY, A.J., P.M. COMOGLIO, and A. HALL, Regulation of scatter factor/hepatocyte growth factor responses by Ras, Rac, and Rho in MDCK cells. *Mol Cell Biol*, **1995**, *15(2)*, 1110–1122.

54 ROYAL, I. and M. PARK, Hepatocyte growth factor-induced scatter of Madin-Darby canine kidney cells requires phosphatidylinositol 3-kinase. *J Biol Chem*, **1995**, *270(46)*, 27780–27787.

55 LAMORTE, L., et al., Crk adapter proteins promote an epithelial-mesenchymal-like transition and are required for HGF-mediated cell spreading and breakdown of epithelial adherens junctions. *Mol Biol Cell*, **2002**, *13(5)*, 1449–1461.

56 BIRCHMEIER, C. and H. BROHMANN, Genes that control the development of migration muscle precursor cells. *Curr Opin Cell Biol*, **2000**, *12(6)*, 725–730.

III

Cell Migration Crucial for Immune Response, Wound Healing, and Tumorigenesis

Cell Migration. Edited by Doris Wedlich
Copyright © 2005 WILEY-VCH Verlag GmbH & Co. KGaA, Weinheim
ISBN: 3-527-30587-4

12
Host–Pathogen Interactions and Cell Motility: Learning from Bacteria

Theresia E. B. Stradal, Silvia Lommel, Jürgen Wehland, and Klemens Rottner

Abstract

Numerous cellular functions and processes depend on dynamic changes of the actin cytoskeleton including morphogenetic movements during embryonic development, chemotactic movement of cells of the immune system and migration of fibroblasts during wound healing. Such changes are also involved in cell shape changes during cell spreading, phagocytosis and intracellular transport events. A large number of actin-binding and regulating proteins have been described involved in polymerization/depolymerization, capping, severing or bundling of actin filaments.

However, the exact signaling pathways and regulatory principles for the tight control of spatially and temporally regulated actin polymerization/depolymerization remained enigmatic for a long time and are still only partially understood. The discovery that similar dynamic changes of the actin cytoskeleton play an essential role in host–pathogen interactions had and still has an enormous impact on this field. Many bacterial pathogens exploit physiological pathways and mechanisms to harness the power of actin polymerization for efficient adhesion to, for getting into or out and for spreading within and between target cells. Moreover, bacterial pathogens evolved mechanisms to interfere with and to manipulate actin-dependent host defense mechanisms such as phagocytosis and specific responses of immune cells.

12.1
Introduction

A large number of microbial pathogens have co-evolved with their host organisms. As a consequence these pathogens have learned to modulate in a highly sophisticated manner cellular functions to ensure their survival, replication and dissemination within the host.

Since evolution has "fine tuned" these interactions, their study provides unique systems and tools which give detailed insights into basic cell biological processes. Here we describe a few eloquent examples of sophisticated strategies bacterial pathogens have evolved in particular to modulate the actin cytoskeleton. These bacteria or some of their virulence factors are now commonly used in many actin laboratories. Among these factors are bacterial toxins that target actin and monomeric GTP-binding proteins to modulate actin dynamics [reviewed in 1]. A fascinating mechanism by which bacteria directly deliver effector proteins into the host cell was discovered in 1994 [2]: the type III secretion system that has been evolved by gram-negative bacteria, which are either extracellular or residing within the phagosome, to inject bacterial proteins into the cytosol of eukaryotic cells. Very often these bacterial proteins precisely mimic host cell factors, enabling these bacteria to influence or even regulate diverse cellular functions [reviewed in 3, 4].

Entero-invasive bacterial pathogens for example can induce their own uptake or phagocytosis into normally nonphagocytic cells such as endothelial or epithelial cells. As a key component in this process the actin cytoskeleton is optimally exploited. Some bacteria expose on their surface proteins that interact with eukaryotic receptors normally involved in cell–cell or cell–matrix adhesion. This interaction triggers signaling cascades within the host cell leading to the recruitment of adaptor and effector proteins that initiate cytoskeletal rearrangements and membrane extensions at the bacterial attachment site culminating in bacterial internalization.

Other bacteria inject effector proteins that modulate the actin cytoskeleton at the attachment site using the type III secretion system. Whereas *Listeria* manipulate host cell surface proteins to gain entry, *Salmonella* and *Shigella* directly activate cytoskeletal components by injecting dedicated bacterial effectors. Some invasive bacterial pathogens can escape from the phagosome and replicate within the cytosol. Here they can recruit cytoskeletal components to induce actin polymerization at one bacterial pole that provides the force to propel these pathogens through the host cell cytoplasm. Because the actin-based intracellular motility of *Listeria* and *Shigella* is the paradigm that has enabled dissecting actin dynamics at the molecular level, this process is described here in more detail. As a third example, the formation of so-called actin pedestals upon attachment of enteropathogenic and enterohaemorrhagic *E. coli* to host cells is discussed. Again the type III secretion system remodels the cytoskeleton underneath the bacteria. We have focused on just a few examples of the various remarkable host–pathogen interactions bacterial pathogens have evolved to exploit the actin cytoskeleton in order to highlight their potential as tools for cell biologists (for overview see Fig. 12.1).

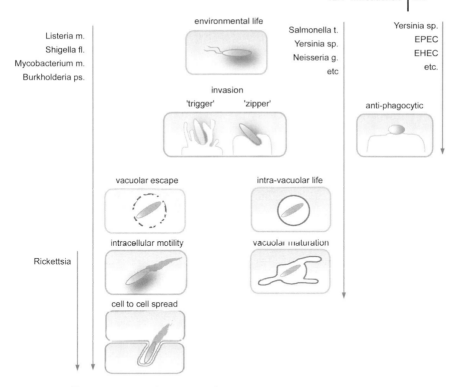

Listeria m.
Shigella fl.
Mycobacterium m.
Burkholderia ps.

environmental life

Salmonella t.
Yersinia sp.
Neisseria g.
etc

Yersinia sp.
EPEC
EHEC
etc.

invasion

'trigger'   'zipper'

anti-phagocytic

vacuolar escape

intra-vacuolar life

Rickettsia

intracellular motility

vacuolar maturation

cell to cell spread

Fig. 12.1 Different strategies of bacterial pathogens to subvert the host cell actin cytoskeleton. Bacterial pathogens usurp the host cell actin cytoskeleton by at least three different means: (i) induced phagocytosis, (ii) intracellular actin-based motility, and (iii) anti-phagocytosis. Entry of bacterial pathogens into their hosts can be achieved by two mechanisms which are roughly divided into the 'trigger' and the 'zipper' mechanism (see section 12.2). Upon entry pathogens are again subdivided into those that escape from the phagosome and subsequently gain actin-based motility in the host cell cytoplasm (see section 12.3), and the ones that reside in the phagosome, which can undergo further maturation (see section 12.2.1). The potential role of the host cell actin system in maturation of pathogen containing vacuoles (e.g. SCVs, see text for details) is under intense investigation. Another group of pathogens actively avoids being engulfed by professional phagocytes. Anti-phagocytosis can be accompanied by actin accumulations as in the case of EPEC and EHEC resulting in the formation of so-called pedestals (see section 12.4) or by inhibition of actin rearrangements in professional phagocytes as in the case of *Yersinia* and macrophages (see section 12.2.2). Note, that *Rickettsiae* are necessarily intracellular. Members of the spotted fever group of these pathogens employ actin-based motility to move within and spread between cells (see section 12.3.3).

12.2
Concepts for Bacterial Entry: 'Trigger' versus 'Zipper'

Many enteric bacterial pathogens have evolved strategies to invade host cells in order to escape the immune system and enhance their survival. To induce their own uptake, the pathogens require to remodel the host cell surface, which explains the large number of virulence factors that amend the host cell actin cytoskeleton. These factors can be roughly subdivided into those directly modifying actin and/or actin-binding proteins [5], and those rendering cellular signaling pathways leading to actin rearrangements [1]. Toxins of the latter group frequently activate, inhibit or modify Rho family GTPases [reviewed in 6], which are key regulatory switches in regulating actin cytoskeletal dynamics in cells [reviewed in 7]. Two major pathogenic strategies to induce internalization into non-professional phagocytes have been described in the past, the 'zipper' and the 'trigger' mechanism, which are categorized based on the host cell morphological changes occurring during bacterial entry [8–12].

12.2.1
Triggered Entry by *Salmonella* and *Shigella*: Variations on a Common Theme

In the 'trigger' mechanism, which is observed with pathogens like *Salmonella typhimurium* or *Shigella flexneri*, massive ruffling occurs, reminiscent of growth factor induced membrane ruffling or macropinocytosis (for overview see Fig. 12.2). Both *Shigella* and *Salmonella* are gram-negative rod-shaped bacteria expressing a type III secretion system (TTSS), which confers the delivery of bacterial virulence factors to the host cell cytoplasm leading to alterations of the actin cytoskeleton [4]. Among these virulence factors are actin-binding, bundling and nucleating factors, but also GEFs (guanine nucleotide exchange factors) and GAPs (GTPase activating proteins) for Rho-family GTPases, which directly induce localized signaling to actin rearrangements at the sites of bacterial attachment and entry [13].

The activities of the *Shigella* invasion plasmid antigens (Ipas) and the related *Salmonella* invasion proteins (Sips) are central to the initiation of the entry process [14, 15].

The TTSS first releases proteins that form a pore in the host cell membrane, which allows further delivery of bacterial effectors. However, the pore components themselves are already prominent modulators of the host cell actin cytoskeleton and absolutely required for bacterial entry [15–17]. For instance, *Salmonella* SipC harbors actin-binding capability in its N terminus and, more excitingly, actin filament nucleation activity in its C terminus [18, 19]. The *Shigella* homolog of SipC, IpaC, is involved only indirectly in actin remodelling by mediating the activation of the

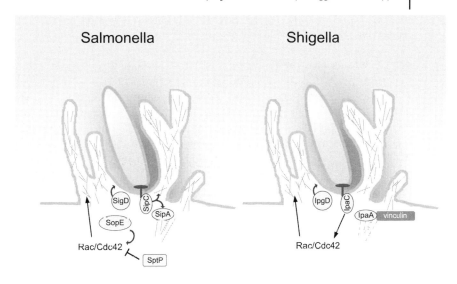

T TTSS

Fig. 12.2 Bacterial entry via the 'trigger' mechanism. The pathogens *Salmonella* and *Shigella* gain access to the host cell by triggering localized actin polymerization at the site of contact. This is achieved by a set of evolutionary related bacterial factors, which however have evolved distinct specificities to manipulate the host cell actin cytoskeleton (for details see text). Both pathogens express a type III secretion system (TTSS) to inject their factors into the host cell. Upon contact, a pore-forming protein complex is secreted, consisting of SipB and SipC for *Salmonella* and IpaB and IpaC for *Shigella*. While SipC directly induces actin accumulation and polymerization, IpaC activates the Rho-GTPases Rac and Cdc42 via an unknown mechanism resulting in localized actin assembly. *Salmonella* induces Rho-GTPase activation via the bacterial Rac- and Cdc42-GEFs SopE1/2, which have no homologs in *Shigella*. The secreted *Salmonella* effector SipA binds to and stabilizes F-actin and cooperates with SipC in the formation of the bacterial entry port. The *Shigella* ortholog IpaA recruits and activates the focal adhesion protein vinculin, which contributes to the formation of the *Shigella* entry focus.

GTPases Rac and Cdc42 [20]. The mechanism through which this occurs is however unclear, since to date no GEF-activity has been measured for IpaC *in vitro* [20]. Recently however, it was demonstrated that host cell tyrosine kinases of the Abl family play an essential role in IpaC-mediated

Cdc42 activation [21]. In the case of *Salmonella*, activation of Rac and Cdc42 is mediated by the SopE1/2 proteins that display classical GEF activity towards these GTPases, despite the absence of sequence homology to any known host cell GEF [22].

One bacterial factor that is delivered by *Shigella* directly into the host cell cytopasm is known as IpaA, which binds tightly to vinculin and recruits it to the site of bacterial entry [23]. This leads to the formation of a focal adhesion-like structure beneath the pathogen, the so-called entry focus [24]. *Shigella* mutants lacking IpaA can induce typical ruffling around the pathogen without efficient engulfment due to impaired 'dragging' into the host cell [23]. More specifically, the IpaA-vinculin complex was shown to depolymerize actin filaments *in vitro*, which is believed to further aid localized invagination of the membrane beneath the bacterium during entry, although the molecular details of such a mechanism are not understood [25].

SipA, the respective *Salmonella* ortholog, differs in its mode of function from IpaA. SipA directly binds to actin, decreases the critical concentration for actin polymerization and inhibits depolymerization by competing off the actin filament severing protein ADF/cofilin. Moreover, the same protein was also reported to be capable of re-annealing gelsolin severed actin filaments *in vitro* [26, 27]. Hence, although not absolutely essential for the uptake process, SipA acts in concert with SipC to positively regulate the formation and stabilization of the actin structures required for *Salmonella* internalization [19].

Together, both *Salmonella* and *Shigella* binding to host cells induce massive localized actin accumulations and only the concerted interplay of direct and indirect modulations of the host cell actin system leads to efficient engulfment and uptake of the respective pathogen [28, 29].

Another common theme of *Salmonella* and *Shigella* entry is the engagement of the phosphoinositide phosphatases IpgD (*Shigella*) [30] and SigD/SopB (*Salmonella*) [31], which are believed to locally influence cytoskeleton/membrane connections either during membrane ruffling [30] or at later steps of pathogen invagination, i.e. when the plasma membrane is resealed. Consequently, membrane fission during *Salmonella* entry was markedly delayed in mutants lacking SigD [31].

Finally, as opposed to *Shigella* (see below), *Salmonella* resides in the infected cell and inside the phagosome accompanied by reverting the cytoskeletal changes after successful entry. This is achieved by downregulation of Rho-GTPase activity through the Rho-GAP SptP [32] and by directed proteasome-dependent degradation of other effectors [33]. This enhances survival of the infected host cell and avoids repeated entry of more *Salmonella* into the same cell [33]. Furthermore, upon phagocytosis, actin rearrangements have also been implicated in maturation of the *Salmonella*-containing vacuole (SCV) [34, 35]. However, the factors and mechanisms involving the actin system in this process are poorly defined.

12.2.2
The 'Zipper' Mechanism of Bacterial Entry

The 'zipper' mechanism is employed e.g. by *Listeria monocytogenes* and *Yersinia* (*pseudotuberculosis* and *enterolitica*) and involves the tight association of bacterial surface proteins with a host cell transmembrane receptor, resulting in a stepwise envelopment of the pathogen. In the case of *Listeria* for instance, Internalin A (InlA) associates with the cell–cell junction protein E-cadherin [36, 37] and, for *Yersinia*, the protein Invasin (Inv) binds to β1-integrin [38–40]. In both cases, host cell adhesion receptors are usurped to mediate this type of entry. Engagement of the respective host cell receptors leads to fulminant signaling which is believed to finally result in acto-myosin-based force generation at the site of receptor clustering and consequently in pulling the pathogen into the cell.

The signaling complex, which is recruited by *Listeria* during InlA-mediated E-cadherin engagement very much resembles that of a cellular adherence junction [41, 42] and is linked to the actin cytoskeleton via α- and β-catenin [43], contains p120 catenin and is linked to myosin VII via vezatin [44, 45]. *Yersinia* engagement of β1-integrins in turn leads to a tyrosine phosphorylation cascade [46] and recruitment of proteins involved in the establishment of focal complexes and focal adhesions [47, 48]. In addition, formation of the *Yersinia*-induced phagocytic cup was found to be accompanied by Rac- and subsequent Arp2/3 activation [49]. In both, *Listeria* and *Yersinia* uptake, the 'zipper' mechanism appears to elicit host cell signaling cascades by mimicking ligands for host cell adhesion receptors without employing additional bacterial factors.

Notably, both pathogens have evolved additional strategies to enter not only their prime targets, epithelial cells, but also other cell types [11, 45]. *Listeria* expresses eight internalins, the function of most of which is unclear [50]. Internalin B (InlB) was identified as a ligand for c-met, the receptor of HGF/SF (hepatocyte growth factor/scatter factor) [51], glycosaminoglycans (GAGs) [52] and the C1q receptor (gC1qR), the first component of the complement cascade [53]. While the function of InlB binding to gC1qR for bacterial entry remains elusive [44], simultaneous interaction of InlB with c-Met and GAGs appears to cooperate in eliciting host cell signaling [52]. InlB interaction with c-met leads to activation of Pi3kinase [54] and of Rho-family GTPases like Rac and Cdc42 [55] and the subsequent engagement of a number of established downstream mediators of actin assembly [reviewed in 44, 45]. It was shown that the InlB pathway is sufficient to induce bacterial uptake in the absence of InlA [56]. However, it is currently believed that this type of uptake does not explain infection of non-epithelial cells during listeriosis *in vivo*, i.e. leading to meningitis, hepatitis or abortus, since the affected tissues may be reached by a combination of both cell to cell spread and professional phagocytes carrying *Listeria* through the blood–brain [57] or to the placental barrier [58].

In contrast to *Listeria*, *Yersinia* cannot survive in professional phago-cytes such as macrophages and therefore has evolved to prohibit its own phagocytosis after breaching the intestinal barrier early in infection. To exert this anti-phagocytic strategy, *Yersinia* delivers inhibitory effectors and toxins (Yops, *Yersinia* outer proteins) [59, 60] to the host cell cyto-plasm via its type III secretion system (TTSS) [4] that first block engulf-ment and subsequently kill the cell. Among these inhibitors are the Rho-GAP YopE, which downregulates Rho, Rac and Cdc42 *in vitro* [61] but ap-pears specific for Rac signaling *in vivo* [62]. Moreover, the toxin YopT is a cystein protease that specifically modifies Rho by removal of the isopre-nylation site in its C terminus resulting in release of the GTPase from the membrane, thereby exerting a dominant negative effect [63]. Another potent *Yersinia* toxin is YopH, a protein tyrosine phosphatase, which tar-gets focal adhesion components like FAK (focal adhesion kinase) and p130Cas [64, 65]. Dephosphorylation of these proteins disrupts not only pre-existing focal adhesions [66], but also interferes with signaling after Inv-induced β1-integrin engagement and therefore with Inv-mediated up-take [67, 68]. Finally the YopJ/P toxin inhibits antiapoptotic signals result-ing in subsequent cell death [reviewed in 59].

An additional aspect of pathogen invasion into host cells is the escape from the membrane-covered phagosome to gain access to the host cell cytoplasm, which can facilitate cell to cell spread of pathogens capable of intracellular actin-based motility. These pathogens, as diverse as *Listeria* and *Shigella* (see also Chapter 13), secrete additional virulence factors that aid lysis of the phagosomal membrane. Subsequent recruitment of parts of the host cell actin polymerization machinery (see below) not only mediates movement through the cytosol, but also the formation of protrusions leading to infection of neighboring cells, thereby avoiding contact with the extracellular space.

12.2.3
Other Mechanisms of Bacterial Invasion

Equally interesting, but less well characterized, are strategies evolved by invasive pathogens to engage caveoli [reviewed in 69] that have been found to be usurped by pathogens such as *Mycobacteria* [70], uropatho-genic FimH expressing *E. coli* [71] or group A *Streptococci* [72], to gain ac-cess to their host cells. Caveolin-mediated endocytosis was suggested to be linked to the actin cytoskeleton based on inhibition of this process by the actin-depolymerizing drug cytochalasin D [73] and more recently by the observation that caveolin-1 can bind to the F-actin cross-linking pro-tein filamin [74]. However, if caveoli-mediated engulfment of pathogens requires actin cytoskeletal rearrangements similar to the 'trigger' or 'zip-per' mechanisms (see above) of bacterial uptake awaits further investiga-tions.

12.3
Intracellular Actin-based Motility

12.3.1
Listeria

Upon gaining access to the host cell cytosol and lysis of the phagosomal vacuole, the gram-positive rod-shaped bacterium *Listeria monocytogenes* is capable of recruiting parts of the host cell actin polymerization machinery required to catalyze the formation of clouds or – more important for efficient motility – of tails built of actin filaments. Hence, *Listeria* and a number additional pathogens (see below) having independently evolved the same strategy to move within the host cell cytoplasm and also into neighboring cells, proved instrumental in gaining insights into the molecular mechanisms driving actin-based motile processes [75–77] (for overview see Fig. 12.3).

In the early 1990s, genetic screens led to the identification of the single listerial virulence factor essential and sufficient to drive the intracellular actin-based motility of this pathogen, the ActA protein [78, 79]. Expression of ActA in *Listeria innocua*, a non-pathogenic and non-motile *Listeria* strain [80] or asymmetric positioning of ActA on *Streptococcus pneumoniae* [81] conferred actin-based motility to these bacteria. Moreover, expression of ActA in eucaryotic cells was accompanied not only by its targeting to mitochondria mediated by its C-terminal membrane anchor, but also by causing focal accumulations of F-actin at these sites [82], proving that ActA by itself is sufficient to recruit host cell factors driving actin assembly. An explosion of subsequent studies led to the recognition of the interaction of ActA with several proteins of the host cell actin polymerization machinery, including actin. The most important discovery however was the elucidation of the capability of ActA to bind and activate the actin filament nucleating complex Arp2/3 [83]. ActA is expressed as a 639 amino acid protein that upon cleavage of the N-terminal secretion signal peptide is anchored in the bacterial membrane at its C terminus leading to the exposure of a 584 residue protein. The amino terminus of the mature protein is homologous to the C termini of WASP/Scar family proteins [84] known as VCA or WA domains, well-established host cell activators of the Arp2/3-complex [reviewed in 85]. More specifically, this part of the protein harbors an N-terminal acidic stretch, two domains with sequence similarity to WASP homology 2 (WH2), domains each capable of binding an actin monomer, and a more C-terminal Arp2/3 binding region [84]. The latter domain of ActA turned out to be essential for recruitment and activation of the actin nucleating activity of Arp2/3, and hence for initiating and maintaining actin tail formation and motility [86, 87]. In summary, the N terminus of ActA – in analogy to WASP/Scar family proteins – is now established as a func-

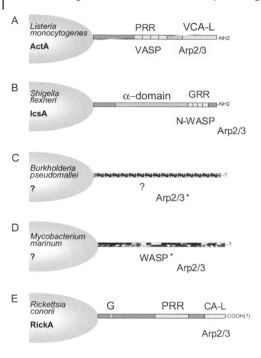

Fig. 12.3 Bacterial virulence factors mediating intracellular actin-based motility.
A) ActA of *Listeria monocytogenes* is anchored to the bacterial surface via its C-terminal membrane anchor. It harbors four central proline-rich repeats (PRR) that recruit Ena/VASP family proteins, which is important, but not essential for bacterial intracellular motility. The N terminus displays sequence similarities with the verprolin-, cofilin-homology, acidic (VCA) domains of WASP family proteins, capable of recruiting and activating the Arp2/3 complex, which is essential for actin comet tail formation (VCA-L: VCA-like). B) *Shigella flexneri* gains intracellular actin-based motility by expression of IcsA (VirG) which is anchored in the outer membrane of the bacterium with the C-terminal so-called beta (ß)-core. The N-terminal alpha (α)-domain is exposed to the host cell cytoplasm and harbors glycine-rich repeats (GRR), capable of specifically recruiting the host cell protein N-WASP, a potent cellular activator of the Arp2/3 complex. How other regions in the α-domain contribute to efficient actin comet tail formation is currently unclear. C) *Burkholderia pseudomallei* was recently shown to form actin comet tails but the bacterial factor(s) conferring this virulence feature are unknown. The Arp2/3 complex was shown to localize to *Burkholderia* actin comets and therefore may drive the motility of this pathogen. WASP family proteins, however, appear to be absent from *Burkholderia* actin tails suggesting a mechanism similar to *Listeria* rather than *Shigella*. D) Recently, *Mycobacterium marinum* was added to the list of bacterial pathogens that are capable of forming actin tails in the host cell cytoplasm. *Mycobacterium* recruits the haematopoietic WASP and the Arp2/3 complex via an unknown mechanism. E) *Rickettsia conorii*, a member of the spotted fever group, expresses the virulence factor RickA, which displays sequence homology to WASP family proteins in its C terminus (CA-L: CA-like, see above). RickA is able to bind and activate the Arp2/3 complex albeit more weakly than ActA of *Listeria*. It is so far unclear whether RickA is anchored to the bacterial surface with its N or C terminus, and it has been suggested that it dimerizes *in vivo*. *Rickettsia* actin tail morphology differs significantly from that of the above pathogens, indicating that alternative mechanisms of Arp2/3 activation are at play to drive the intracellular motility of this pathogen (for details see text).

tional unit driving Arp2/3-mediated catalysis of actin filament assembly by (i) binding Arp2/3 and mediating a conformational change allowing Arp2/3 activation and by (ii) recruiting monomeric actin allowing efficient actin filament nucleation.

A more central proline-rich stretch of ActA harboring 4 copies of the D/EFPPPPXE/D consensus sequence is essential for interacting with the Ena/VASP homology 1 (EVH1) domains of Ena/VASP family proteins, additional prominent regulators of the actin cytoskeleton. This protein family in mammals comprises three members: the vasodilator-stimulated phosphoprotein, VASP; mammalian Enabled (Mena), the ortholog of the *Drosophila* protein Ena; and the Ena-VASP-like protein Evl. All members of this protein family harbor the aforementioned N-terminal EVH1 domain, more central proline-rich stretches, and a C-terminal Ena/VASP homology 2 (EVH2) domain [reviewed in 88]. The proline-rich stretches with the consensus GPPPPP interact directly with the actin monomer-binding protein profilin [89, 90] and sequences within the EVH2 domain were implicated in both F-actin and G-actin binding and in multimerization of Ena/VASP proteins [91, 92]. By a variety of means, the recruitment of Ena/VASP proteins to *Listeria* surfaces was demonstrated to enhance actin polymerization and hence motility, although it has emerged that they are not absolutely essential [93–96]. First, *Listeria* mutants expressing ActA lacking the central proline-rich EVH1-interaction domain moved with significantly slower rates [93, 94]. Furthermore, MVD7, a cell line which is genetically deficient for Mena and VASP [97], supported efficient *Listeria* motility only upon re-expression of either protein in a dose-dependent manner [96]. It was proposed that Ena/VASP function to drive *Listeria* motility is mediated by Ena/VASP-dependent recruitment of profilin–actin leading to efficient shuttling of actin monomers onto the actin filaments abutting the bacterial surface. In line with this model, *Listeria* speed directly correlated with profilin recruitment to the bacterial surface [98] and injection of cross-linked profilin–actin complexes into cells severely compromised bacterial motility [99]. *In vitro* reconstitution of *Listeria* motility using a minimal set of proteins not only showed that Ena/VASP proteins and profilin can significantly enhance bacterial speed, but also that Ena/VASP proteins are able to promote motility in the absence of profilin [95]. Recent motility assays employing ActA-coated beads for instance indicated that VASP increases the rate of filament dissociation, thereby supporting efficient elongation and motility [100]. Finally, the indirect recruitment of profilin:actin via the GPPPPP motifs of Ena/VASP proteins could also overcome the lack of an actin monomer binding region within the N terminus of ActA [101], suggesting that by providing actin monomers, Ena/VASP proteins can also contribute directly to the Arp2/3-dependent nucleation process, and not just to filament elongation.

Together, besides highlighting the fundamental role of the Arp2/3-complex in certain actin-based processes, *Listeria* have also helped to improve our understanding of the actin polymerization promoting function of Ena/VASP family proteins, although more recently it has been suggested that this protein family regulates the host cell actin cytoskeleton by additional and different means [88, 100, 102].

12.3.2
Shigella

Gram-negative bacteria of the genus *Shigella* have evolved a mechanism to recruit the host cell actin polymerization machinery that is different from *Listeria monocytogenes*. The sole bacterial virulence factor required for actin tail assembly and movement through the cytosol is termed IcsA, also known as VirG [103]. IcsA is a 1102 aminoacid protein, of which approximately 750 N-terminal residues, comprising the so-called *a*-domain, are exposed on the bacterial surface [see 77]. Interestingly, in contrast to listerial ActA [104], IcsA is targeted to and inserted into one pole of the bacterium, which is where actin assembly is initiated [105]. Translocation across the outer membrane is mediated by the smaller C-terminal *β*-domain [reviewed in 106]. *In vitro* experiments have shown that IcsA is not only necessary, but also sufficient for *Shigella* motility [80, 107].

The *a*-domain of IcsA was initially shown to interact directly with the focal adhesion protein vinculin [108]. Microinjection of the 95-kDa head domain of vinculin, that harbors a proline-rich region capable of interaction with EVH1 domains of Ena/VASP family proteins [94], was reported to accelerate *Shigella* motility [109]. These observations have led to the suggestion that IcsA – in analogy to listerial ActA – may be able via vinculin to recruit proteins of the Ena/VASP family that would then mediate the recruitment of profilin–actin complexes, promoting bacterial motility. However, as opposed to *Listeria*, the relevance of Ena/VASP proteins for *Shigella* motility may be minimal, since MV_{D7} cells supported *Shigella* movement equally well as MV_{D7} cells re-expressing EGFP-tagged Mena [110]. Moreover, cells expressing a truncated vinculin were not defective in *Shigella* movement [111] and, more significantly, in *in vitro* experiments employing *E. coli* expressing IcsA, actin-based motility was reconstituted in the absence of vinculin [95], strongly supportive of the view that vinculin is dispensable for *Shigella* movement. Interestingly, *in vitro*, the presence of profilin was shown to enhance both IcsA-mediated *E. coli* and *Listeria* motility up to three times, although it was also not essential [95]. These data indicate that while profilin promotes *Shigella* movement, vinculin and Ena/VASP proteins are not relevant for the intracellular actin-based movement of this pathogen.

Shigella motility requires N-WASP (neural Wiskott-Aldrich Syndrome protein) [95, 112–115], a member of the WASP/Scar family of proteins. In mammals, this protein family comprises the haematopoietic WASP, the ubiquitous N-WASP, the expression of which is strongest in brain and three WAVE (WASP family verprolin homologous) isoforms, the *Dictyostelium* ortholog of which is known as Scar (suppressor of cAMP receptor). In analogy to listerial ActA (see above), these proteins share a WA domain, located at their C termini, which is essential for Arp2/3 binding and activation. Hence, WASP/Scar proteins have emerged as key mediators of cellular Arp2/3 function [85].

IcsA directly binds N-WASP via its glycine-rich repeats and recruits it to the *Shigella* surface [113–116]. Significantly, *Xenopus* cytoplasmic extracts immunodepleted for N-WASP, or platelet extracts naturally lacking N-WASP, supported significant actin tail formation by *E. coli* expressing IcsA only after re-addition of or pre-incubation of the bacteria with N-WASP [113]. Likewise, cell lines genetically deficient for N-WASP were capable of engulfing *Shigella*, but did not allow actin tail formation and movement of this pathogen unless the protein was re-expressed in these cells [114, 115]. This effect was specific to *Shigella*, since *Listeria*-induced actin tail assembly appeared normal in the absence of N-WASP [114]. *Shigella* actin tail formation coincides with incorporation of this complex into the actin comet tails. Depletion of the Arp2/3-complex from platelet extracts caused the loss of IcsA-expressing *E. coli* motility, which was again restored by addition of purified Arp2/3 [113] and *in vitro* reconstitution of *Shigella* movement required Arp2/3 [95]. Finally, detailed analysis of the domain requirements of N-WASP for reconstitution of *Shigella* motility in N-WASP-defective cells revealed the absolute dependence on the C terminal Arp2/3-activating WA domain [114]. Together, all these observations unambiguously prove that *Shigella* movement requires recruitment and activation of the Arp2/3-complex by the widely expressed WASP/Scar family member N-WASP.

The molecular details of N-WASP recruitment and activation in *Shigella* motility are less well understood. N-WASP and WASP share a CRIB (Cdc42/Rac interactive binding) domain mediating binding to the Rho family GTPase Cdc42 [117, 118]. Rho family GTPases have long been recognized as key signaling switches in the sequence of events leading to actin cytoskeleton reorganization [reviewed in 7]. Cdc42, which is prominent for driving the formation of thin cellular protrusions filled with densely packed actin filaments known as filopodia [for review see 119], was also implicated in N-WASP-mediated actin tail assembly at the *Shigella* surface. Indeed, it was suggested that IcsA might form a ternary complex with N-WASP and Cdc42 allowing efficient initiation of actin nucleation [for review see 76]. However, a significant body of evidence now indicates a much less prominent role for Cdc42 in this process. Initially, *Shigella*-induced actin tail formation was not significantly blocked by *Clostridium difficile* exotoxin TcdB-10436, which inactivates most Rho-GTPases including Cdc42 [13]. And there was no evidence for the presence of Cdc42 in the reconstituted motility system supporting the movement of IcsA-expressing *E. coli* [95]. Consistent with the above observations, actin tail formation was restored equally well by wild-type N-WASP or N-WASP mutant proteins incapable of interaction with Cdc42 [114] and finally, *Shigella* were reported to move in the cytosol of Cdc42-deficient cells [120]. Hence, Cdc42 appears to have no significant role in *Shigella* actin tail formation and motility.

The central recruitment domain of N-WASP to the *Shigella* surface could be confined to approx. 50 residues C-terminal to the CRIB domain [114], the latter of which is highly conserved between N-WASP and the haematopoietic WASP and mediating interaction with Cdc42 for both proteins. The fact that the CRIB domain is not involved in mediating N-WASP recruitment to *Shigella* [114] is therefore not a contradiction to the observation that the motility of this pathogen is apparently specific for N-WASP, since expression of WASP in N-WASP –/– cells could not restore *Shigella*-induced actin tail formation [115].

The WH1 (WASP homology 1) domain of N-WASP can interact with a family of proteins known as WIP (WASP interacting proteins) and the more central polyproline-rich domain has interaction surfaces for multiple proteins including profilin and SH3-adaptor proteins such as Nck [reviewed in 85]. The latter protein can also bind WIP and it was proposed therefore that N-WASP may be recruited to the *Shigella* surface in tripartite complex with WIP and Nck, although the relevance of such a mechanism remained elusive [76].

N-WASP is also essential for the actin-based motility of vaccinia virus, a member of the poxvirus family of large enveloped double-stranded DNA viruses, which have evolved to mimic a receptor tyrosine kinase signaling pathway. Thus, engagement of the actin polymerization machinery that is mediated by phosphotyrosine residues on a virally encoded protein at the particle surface involves SH2-SH3 adaptor proteins Nck [121] and/or Grb2 [122] as well as WIP upstream of N-WASP, and not downstream, as in *Shigella* [reviewed in 76].

Therefore, these two pathogens represent remarkable examples of convergent evolution to usurp the same protein, N-WASP, in order to exploit the actin-based motility to spread infection.

12.3.3
Rickettsia

Coccobacillary gram-negative bacteria of the genus *Rickettsia* are obligate intracellular pathogens. Actin tail formation and motility within and in between host cells has been observed for the spotted fever group *Rickettsiae* and for *R. typhi*, but not for the typhus group species *R. prowazekii* [for review see 77]. Although intracellular movements of *Rickettsiae* have been known for a long time [123], it was recognized much later that these movements were associated with the formation of actin tails [124,

125]. Teysseire et al. [124] reported actin tails associated with the frequently studied species *R. conorii*, and, to a much lesser extent, with *R. typhi*. In addition, Heinzen et al. [125] blocked the release of *R. rickettsii* into the tissue culture medium by inhibiting actin polymerization with Cytochalasin D. Video microscopy showed that this type of actin-based motility is about 2.5–3 times slower than that observed for *Shigella* and *Listeria* [126, 127].

Interestingly, the morphology of the actin tails formed by these bacteria are quite different from the ones observed with *Shigella* or *Listeria*. While the latter form tails composed of densely packed and cross-linked arrays of relatively short actin filaments, *Rickettsia* tails consist of long filaments that are usually arranged in a parallel manner [126, 128]. In addition to these morphological differences, a number of proteins significant for *Listeria* or *Shigella* motility, such as cofilin or Arp3, have been reported to be absent from *Rickettsia* tails [126].

The above observations have led to the hypothesis that *Rickettsia* actin-based motility reflects the only exception to the otherwise general rule that rapid actin-based motility, as observed with the aforementioned bacterial or viral pathogens and also with endosomal vesicles [129, 130], requires filament nucleation by Arp2/3. This hypothesis was further corroborated by the observation that, although *R. rickettsii* apparently recruited GFP-tagged Cdc42, overexpression of dominant-negative N-WASP constructs did not significantly interfere with the motility of this species [131, 132]. The potential role of Cdc42 in this process is elusive, while the conclusion that *Rickettsia* motility occurs independently of Arp2/3 function is significantly challenged by a more recent report. Excitingly, Gouin et al. [133] – by comparison of the genome of *R. conorii* with that of the non-motile *R. prowazekii* species – identified a 517-aminoacid protein harboring not only a potential N-terminal G-actin binding domain and a central proline-rich domain, but more significantly, a C-terminal CA domain with sequence homology to the same domains in WASP/Scar family proteins. Consequently, the same protein, which was termed RickA, was capable of activating the Arp2/3-complex and was shown to co-localize with the Arp2/3-complex at the surface of *R. conorii*. As the genetic manipulation of *Rickettsia* is not yet established, the relevance of RickA for motility is not formally proven, but it is an attractive hypothesis, that this type of actin-based motility does indeed involve Arp2/3-mediated nucleation of actin filaments at the *Rickettsia* surface, as with *Listeria* or *Shigella*, although the complex does not stay incorporated into *Rickettsia*-induced actin tails.

In this context, it is worth mentioning that by employing elegant perfusion chamber experiments, it was shown recently that the propulsion of *Listeria* can continue, at least for some time, in the absence of Arp2/3 activity if *Listeria*-attached actin filaments were nucleated by Arp2/3 before its inhibition [134]. Significantly, in these experiments, actin tail morphologies varied significantly between Arp2/3-dependent and -independent phases of *Listeria* propulsion, with tail morphologies upon Arp inhibition being highly reminiscent of *Rickettsia* tails [134]. It is therefore tempting to speculate that *Rickettsiae* may have evolved a mechanism of infrequent or transient Arp2/3-dependent nucleation at their surfaces, which is followed by rapid filament debranching, bundling and elongation similar to the experimentally-induced Arp2/3-independent propulsion of *Listeria*. Indeed, the latter process required bundling activity of proteins such as fascin [134], which was also shown to be prominently associated to *Rickettsia conorii* tails [133].

It will be exciting to learn if RickA is capable by itself of supporting actin-based motility of otherwise non-motile bacteria or of beads coated with purified protein, as demonstrated for ActA and IcsA, or if additional factors are required. Moreover, in the light of these novel findings, it will be important to test for *Rickettsia* motility in extracts depleted of the Arp2/3-complex, since – as opposed to earlier results [131] obtained with *R. rickettsii* – sequestration of the Arp2/3-complex by the WA domain of WASP/Scar proteins seemed to block actin tail formation at least of *Rickettsia conorii* [133]. The reason for these differences is currently unclear, but it will be exciting to see if engagement of Arp2/3 is common to all *Rickettsiae* including *R. typhi*.

12.3.4
Intracellular Motility of less-well Characterized Bacterial Pathogens

The list of intracellular bacterial pathogens surfing through the host cell cytosol by recruitment of the host cell actin polymerization machinery is continuously growing. Examples additional to the pathogens mentioned above are the gram-negative rod bacterium *Burkholderia pseudomallei* and the gram-positive *Mycobacterium marinum*.

Infections of humans with *Burkholderia pseudomallei* lead to development of a disease known as melioidosis. *B. pseudomallei* is a facultative intracellular organism, which invades professional phagocytes and non-phagocytic cells. Kespichayawattana et al. [135] were the first to report the formation of actin tails associated with intracellular *B. pseudomallei* and with peripheral membrane protrusions. The mechanism of actin assembly induced by this pathogen is largely unknown, except that – as with *Shigella* and *Listeria* – actin tail formation was accompanied by incorporation of the Arp2/3-complex into the tails and by recruitment of the actin-bundling protein a-actinin. As opposed to *Shigella* and *Listeria*, respectively, actin tail formation required neither N-WASP nor Ena/VASP family proteins [136]. Although the prominent Arp2/3 association is indicative of its involvement in motility, data supporting its essential function as well as a potential mechanism for its recruitment and activation are still missing.

Mycobacterium marinum is the causative agent of a systemic tuberculosis-like disease in fish and frogs and of localized skin infections in humans. *M. marinum* was shown recently not only to escape the phagosome thereby entering the cytoplasm of infected macrophages, but also to move within these cells with rates highly similar to those observed with *Shigella* and *Listeria*. In addition, the morphology of the actin tails observed on electron micrographs is similar to those formed by *Listeria* or *Shigella*. The prominent actin tails recruited both Ena/VASP proteins and Arp2/3-complex as well as WASP at the bacterium–tail interface [137], suggestive of a mechanism of recruitment of the actin polymerization machinery similar to, but distinct from, *Shigella*, since the latter pathogen is unable to employ the haematopoietic WASP [115].

12.4
Actin Pedestal Formation by Pathogenic *E. coli*:
Directing Cytoskeletal Organization across the Plasma Membrane

In contrast to other intracellular pathogens that impinge on the actin cytoskeleton from inside the host cell cytoplasm, the gram-negative diarrheagenic bacterial pathogen enteropathogenic *Escherichia coli* (EPEC) and its Shiga-toxin producing relative, enterohaemorrhagic *E. coli* (EHEC), direct rearrangement of the actin cytoskeleton without entering their host cells. They have emerged as paradigms for the analysis of host–pathogen interactions and cytoskeletal rearrangements that occur at the host cell membrane. EPEC and EHEC are two closely related members of the 'attaching and effacing' (A/E) family of enteropathogens, which cause severe diarrheal disease in humans and animals and present a significant risk to human health worldwide [reviewed in 138]. Upon colonization of the intestinal mucosa, EPEC and EHEC induce the formation of unique histopathological lesions termed attaching and effacing (A/E) lesions on the apical surfaces of gut epithelial cells central to pathogenesis. A/E lesions are characterized by a localized loss of microvilli (effacement), intimate attachment of the bacteria to the cell surface followed by formation of pseudopod-like filamentous actin-based structures underneath the adherent bacteria, termed actin pedestals [139–141].

These phenotypes depend on a 35kb chromosomal pathogenicity island, termed the locus of enterocyte effacement (LEE), which is highly conserved among A/E family enteropathogens [142–145]. Interestingly however, although the LEE of EHEC contains all the genes found in the EPEC LEE in the same organisation, only the LEE of EPEC is able to confer the A/E phenotype upon *E. coli* K12 [143, 146] suggesting functional and/or regulatory differences between EPEC and EHEC.

The LEE encodes a type III protein secretion apparatus for the delivery of bacterial proteins into the underlying host cell, an adhesin (intimin), regulatory proteins, as well as translocated effector proteins, termed *E. coli*-secreted proteins (Esp) and their chaperons.

Of the secreted proteins encoded by the LEE, six (EspB, EspF, EspG, Map, EspH and Tir (EspE)) have been described as modulators of the host cytoskeleton [for review see 147, 148], with the function of Tir being best understood. EspB has been implicated in the downregulation of stress fiber formation in addition to its role as part of the type III translocon [149]. EspF disrupts tight junction integrity and intestinal barrier function and can induce host cell death by as yet unknown mechanisms [150, 151]. EspG is homologous with the *Shigella flexneri* protein VirA [152], which has been implicated in destabilization of microtubules [153]. Whether EspG has a similar function in EPEC and EHEC infection remains to be seen. Map (Mitochondrial-*a*ssociated *p*rotein) is targeted to mitochondria where it appears to interfere with the mitochondrial membrane potential [154]. In addition and independently from its targeting to mitochondria, Map was reported to induce the formation of filopodia via the Rho GTPase Cdc42, an early and transient event in A/E lesion formation [155]. EspH was described as a modulator of the host actin cytoskeleton by repressing filopodia formation and enhancing actin pedestal formation [156].

Of special importance is Tir (translocated intimin receptor, EspE) which serves multiple functions in A/E lesion formation. Upon translocation and modification by host kinases, Tir becomes inserted into the host cell plasma membrane in a hairpin-like conformation where it acts as the bacterial receptor by binding to the bacterial outer surface protein intimin [157, 158]. This interaction allows the bacteria to intimately attach to the host cell surface and in addition triggers signaling events leading to pedestal formation [157, 158, 159]. Thus, upon binding to intimin, Tir interacts on the cytoplasmic side with cytoskeletal and signaling components and orchestrates the massive cytoskeletal rearrangements beneath adherent bacteria that ultimately lead to pedestal formation [reviewed in 160].

Although EPEC and EHEC produce lesions in the gut which are apparently morphologically identical, EPEC colonizing the small intestine and EHEC colonizing the large intestine [reviewed in 161], studies of actin pedestal formation induced by EPEC and EHEC in cultured cells have revealed that pedestal composition and the molecular mechanisms of pedestal formation differ [reviewed in 160].

Interestingly, the Tir proteins of EPEC and EHEC are not functionally identical. Upon binding to intimin, EPEC Tir becomes phosphorylated on tyrosine 474 [162]. This phosphorylation is critical for EPEC-induced actin pedestal formation as it allows recruitment of the cellular signaling adaptor protein Nck, thereby exploiting cellular signaling cascades leading to actin assembly [163–165]. In contrast, EHEC-induced pedestals form independently of Tir tyrosine phosphorylation [158, 166] and Nck [163]. The fact that EHEC Tir does not fully complement an EPEC Tir deletion mutant with regard to pedestal formation [167], suggests that pedestal formation by EHEC requires the type III delivery of additional EHEC factors into the host cell.

Despite these differences, both pathogens trigger recruitment and activation of N-WASP, which is absolutely essential to drive Arp2/3 complex-mediated actin polymerization [114, 168, 169] (see Fig. 12.4).

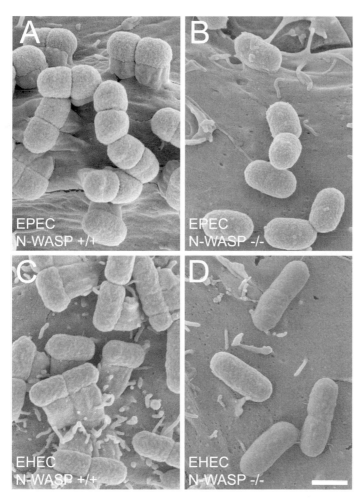

Fig. 12.4 Actin pedestal formation by EPEC and EHEC depends on N-WASP. Scanning electron micrographs showing cells infected with EPEC (A, B) or EHEC (C, D). Both EPEC and EHEC induce the formation of actin pedestals on the surface of N-WASP-expressing cells (labeled N-WASP +/+ in (A, C)). In N-WASP-defective cells (labeled N-WASP–/– in (B, D)), EPEC and EHEC readily adhere to the cell surface, but are unable to induce the formation of actin pedestals. Bar in D equals 1 μm and is valid for (A–D).

How the two pathogens achieve recruitment of N-WASP is still not clear (see Fig. 12.5). Nck, which prominently targets to EPEC attachment sites [160, 163, 169] is capable of physically interacting with and activating N-WASP *in vitro* via its SH3 domains which bind to the polyproline domain of N-WASP [170]. However, recruitment of N-WASP to EPEC attachment sites requires the amino-terminal half of the protein [114], whereas the polyproline domain of N-WASP by itself fails to target to sites of EPEC attachment. In analogy to the actin-based motility of vaccinia virus, it was therefore suggested that N-WASP is recruited downstream of Nck not by direct interaction, but indirectly via the Nck- and

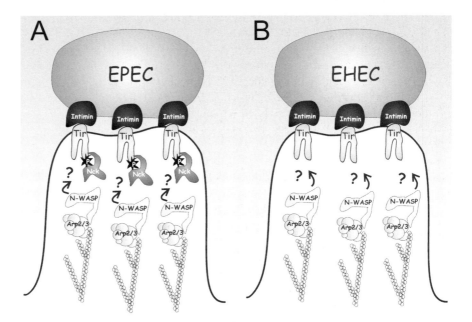

Fig. 12.5 Model of actin signaling by EPEC and EHEC during pedestal formation. Schematic representation of the signaling events leading to EPEC- versus EHEC-induced actin pedestal formation after insertion of Tir into the host cell membrane. A) Upon interaction with intimin, tyrosine phosphorylation of EPEC Tir provides a docking site for the cellular adaptor protein Nck. Downstream of Nck, N-WASP is recruited by a so far unidentified factor via sequences in its amino-terminal half. Recruitment and activation of N-WASP then leads to Arp2/3-mediated actin assembly driving pedestal formation. B) EHEC Tir is not tyrosine phosphorylated. Together with so far unidentified factors, EHEC Tir triggers recruitment of N-WASP via amino acid residues 226–274. As in (A), recruitment and activation of N-WASP then leads to Arp2/3-mediated actin assembly.

N-WASP-interacting protein WIP [171]. However, direct interaction of N-WASP with WIP is not required for pedestal formation [169], arguing against such a sequence of events. Thus, the factor(s) that bring about recruitment of N-WASP to sites of EPEC attachment via interaction with its amino terminus still await identification.

In EHEC, aminoacid residues 226–274 of N-WASP are sufficient to target N-WASP to sites of bacterial attachment [169]. Interestingly, the same residues also constitute the central recruitment domain used by *Shigella flexneri* for targeting N-WASP to the bacterial surface ([114] and see above). Whether EHEC possesses a bacterial factor capable of direct interaction with residues 226–274 of N-WASP to bring about its recruitment, in analogy to the interaction of N-WASP with *Shigella* IcsA, or whether a so far unidentified cellular protein is employed remains to be seen. As stated above, the fact that EHEC Tir is unable to compensate for EPEC Tir suggests that pedestal formation by EHEC requires additional EHEC effectors. On the other hand, the finding that WASP can compensate for N-WASP in EHEC-induced actin pedestal formation, yet is unable to do so in actin-based motility of *Shigella flexneri*, although both pathogens target the same recruitment domain, is rather supportive of EHEC utilizing a cellular factor, which has evolved to interact with both N-WASP/WASP, to mediate recruitment.

In addition to N-WASP and the Arp2/3 complex, numerous other cellular proteins have been identified that localize to pedestals induced by EPEC and EHEC, the role of most in pedestal formation by EPEC and EHEC being much less clear [reviewed in 160]. Thus, to list just a few of them, actin-associated proteins such as cofilin, gelsolin, VASP, cortactin, as well as the focal adhesion proteins α-actinin, talin and vinculin have been reported to localize to pedestals [172–177]. α-actinin, talin and vinculin were further described to be able to interact directly with Tir *in vitro* [173–175, 178] and these interactions were suggested to provide tight anchorage of the bacteria to the host actin cytoskeleton. In addition, a recent study suggested that Tir also connects to the intermediate filament system by direct interaction with cytokeratin-18 [179]. Nevertheless, pedestals are highly dynamic and can elongate up to 10 μm above the cell surface, shorten, bend and as was shown for EPEC using cultured cells, even translocate along the host cell surface, driven by the polymerization of actin at the pedestal tip adjacent to the bacterium [180]. In this respect, EPEC motility on host cells resembles the motility of vaccinia virus, taking place at the cell membrane and being dependent on tyrosine phosphorylation of a pathogen-derived membrane protein (A36R) as well as on N-WASP and the Arp2/3 complex [reviewed in 76].

Although correlated with disease, the functional relevance of actin pedestal formation during EPEC and EHEC infection has yet to be established. What has become clear is that EPEC and EHEC have independently evolved distinct mechanisms to recruit the cellular nucleation promoting factor N-WASP to drive actin assembly. Identification of the factor(s) directly upstream of N-WASP in EPEC- and EHEC-induced pedestal formation will thus not only broaden our knowledge about the pathogenesis of EPEC and EHEC, but will provide valuable insights into the signaling pathways regulating cellular actin dynamics.

12.5
Conclusions

Undoubtedly, during the last few years pathogen-elicited actin assembly has contributed enormously to our understanding of the molecular mechanisms of actin reorganisations in mammalian cells. For instance, bacterial pathogens have been providing an ever growing list of extremely useful biological probes, for instance toxins targeting Rho-family GTPases. Furthermore, *in vitro* experiments on the actin-based motility of *Listeria* and *Shigella* using cell-free systems, which culminated in the reconstitution of this type of actin-based motility using pure proteins, have unraveled a number of key features of the phenomenon of actin-based motility. These experiments have not only demonstrated for the first time the key function of the Arp2/3-complex in actin filament nucleation, but have also highlighted the need for actin filament capping (to spatially restrict polymerization and force generation to the pathogen pole) and for depolymerizing factors (to maintain turnover and therefore continuous motility). Last but not least, by studying the variety of strategies employed by pathogens to manipulate actin reorganisations at the plasma membrane we are just beginning to grasp the complexity of signaling pathways regulating the formation of diverse cell surface protrusions.

12.6
Acknowledgments

This work was supported by the Deutsche Forschungsgemeinschaft and the Fonds der Chemischen Industrie. The authors thank Manfred Rohde for electron microscopy and the laboratory members who have over the years contributed to the different topics.

12.7
References

1 BARBIERI, J.T., M. J. RIESE, and K. AKTORIES, *Annu Rev Cell Dev Biol*, 2002, **18**, 315–344.

2 ROSQVIST, R., K.E. MAGNUSSON, and H. WOLF-WATZ, *EMBO J*, 1994, **13**(4), 964–972.

3 HUECK, C.J., *Microbiol Mol Biol Rev*, 1998, **62**(2), 379–433.

4 ZAHARIK, M.L., S. GRUENHEID, A.J. PERRIN, and B.B. FINLAY, *Int J Med Microbiol*, 2002, **291**(8), 593–603.

5 HAYWARD, R.D. and V. KORONAKISS, *Trends Cell Biol*, 2002, **12**(1), 15–20.

6 FIORENTINI, C., L. FALZANO, S. TRAVAGLIONE, and A. FABBRI, *Cell Death Differ*, 2003, **10**(2), 147–152.

7 HALL, A., *Science*, 1998, **279**(5350), 509–514.

8 SWANSON, J.A. and S.C. BAER, *Trends Cell Biol*, 1995, **5**(3), 89–93.

9 DRAMSI, S. and P. COSSART, *Annu Rev Cell Dev Biol*, 1998, **14**, 137–166.

10 DONNENBERG, M.S., *Nature*, 2000, **406**(6797), 768–774.

11 SANSONETTI, P., *Semin Immunol*, 2001, **13**(6), 381–390.

12 GRUENHEID, S. and B.B. FINLAY, *Nature*, 2003, **422**(6933), 775–781.

13 MOUNIER, J., V. LAURENT, A. HALL, P. FORT, M.F. CARLIER, P.J. SANSONETTI, and C. EGILE, *J Cell Sci*, 1999, **112** (Pt 13), 2069–2080.

14 KANIGA, K., D. TROLLINGER, and J.E. GALAN, *J Bacteriol*, 1995, **177**(24), 7078–7085.

15 KANIGA, K., S. TUCKER, D. TROLLINGER, and J.E. GALAN, *J Bacteriol*, 1995, **177**(14), 3965–3971.

16 HAYWARD, R.D., E.J. MCGHIE, and V. KORONAKIS, *Mol Microbiol*, 2000, **37**(4), 727–739.

17 HUME, P.J., E.J. MCGHIE, R.D. HAYWARD, and V. KORONAKIS, *Mol Microbiol*, 2003, **49**(2), 425–439.

18 HAYWARD, R.D. and V. KORONAKIS, *EMBO J*, 1999, **18**(18), 4926–4934.

19 MCGHIE, E.J., R.D. HAYWARD, and V. KORONAKIS, *EMBO J*, 2001, **20**(9), 2131–2139.

20 TRAN VAN NHIEU, G., E. CARON, A. HALL, and P.J. SANSONETTI, *EMBO J*, 1999, **18**(12), 3249–3262.

21 BURTON, E.A., R. PLATTNER, and A.M. PENDERGAST, *EMBO J*, 2003, **22**(20), 5471–5479.

22 HARDT, W.D., L.M. CHEN, K.E. SCHUEBEL, X.R. BUSTELO, and J.E. GALAN, *Cell*, 1998, **93**(5), 815–826.

23 BOURDET-SICARD, R. and G. TRAN VAN NHIEU, *Trends Microbiol*, 1999, **7**(8), 309–310.

24 TRAN VAN NHIEU, G., R. BOURDET-SICARD, G. DUMENIL, A. BLOCKER, and P.J. SANSONETTI, *Cell Microbiol*, 2000, **2**(3), 187–193.

25 BOURDET-SICARD, R., M. RUDIGER, B.M. JOCKUSCH, P. GOUNON, P.J. SANSONETTI, and G.T. NHIEU, *EMBO J*, 1999, **18**(21), 5853–5862.

26 ZHOU, D., M.S. MOOSEKER, and J.E. GALAN, *Science*, 1999, **283**(5410), 2092–2095.

27 McGhie, E.J., R.D. Hayward, and V. Koronakis, *Mol Cell*, 2004, **13**(4), 497–510.

28 Bourdet-Sicard, R., C. Egile, P.J. Sansonetti, and G. Tran Van Nhieu, *Microbes Infect*, 2000, **2**(7), 813–819.

29 Lesser, C.F., C.A. Scherer, and S.I. Miller, *Trends Microbiol*, 2000, **8**(4), 151–152.

30 Niebuhr, K., S. Giuriato, T. Pedron, D.J. Philpott, F. Gaits, J. Sable, M.P. Sheetz, C. Parsot, P.J. Sansonetti, and B. Payrastre, *EMBO J*, 2002, **21**(19), 5069–5078.

31 Terebiznik, M.R., O.V. Vieira, S.L. Marcus, A. Slade, C.M. Yip, W.S. Trimble, T. Meyer, B.B. Finlay, and S. Grinstein, *Nat Cell Biol*, 2002, **4**(10), 766–773.

32 Fu, Y. and J.E. Galan, *Nature*, 1999, **401**(6750), 293–297.

33 Kubori, T. and J.E. Galan, *Cell*, 2003, **115**(3), 333–342.

34 Meresse, S., O. Steele-Mortimer, E. Moreno, M. Desjardins, B. Finlay, and J.P. Gorvel, *Nat Cell Biol*, 1999, **1**(7), E183–188.

35 Meresse, S., K.E. Unsworth, A. Habermann, G. Griffiths, F. Fang, M.J. Martinez-Lorenzo, S.R. Waterman, J.P. Gorvel, and D.W. Holden, *Cell Microbiol*, 2001, **3**(8), 567–577.

36 Mengaud, J., H. Ohayon, P. Gounon, R.M. Mege, and P. Cossart, *Cell*, 1996, **84**(6), 923–932.

37 Schubert, W.D., C. Urbanke, T. Ziehm, V. Beier, M.P. Machner, E. Domann, J. Wehland, T. Chakraborty, and D.W. Heinz, *Cell*, 2002, **111**(6), 825–836.

38 Isberg, R.R., D.L. Voorhis, and S. Falkow, *Cell*, 1987, **50**(5), 769–778.

39 Isberg, R.R. and J.M. Leong, *Cell*, 1990, **60**(5), 861–871.

40 Rankin, S., R.R. Isberg, and J.M. Leong, *Infect Immun*, 1992, **60**(9), 3909–3912.

41 Nagafuchi, A., *Curr Opin Cell Biol*, 2001, **13**(5), 600–603.

42 Vasioukhin, V. and E. Fuchs, *Curr Opin Cell Biol*, 2001, **13**(1), 76–84.

43 Lecuit, M., R. Hurme, J. Pizarro-Cerda, H. Ohayon, B. Geiger, and P. Cossart, *Proc Natl Acad Sci USA*, 2000, **97**(18), 10008–10013.

44 Cossart, P., J. Pizarro-Cerda, and M. Lecuit, *Trends Cell Biol*, 2003, **13**(1), 23–31.

45 Cossart, P. and P.J. Sansonetti, *Science*, 2004, **304**(5668), 242–248.

46 Alrutz, M.A. and R.R. Isberg, *Proc Natl Acad Sci USA*, 1998, **95**(23), 13658–13663.

47 Weidow, C.L., D.S. Black, J.B. Bliska, and A.H. Bouton, *Cell Microbiol*, 2000, **2**(6), 549–560.

48 Bruce-Staskal, P.J., C.L. Weidow, J.J. Gibson, and A.H. Bouton, *J Cell Sci*, 2002, **115**(Pt 13), 2689–2700.

49 Alrutz, M.A., A. Srivastava, K.W. Wong, C. D'Souza-Schorey, M. Tang, L.E. Ch'Ng, S.B. Snapper, and R.R. Isberg, *Mol Microbiol*, 2001, **42**(3), 689–703.

50 Vazquez-Boland, J.A., G. Dominguez-Bernal, B. Gonzalez-Zorn, J. Kreft, and W. Goebel, *Microbes Infect*, 2001, **3**(7), 571–584.

51 Shen, Y., M. Naujokas, M. Park, and K. Ireton, *Cell*, 2000, **103**(3), 501–510.

52 JONQUIERES, R., J. PIZARRO-CERDA, and P. COSSART, *Mol Microbiol*, 2001, **42**(4), 955–965.

53 BRAUN, L., B. GHEBREHIWET, and P. COSSART, *EMBO J*, 2000, **19**(7), 1458–1466.

54 IRETON, K., B. PAYRASTRE, H. CHAP, W. OGAWA, H. SAKAUE, M. KASUGA, and P. COSSART, *Science*, 1996, **274**(5288), 780–782.

55 BIERNE, H., E. GOUIN, P. ROUX, P. CARONI, H. L. YIN, and P. COSSART, *J Cell Biol*, 2001, **155**(1), 101–112.

56 BRAUN, L., H. OHAYON, and P. COSSART, *Mol Microbiol*, 1998, **27**(5), 1077–1087.

57 DREVETS, D. A., P. J. LEENEN, and R. A. GREENFIELD, *Clin Microbiol Rev*, 2004, **17**(2), 323–347.

58 LECUIT, M., D. M. NELSON, S. D. SMITH, H. KHUN, M. HUERRE, M. C. VACHER-LAVENU, J. I. GORDON, and P. COSSART, *Proc Natl Acad Sci USA*, 2004, **101**(16), 6152–6157.

59 CORNELIS, G. R., *Nat Rev Mol Cell Biol*, 2002, **3**(10), 742–752.

60 CORNELIS, G. R., *Int J Med Microbiol*, 2002, **291**(6–7), 455–462.

61 VON PAWEL-RAMMINGEN, U., M. V. TELEPNEV, G. SCHMIDT, K. AKTORIES, H . WOLF-WATZ, and R. ROSQVIST, *Mol Microbiol*, 2000, **36**(3), 737–748.

62 ANDOR, A., K. TRULZSCH, M. ESSLER, A. ROGGENKAMP, A. WIEDEMANN, J. HEESEMANN, and M. AEPFELBACHER, *Cell Microbiol*, 2001, **3**(5), 301–310.

63 SHAO, F. and J. E. DIXON, *Adv Exp Med Biol*, 2003, **529**, 79–84.

64 PERSSON, C., N. CARBALLEIRA, H. WOLF-WATZ, and M. FALLMAN, E *EMBO J* J, 1997, **16**(9), 2307–2318.

65 HAMID, N., A. GUSTAVSSON, K. ANDERSSON, K. McGEE, C. PERSSON, C. E. RUDD, and M. FALLMAN, *Microb Pathog*, 1999, **27**(4), 231–242.

66 PERSSON, C., R. NORDFELTH, K. ANDERSSON, A. FORSBERG, H. WOLF-WATZ, and M. FALLMAN, *Mol Microbiol*, 1999, **33**(4), 828–838.

67 ERNST, J. D., *Cell Microbiol*, 2000, **2**(5), 379–386.

68 FALLMAN, M., F. DELEUIL, and K. McGEE, *Int J Med Microbiol*, 2002, **291**(6–7), 501–509.

69 DUNCAN, M. J., J. S. SHIN, and S. N. ABRAHAM, *Cell Microbiol*, 2002, **4**(12), 783–791.

70 GATFIELD, J. and J. PIETERS, *Science*, 2000, **288**(5471), 1647–1650.

71 BAORTO, D. M., Z. GAO, R. MALAVIYA, M. L. DUSTIN, A. VAN DER MERWE, D. M. LUBLIN, and S. N. ABRAHAM, *Nature*, 1997, **389**(6651), 636–639.

72 ROHDE, M., E. MULLER, G. S. CHHATWAL, and S. R. TALAY, *Cell Microbiol*, 2003, **5**(5), 323–342.

73 PARTON, R. G., B. JOGGERST, and K. SIMONS, *J Cell Biol*, 1994, **127**(5), 1199–1215.

74 STAHLHUT, M. and B. VAN DEURS, *Mol Biol Cell*, 2000, **11**(1), 325–337.

75 CAMERON, L. A., P. A. GIARDINI, F. S. SOO, and J. A. THERIOT, *Nat Rev Mol Cell Biol*, 2000, **1**(2), 110–119.

76 FRISCHKNECHT, F. and M. WAY, *Trends Cell Biol*, 2001, **11**(1), 30–38.

77 GOLDBERG, M. B., *Microbiol Mol Biol Rev*, 2001, **65**(4), 595–626, table of contents.

78 DOMANN, E., M. LEIMEISTER-WACHTER, W. GOEBEL, and T. CHAKRABORTY, *Infect Immun*, 1991, **59**(1), 65–72.

79 Kocks, C., E. Gouin, M. Tabouret, P. Berche, H. Ohayon, and P. Cossart, *Cell*, 1992, **68**(3), 521–531.

80 Kocks, C., J. B. Marchand, E. Gouin, H. d'Hauteville, P. J. Sansonetti, M. F. Carlier, and P. Cossart, *Mol Microbiol*, 1995, **18**(3), 413–423.

81 Smith, G. A., D. A. Portnoy, and J. A. Theriot, *Mol Microbiol*, 1995, **17**(5), 945–951.

82 Pistor, S., T. Chakraborty, K. Niebuhr, E. Domann, and J. Wehland, *EMBO J*, 1994, **13**(4), 758–763.

83 Welch, M. D., A. H. DePace, S. Verma, A. Iwamatsu, and T. J. Mitchison, *J Cell Biol*, 1997, **138**(2), 375–384.

84 Zalevsky, J., I. Grigorova, and R. D. Mullins, *J Biol Chem*, 2001, **276**(5), 3468–3475.

85 Stradal, T. E. B., K. Rottner, A. Disanza, S. Confalioneri, M. Innocenti, and G. Scita, *Trends Cell Biol*, 2004, in press.

86 Welch, M. D., J. Rosenblatt, J. Skoble, D. A. Portnoy, and T. J. Mitchison, *Science*, 1998, **281**(5373), 105–108.

87 May, R. C., M. E. Hall, H. N. Higgs, T. D. Pollard, T. Chakraborty, J. Wehland, L. M. Machesky, and A. S. Sechi, *Curr Biol*, 1999, **9**(14), 759–762.

88 Sechi, A. S. and J. Wehland, *Front Biosci*, 2004, **9**, 1294–1310.

89 Reinhard, M., K. Giehl, K. Abel, C. Haffner, T. Jarchau, V. Hoppe, B. M. Jockusch, and U. Walter, *EMBO J*, 1995, **14**(8), 1583–1589.

90 Gertler, F. B., K. Niebuhr, M. Reinhard, J. Wehland, and P. Soriano, *Cell*, 1996, **87**(2), 227–239.

91 Bachmann, C., L. Fischer, U. Walter, and M. Reinhard, *J Biol Chem*, 1999, **274**(33), 23549–23557.

92 Walders-Harbeck, B., S. Y. Khaitlina, H. Hinssen, B. M. Jockusch, and S. Illenberger, *FEBS Lett*, 2002, **529**(2–3), 275–280.

93 Smith, G. A., J. A. Theriot, and D. A. Portnoy, *J Cell Biol*, 1996, **135**(3), 647–660.

94 Niebuhr, K., F. Ebel, R. Frank, M. Reinhard, E. Domann, U. D. Carl, U. Walter, F. B. Gertler, J. Wehland, and T. Chakraborty, *EMBO J*, 1997, **16**(17), 5433–5444.

95 Loisel, T. P., R. Boujemaa, D. Pantaloni, and M. F. Carlier, *Nature*, 1999, **401**(6753), 613–616.

96 Geese, M., J. J. Loureiro, J. E. Bear, J. Wehland, F. B. Gertler, and A. S. Sechi, *Mol Biol Cell*, 2002, **13**(7), 2383–2396.

97 Bear, J. E., J. J. Loureiro, I. Libova, R. Fassler, J. Wehland, and F. B. Gertler, *Cell*, 2000, **101**(7), 717–728.

98 Geese, M., K. Schluter, M. Rothkegel, B. M. Jockusch, J. Wehland, and A. S. Sechi, *J Cell Sci*, 2000, **113 (Pt 8)**, 1415–1426.

99 Grenklo, S., M. Geese, U. Lindberg, J. Wehland, R. Karlsson, and A. S. Sechi, *EMBO Rep*, 2003, **4**(5), 523–529.

100 Samarin, S., S. Romero, C. Kocks, D. Didry, D. Pantaloni, and M. F. Carlier, *J Cell Biol*, 2003, **163**(1), 131–142.

101 Skoble, J., V. Auerbuch, E. D. Goley, M. D. Welch, and D. A. Portnoy, *J Cell Biol*, 2001, **155**(1), 89–100.

102 Bear, J. E., T. M. Svitkina, M. Krause, D. A. Schafer, J. J. Loureiro, G. A. Strasser, I. V. Maly, O. Y. Chaga, J. A. Cooper, G. G. Borisy, and F. B. Gertler, *Cell*, 2002, **109**(4), 509–521.

103 LETT, M.C., C. SASAKAWA, N. OKADA, T. SAKAI, S. MAKINO, M. YAMADA, K. KOMATSU, and M. YOSHIKAWA, *J Bacteriol*, 1989, **171**(1), 353–359.

104 NIEBUHR, K., T. CHAKRABORTY, M. ROHDE, T. GAZLIG, B. JANSEN, P. KOLLNER, and J. WEHLAND, *Infect Immun*, 1993, **61**(7), 2793–2802.

105 GOLDBERG, M.B., O. BARZU, C. PARSOT, and P.J. SANSONETTI, *J Bacteriol*, 1993, **175**(8), 2189–2196.

106 HENDERSON, I.R., F. NAVARRO-GARCIA, and J.P. NATARO, *Trends Microbiol*, 1998, **6**(9), 370–378.

107 GOLDBERG, M.B. and J.A. THERIOT, *Proc Natl Acad Sci USA*, 1995, **92**(14), 6572–6576.

108 SUZUKI, T., S. SAGA, and C. SASAKAWA, *J Biol Chem*, 1996, **271**(36), 21878–21885.

109 LAINE, R.O., W. ZEILE, F. KANG, D.L. PURICH, and F.S. SOUTHWICK, *J Cell Biol*, 1997, **138**(6), 1255–1264.

110 ALLY, S., N.J. SAUER, J.J. LOUREIRO, S.B. SNAPPER, F.B. GERTLER, and M.B. GOLDBERG, *Cell Microbiol*, 2004, **6**(4), 355–366.

111 GOLDBERG, M.B., *Cell Motil Cytoskeleton*, 1997, **37**(1), 44–53.

112 SUZUKI, T., H. MIKI, T. TAKENAWA, and C. SASAKAWA, *EMBO J*, 1998, **17**(10), 2767–2776.

113 EGILE, C., T.P. LOISEL, V. LAURENT, R. LI, D. PANTALONI, P.J. SANSONETTI, and M.F. CARLIER, *J Cell Biol*, 1999, **146**(6), 1319–1332.

114 LOMMEL, S., S. BENESCH, K. ROTTNER, T. FRANZ, J. WEHLAND, and R. KUHN, *EMBO Rep*, 2001, **2**(9), 850–857.

115 SNAPPER, S.B., F. TAKESHIMA, I. ANTON, C.H. LIU, S.M. THOMAS, D. NGUYEN, D. DUDLEY, H. FRASER, D. PURICH, M. LOPEZ-ILASACA, C. KLEIN, L. DAVIDSON, R. BRONSON, R.C. MULLIGAN, F. SOUTHWICK, R. GEHA, M.B. GOLDBERG, F.S. ROSEN, J.H. HARTWIG, and F.W. ALT, *Nat Cell Biol*, 2001, **3**(10), 897–904.

116 MOREAU, V., F. FRISCHKNECHT, I. RECKMANN, R. VINCENTELLI, G. RABUT, D. STEWART, and M. WAY, *Nat Cell Biol*, 2000, **2**(7), 441–448.

117 ASPENSTROM, P., U. LINDBERG, and A. HALL, *Curr Biol*, 1996, **6**(1), 70–75.

118 MIKI, H., T. SASAKI, Y. TAKAI, and T. TAKENAWA, *Nature*, 1998, **391**(6662), 93–96.

119 SMALL, J.V., T. STRADAL, E. VIGNAL, and K. ROTTNER, *Trends Cell Biol*, 2002, **12**(3), 112–120.

120 SHIBATA, T., F. TAKESHIMA, F. CHEN, F.W. ALT, and S.B. SNAPPER, *Curr Biol*, 2002, **12**(4), 341–345.

121 FRISCHKNECHT, F., V. MOREAU, S. ROTTGER, S. GONFIONI, I. RECKMANN, G. SUPERTI-FURGA, and M. WAY, *Nature*, 1999, **401**(6756), 926–929.

122 SCAPLEHORN, N., A. HOLMSTROM, V. MOREAU, F. FRISCHKNECHT, I. RECKMANN, and M. WAY, *Curr Biol*, 2002, **12**(9), 740–745.

123 SCHAECHTER, M., F.M. BOZEMAN, and J.E. SMADEL, *Virology*, 1957, **3**(1), 160–172.

124 TEYSSEIRE, N., C. CHICHE-PORTICHE, and D. RAOULT, *Res Microbiol*, 1992, **143**(9), 821–829.

125 HEINZEN, R.A., S.F. HAYES, M.G. PEACOCK, and T. HACKSTADT, *Infect Immun*, 1993, **61**(5), 1926–1935.

126 Gouin, E., H. Gantelet, C. Egile, I. Lasa, H. Ohayon, V. Villiers, P. Gounon, P. J. Sansonetti, and P. Cossart, *J Cell Sci*, 1999, **112 (Pt 11)**, 1697–1708.

127 Heinzen, R. A., S. S. Grieshaber, L. S. Van Kirk, and C. J. Devin, *Infect Immun*, 1999, **67**(8), 4201–4207.

128 Van Kirk, L. S., S. F. Hayes, and R. A. Heinzen, *Infect Immun*, 2000, **68**(8), 4706–4713.

129 Taunton, J., B. A. Rowning, M. L. Coughlin, M. Wu, R. T. Moon, T. J. Mitchison, and C. A. Larabell, *J Cell Biol*, 2000, **148**(3), 519–530.

130 Benesch, S., S. Lommel, A. Steffen, T. E. Stradal, N. Scaplehorn, M. Way, J. Wehland, and K. Rottner, *J Biol Chem*, 2002, **277**(40), 37771–37776.

131 Harlander, R. S., M. Way, Q. Ren, D. Howe, S. S. Grieshaber, and R. A. Heinzen, *Infect Immun*, 2003, **71**(3), 1551–1556.

132 Heinzen, R. A., *Ann N Y Acad Sci*, 2003, **990**, 535–547.

133 Gouin, E., C. Egile, P. Dehoux, V. Villiers, J. Adams, F. Gertler, R. Li, and P. Cossart, *Nature*, 2004, **427**(6973), 457–461.

134 Brieher, W. M., M. Coughlin, and T. J. Mitchison, *J Cell Biol*, 2004, **165**(2), 233–242.

135 Kespichayawattana, W., S. Rattanachetkul, T. Wanun, P. Utaisincharoen, and S. Sirisinha, *Infect Immun*, 2000, **68**(9), 5377–5384.

136 Breitbach, K., K. Rottner, S. Klocke, M. Rohde, A. Jenzora, J. Wehland, and I. Steinmetz, *Cell Microbiol*, 2003, **5**(6), 385–393.

137 Stamm, L. M., J. H. Morisaki, L. Y. Gao, R. L. Jeng, K. L. McDonald, R. Roth, S. Takeshita, J. Heuser, M. D. Welch, and E. J. Brown, *J Exp Med*, 2003, **198**(9), 1361–1368.

138 Nataro, J. P. and J. B. Kaper, *Clin Microbiol Rev*, 1998, **11**(1), 142–201.

139 Ulshen, M. H. and J. L. Rollo, *N Engl J Med*, 1980, **302**(2), 99–101.

140 Moon, H. W., S. C. Whipp, R. A. Argenzio, M. M. Levine, and R. A. Giannella, *Infect Immun*, 1983, **41**(3), 1340–1351.

141 Tzipori, S., I. K. Wachsmuth, C. Chapman, R. Birden, J. Brittingham, C. Jackson, and J. Hogg, *J Infect Dis*, 1986, **154**(4), 712–716.

142 McDaniel, T. K., K. G. Jarvis, M. S. Donnenberg, and J. B. Kaper, *Proc Natl Acad Sci USA*, 1995, **92**(5), 1664–1668.

143 McDaniel, T. K. and J. B. Kaper, *Mol Microbiol*, 1997, **23**(2), 399–407.

144 Elliott, S. J., L. A. Wainwright, T. K. McDaniel, K. G. Jarvis, Y. K. Deng, L. C. Lai, B. P. McNamara, M. S. Donnenberg, and J. B. Kaper, *Mol Microbiol*, 1998, **28**(1), 1–4.

145 Perna, N. T., G. F. Mayhew, G. Posfai, S. Elliott, M. S. Donnenberg, J. B. Kaper, and F. R. Blattner, *Infect Immun*, 1998, **66**(8), 3810–3817.

146 Elliott, S. J., J. Yu, and J. B. Kaper, *Infect Immun*, 1999, **67**(8), 4260–4263.

147 Kenny, B., *Int J Med Microbiol*, 2002, **291**(6–7), 469–477.

148 Clarke, S. C., R. D. Haigh, P. P. Freestone, and P. H. Williams, *Clin Microbiol Rev*, 2003, **16**(3), 365–378.

149 TAYLOR, K.A., P.W. LUTHER, and M.S. DONNENBERG, *Infect Immun*, 1999, **67**(1), 120–125.

150 CRANE, J.K., B.P. McNAMARA, and M.S. DONNENBERG, *Cell Microbiol*, 2001, **3**(4), 197–211.

151 McNAMARA, B.P., A. KOUTSOURIS, C.B. O'CONNELL, J.P. NOUGAYREDE, M.S. DONNENBERG, and G. HECHT, *J Clin Invest*, 2001, **107**(5), 621–629.

152 ELLIOTT, S.J., E.O. KREJANY, J.L. MELLIES, R.M. ROBINS-BROWNE, C. SASAKAWA, and J.B. KAPER, *Infect Immun*, 2001, **69**(6), 4027–4033.

153 YOSHIDA, S., E. KATAYAMA, A. KUWAE, H. MIMURO, T. SUZUKI, and C. SASAKAWA, *EMBO J*, 2002, **21**(12), 2923–2935.

154 KENNY, B. and M. JEPSON, *Cell Microbiol*, 2000, **2**(6), 579–590.

155 KENNY, B., S. ELLIS, A.D. LEARD, J. WARAWA, H. MELLOR, and M.A. JEPSON, *Mol Microbiol*, 2002, **44**(4), 1095–1107.

156 TU, X., I. NISAN, C. YONA, E. HANSKI, and I. ROSENSHINE, *Mol Microbiol*, 2003, **47**(3), 595–606.

157 KENNY, B., R. DeVINNEY, M. STEIN, D.J. REINSCHEID, E.A. FREY, and B.B. FINLAY, *Cell*, 1997. **91**(4), 511–520.

158 DEIBEL, C., S. KRAMER, T. CHAKRABORTY, and F. EBEL, *Mol Microbiol*, 1998, **28**(3), 463–474.

159 ROSENSHINE, I., S. RUSCHKOWSKI, M. STEIN, D.J. REINSCHEID, S.D. MILLS, and B.B. FINLAY, *EMBO J*, 1996, **15**(11), 2613–2624.

160 CAMPELLONE, K.G. and J.M. LEONG, *Curr Opin Microbiol*, 2003, **6**(1), 82–90.

161 HART, C.A., R.M. BATT, and J.R. SAUNDERS, *Ann Trop Paediatr*, 1993, **13**(2), 121–131.

162 KENNY, B., *Mol Microbiol*, 1999, **31**(4), 1229–1241.

163 GRUENHEID, S., R. DeVINNEY, F. BLADT, D. GOOSNEY, S. GELKOP, G.D. GISH, T. PAWSON, and B.B. FINLAY, *Nat Cell Biol*, 2001, **3**(9), 856–859.

164 CAMPELLONE, K.G., A. GIESE, D.J. TIPPER, and J.M. LEONG, *Mol Microbiol*, 2002, **43**(5), 1227–1241.

165 CAMPELLONE, K.G., S. RANKIN, T. PAWSON, M.W. KIRSCHNER, D.J. TIPPER, and J.M. LEONG, *J Cell Biol*, 2004, **164**(3), 407–416.

166 DeVINNEY, R., M. STEIN, D. REINSCHEID, A. ABE, S. RUSCHKOWSKI, and B.B. FINLAY, *Infect Immun*, 1999, **67**(5), 2389–2398.

167 DeVINNEY, R., J.L. PUENTE, A. GAUTHIER, D. GOOSNEY, and B.B. FINLAY, *Mol Microbiol*, 2001, **41**(6), 1445–1458.

168 KALMAN, D., O.D. WEINER, D.L. GOOSNEY, J.W. SEDAT, B.B. FINLAY, A. ABO, and J.M. BISHOP, *Nat Cell Biol*, 1999, **1**(6), 389–391.

169 LOMMEL, S., S. BENESCH, M. ROHDE, J. WEHLAND, and K. ROTTNER, *Cell Microbiol*, 2004, **6**(3), 243–254.

170 ROHATGI, R., P. NOLLAU, H.-Y.H. HO, M.W. KIRSCHNER, and B.J. MAYER, *J Biol Chem*, 2001, **276**(28), 26448–26452.

171 WELCH, M.D. and R.D. MULLINS, *Annu Rev Cell Dev Biol*, 2002, **18**, 247–288.

172 CANTARELLI, V.V., A. TAKAHASHI, Y. AKEDA, K. NAGAYAMA, and T. HONDA, *Infect Immun*, 2000, **68**(1), 382–386.

173 FREEMAN, N.L., D.V. ZURAWSKI, P. CHOWRASHI, J.C. AYOOB, L. HUANG, B. MITTAL, J.M. SANGER, and J.W. SANGER, *Cell Motil Cytoskeleton*, 2000, **47**(4), 307–318.

174 GOOSNEY, D.L., R. DeVINNEY, R.A. PFUETZNER, E.A. FREY, N.C. STRYNADKA, and B.B. FINLAY, *Curr Biol*, 2000, **10**(12), 735–738.

175 CANTARELLI, V.V., A. TAKAHASHI, I. YANAGIHARA, Y. AKEDA, K. IMURA, T. KODAMA, G. KONO, Y. SATO, and T. HONDA, *Cell Microbiol*, 2001, **3**(11), 745–751.

176 GOOSNEY, D.L., R. DeVINNEY, and B.B. FINLAY, *Infect Immun*, 2001, **69**(5), 3315–3322.

177 CANTARELLI, V.V., A. TAKAHASHI, I. YANAGIHARA, Y. AKEDA, K. IMURA, T. KODAMA, G. KONO, Y. SATO, T. IIDA, and T. HONDA, *Infect Immun*, 2002, **70**(4), 2206–2209.

178 HUANG, L., B. MITTAL, J.W. SANGER, and J.M. SANGER, *Cell Motil Cytoskeleton*, 2002, **52**(4), 255–265.

179 BATCHELOR, M., J. GUIGNOT, A. PATEL, N. CUMMINGS, J. CLEARY, S. KNUTTON, D.W. HOLDEN, I. CONNERTON, and G. FRANKEL, *EMBO Rep*, 2004, **5**(1), 104–110.

180 SANGER, J.M., R. CHANG, F. ASHTON, J.B. KAPER, and J.W. SANGER, *Cell Motil Cytoskeleton*, 1996, **34**(4), 279–287.

13
Transendothelial Migration of Leukocytes

William A. Muller and Alan R. Schenkel

13.1
Sequential Adhesive Events in Leukocyte Emigration

Leukocyte emigration from the bloodstream into inflamed tissues occurs within minutes *in vivo* and involves a series of adhesive interactions of increasing strength between molecules on the leukocyte surface and those on the endothelial cells lining the blood vessel. In general each successive step is a prerequisite for the next one, and each is governed by a different class of adhesion molecules on the leukocytes or endothelium (see Fig. 13.1).

We recognize these distinct steps because we have reagents that can block each one. As we shall show in this chapter, the discovery of new molecular interactions between leukocytes and endothelial cells has recently led to the identification of an additional step (locomotion) and the dissection of the diapedesis process into two distinct molecularly-defined interactions. This section will provide a brief overview of the entire process of leukocyte emigration. We will then describe the locomotion and diapedesis stages that have been the focus of our recent studies.

The studies of leukocyte emigration described in this chapter have been carried out predominantly on monocytes and neutrophils. In general, where leukocyte subsets have been compared, the molecular interactions governing leukocyte-endothelial cell interactions are similar. The degree to which distinct steps in the emigration process rely on a specific adhesion molecule may vary among cell types due to the expression on some cells of alternative molecules that can carry out that particular step. In other words, although the process of emigration is similar for all types of leukocytes, the exact adhesion molecules that they use to carry out the rolling, adhesion, and transmigration steps may vary among the leukocyte types. In addition, different adhesion molecules may predominate in different tissues or under different inflammatory conditions. Finally, there is a great deal of redundancy in the control of this process. Several different molecules can carry out the same step of leukocyte emigration (Fig. 13.1). This has made analysis of the process difficult, since inhibit-

Cell Migration. Edited by Doris Wedlich
Copyright © 2005 WILEY-VCH Verlag GmbH & Co. KGaA, Weinheim
ISBN: 3-527-30587-4

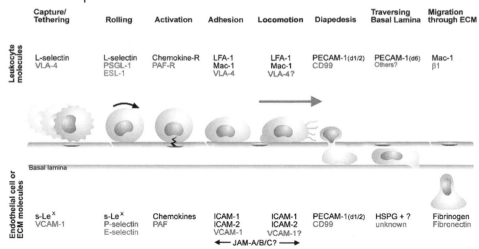

	Capture/ Tethering	Rolling	Activation	Adhesion	Locomotion	Diapedesis	Traversing Basal Lamina	Migration through ECM
Leukocyte molecules	L-selectin VLA-4	L-selectin PSGL-1 ESL-1	Chemokine-R PAF-R	LFA-1 Mac-1 VLA-4	LFA-1 Mac-1 VLA-4?	PECAM-1(d1/2) CD99	PECAM-1(d6) Others?	Mac-1 β1
Endothelial cell or ECM molecules	s-Le^x VCAM-1	s-Le^x P-selectin E-selectin	Chemokines PAF	ICAM-1 ICAM-2 VCAM-1	ICAM-1 ICAM-2 VCAM-1?	PECAM-1(d1/2) CD99	HSPG + ? unknown	Fibrinogen Fibronectin

◄─── JAM-A/B/C? ───►

Fig. 13.1 Sequential steps in leukocyte emigration are controlled by specific adhesion molecules on leukocytes and endothelial cells. The various steps of leukocyte emigration described in the text are depicted schematically here. For each step the interacting pairs of adhesion molecules, ligands, or counter-receptors expressed by the leukocytes and endothelial cells or extracellular matrix are shown in the same color. This diagram is not inclusive, and other molecules may mediate each of these events for distinct leukocyte types under different inflammatory conditions. For the Capture/Tethering step, the protrusions on the leukocyte surface are meant to represent the microvilli that bear L-selectin or VLA-4. The lightning bolt at the Activation step represents the triggering of inside-out activation of leukocyte integrins by signals from the endothelium and endothelial surface via G protein coupled receptors. The arrow under Locomotion indicates translational movement across the endothelial surface. Question marks next to some molecules indicate that their role in this step is postulated, but not documented. The basal lamina is depicted as separate from the remainder of the extracellular matrix (ECM) since migration across the subendothelial basal lamina appears to be a separate step controlled by distinct molecules [34, 50]. However, these steps may not be molecularly distinct (see text.) Since the exact β_1 integrin(s) that are involved in migration through ECM via fibronectin interactions have not been defined, the leukocyte molecules are designated simply as β_1. Abbreviations: s-Le^x = sialyl-Lewis^x carbohydrate antigen; PSGL-1 = P-selectin glycoprotein ligand 1; ESL-1 = E-selectin ligand 1; PAF = platelet activating factor; PAF R = PAF receptor; PECAM-1 (d1/2) = interaction involves immunoglobulin domains 1 and/or 2 of PECAM-1; PECAM-1 (d6) = interaction involves immunoglobulin domain 6 of PECAM-1; HSPG = heparan sulfate proteoglycan. (This figure also appears on the color plates.)

ing the function of a molecule that really plays a role in leukocyte emigration may not lead to an observable phenotype if other molecules take over for it. This chapter will deal with the molecules and events that we have been studying in our laboratory. There is evidence that members of the junctional adhesion molecule (JAM) family play a role in these processes as well. This will not be covered for lack of space. For a more extensive overview of the process of leukocyte emigration in general or diapedesis in particular, the reader is referred to references [1] or [2], respectively.

13.1.1
Tethering and Rolling

Since leukocytes cannot swim, they must first attach to the endothelium lining the blood vessel wall before they can pass through into the inflamed tissues. These initial capturing interactions are generally mediated by the selectin family of adhesion molecules and their sulfated, sialylated, fucosylated, glycoprotein ligands [3–5]. These are generally of the configuration of the blood group carbohydrate sialyl-Lewis^x. For optimal P-selectin and L-selectin binding, tyrosine residues on the amino terminus of the protein bearing sialyl-Lewis^x must be sulfated [5–7]. The integrin $a4\beta1$ can also mediate rolling of T-cells [8] and neutrophils [9].

L-selectin on leukocytes is enriched on microvillous projections of the cell membrane, while the integrin molecules that mediate the next step of emigration (see below) are restricted to the body of the leukocyte between microvilli [10]. In this way, the molecules mediating the initial tethering of leukocytes contact the endothelium first. Soon after this initial contact, the microvilli retract, allowing contact of the integrins with their counter-receptors on the endothelium. The integrin $a_4\beta_1$ on leukocytes interacting with vascular cell adhesion molecule-1 (VCAM-1) on inflamed endothelium may be able to play a similar role [9]. In such cases, $a_4\beta_1$ has been shown to be enriched in microvillous projections, similar to L-selectin [8].

13.1.2
Arrest and Adhesion

In order to stop rolling, the low affinity transient interactions of rolling must be replaced by high affinity adhesion between the leukocyte and the endothelial surface. Such high affinity interactions are triggered by stimuli presented to the leukocyte on the endothelial surface. These may be intrinsic endothelial cell surface molecules [11] or molecules such as chemokines from the inflammatory site that are bound to endothelial surface glycosaminoglycans and presented to the leukocytes [12]. The common denominator is that interaction with these molecules activates the leukocyte resulting in adhesion by yet another family of adhesion molecules, the integrins.

Leukocytes possess a unique family of integrins, the β_2 or leukocyte-specific integrins (also known as CD18), that are primarily involved in adhesion to other cells. Of the four identified α chains that pair with CD18, α_L (CD11a) and α_M (CD11b) are the most important for the adhesion of leukocytes to endothelium. The integrins formed, $\alpha_L\beta_2$ (also known as CD11a/CD18 and leukocyte function associated antigen 1 (LFA-1)) and $\alpha_M\beta_2$ (also known as CD11b/CD18 and Mac-1) are expressed to different extents on the various leukocyte types. In general, they can substitute for each other, and to effectively inhibit leukocyte adhesion therapeutically in most models, both integrins (or the common β chain) must be blocked. In addition many leukocytes express the β_1-integrin $\alpha4\beta1$ (very late antigen 4, VLA-4). This can adhere to vascular cell adhesion molecule-1 (VCAM-1), a molecule expressed on the surface of inflamed endothelium that is distinct from the counter-receptor for the β_2-integrins (see below.)

Integrins exist in a low and high affinity binding state with the low affinity state favored. They must be activated to favor the high affinity state and efficiently bind their ligands. This can be done in two ways. Avidity modulation clusters the integrins in the plane of the membrane. A higher density of integrin molecules increases the probability that a molecule in that cluster will be in the active conformation at any given time. Affinity modulation involves a conformational change in the integrin chains that favors their assumption of the active state. Signals that trigger affinity modulation can be transduced by a variety of receptors for inflammatory mediators. Recent research has highlighted the role of heptahelical receptors for chemokines and formylated peptides, all coupled to heterotrimeric G-proteins, in the transmission of signals that activate leukocyte integrins.

When activated, the leukocyte integrins bind to their ligands or counter-receptors on endothelium. These molecules, intercellular adhesion molecule (ICAM) -1 and -2 and VCAM-1 make high affinity adhesions with the leukocyte integrins, allowing them to stop rolling and begin crawling on the endothelial cell surface – the next step in their migration. The avidity of the counter-receptors can be regulated as well. The active form of ICAM-1 functions as a dimer [13, 14].

13.1.3
Locomotion: Migration on the Endothelial Surface

While the leukocyte may roll for several seconds over many endothelial cells before finally arresting, once tightly adherent to the endothelial cell surface, the leukocyte appears to migrate into the tissues by crossing the border of the endothelial cell to which it is adherent. Migration on the surface of endothelium toward the junction requires leukocytes to maintain adhesion at the front end of the cell while letting go at the rear. Until recently, little was known about the regulation of these cycles. The distance traveled is not likely to be too great. Some studies show that under conditions resembling blood flow most neutrophils arrest within one cell diameter of the junction they will migrate across [15].

Leukocyte migration on the apical surface of endothelium, which we term "locomotion", appears to be mediated largely by interactions between β_2 integrins of the leukocyte and intercellular adhesion molecules (ICAMs) on the endothelial surface [16]. This will be discussed in detail in a subsequent section.

13.1.4
Diapedesis

During diapedesis, when the leukocyte squeezes in ameboid fashion between the tightly apposed endothelial cell borders, there is a fundamental change in the interactions between the leukocytes and endothelial cells. In the preceding steps leukocytes interact with the apical surface of endothelial cells as if adhering to a two dimensional plane. Beginning with diapedesis, adhesive interactions take place in three dimensions. The earlier steps involve molecules of one molecular family on the leukocyte engaging molecules of a distinct molecular family on the endothelium (heterophilic interaction). In contrast, diapedesis requires distinct molecules on the leukocyte engaging the same molecule expressed by the endothelial cell (homophilic interaction).

Once leukocytes arrest tightly on the endothelial surface and begin to transmigrate, it is generally believed that the subsequent process of transendothelial migration is independent of fluid shear [17]. However, some papers have appeared recently with data demonstrating more efficient adhesion [18, 19] or transmigration [20] under conditions of fluid shear. While monocytes and neutrophils can transmigrate efficiently in the absence of flow in many *in vitro* models [21, 22], T-cell transmigration was remarkably more efficient in the presence of shear stress [20].

Platelet/endothelial cell adhesion molecule-1 (PECAM, CD31) and CD99 are both molecules expressed by leukocytes and concentrated at the borders of endothelial cells. Homophilic interaction of these molecules is required for leukocyte diapedesis [23–25]. In our experience, these are the major molecules regulating diapedesis of monocytes and neutrophils. Antibodies against other molecules that have been reported by others to block diapedesis appear to block earlier steps of tight adhesion or the recently-discovered step of locomotion (see below) and not diapedesis *per se*. However, there are clearly molecules in addition to CD31 and CD99 that can support diapedesis, and the molecules controlling this step may be different for different leukocytes, vascular beds, or inflammatory stimuli.

13.2
Locomotion

Cell migration requires that leukocytes polarize with cycles of adhesion at the front and de-adhesion at the rear. Membrane protrusion/extension at the leading edge provides a mechanism to move the cell forward. At the same time, membrane is removed from the trailing edge and reincorporated into the main cell body. There is evidence that at least part of cell movement involves internalization of integrin-bearing membrane at the rear by endocytosis and recycling of that membrane to the front of the cell [26]. While these studies were performed on cells migrating on vitronectin on coated glass, not on endothelial cells, a recent report on monocyte transmigration across endothelial cells supports the generalization of this idea. In that study, transmigration was blocked by inhibition of the small GTPase RhoA. The investigators found that β_2 integrins remained localized in the unretracted tail of the monocyte [27].

We recently studied the molecules used by leukocytes to move from their point of firm attachment on the apical surface of the endothelial cell to the cell borders across which they would diapedese [16]. Since firm adhesion to the apical surface of endothelial cells is a prerequisite for migration to the cell border, we designed an assay system in which we could study the role of a given molecule in this migration process independent of its role in firm adhesion. We took advantage of the fact that monocytes can adhere to the apical surface of endothelial cells via β_2-integrins binding to ICAMs or the β_1-integrin VLA-4 binding to VCAM-1 (Fig. 13.1). By blocking the function of a molecule in one of these pairs using monoclonal antibodies, we were able to study the role of that molecule in locomotion while the monocyte remained tightly adherent via the other undisturbed molecular interactions.

We visualized leukocyte–endothelial cell interactions in real time to observe the effects of these antibodies on leukocyte transmigration. In the absence of antibody blockade, leukocytes polarized, spread, and moved to a nearby cell border shortly after tight adhesion to the endothelial cell. Within minutes, they would diapedese across this cell border. This activity was particularly rapid when the endothelial cells had been activated by the proinflammatory cytokines interleukin 1β (IL-1β) or tumor necrosis factor α (TNFα). Blocking CD11a and CD11b or their common β chain CD18 had a profound effect on leukocyte migration along the apical surface of the endothelial cell. Similarly, blocking the β_2-integrin counter-receptors ICAM-1 or ICAM-2 (and especially when both were blocked together) greatly inhibited leukocyte emigration without affecting the total number of adherent monocytes. In both cases, leukocytes would adhere to the apical surface of the cell and exhibit one or both of two aberrant migration phenotypes [16]. Cells would perform "search pirouettes" in which they would rotate around their trailing edge (uropod) for several minutes, sending filopodia out in alternating directions, as if looking for a migration signal, but unable to find it. Search pirouettes were especially long and protracted when the leukocyte β_2-integrin was blocked [16]. In addition, cells would often travel for long distances ("long walks" >40 μm) crossing over cell borders onto adjacent cells without stopping, as if they did not recognize that they had reached the cell border.

Blocking $a_4\beta_1$ (VLA-4) and/or its counter-receptor VCAM-1 primarily served to decrease adhesion of monocytes to the apical surface. They did not seem to play a significant role in migration of monocytes from the point of attachment to the cell border [16]. Likewise, blocking PECAM-1 (CD31) had no effect on adhesion or locomotion, but profoundly blocked monocyte diapedesis (see below).

It will be important to follow up these data with studies of locomotion performed under conditions of physiologic fluid shear. We believe that by performing these experiments under static conditions we may have made cellular interactions that would be difficult to observe under shear stress more obvious. However, flowing blood could potentially affect the behavior of leukocytes as they migrate across the surface of the endothelium. For example, fluid shear might influence directionality of movement. Morever, JAM family members or molecules such as selectins, which are designed to function under such conditions, may play a role in this process as well.

13.3
Diapedesis

Two molecules have been described that appear to function selectively in diapedesis: PECAM and CD99. As mentioned above, these molecules are expressed by both leukocytes and endothelium, and homophilic interaction between them is required for transendothelial migration to proceed. Blocking the function of either molecule selectively blocks diapedesis; there is no effect on the earlier steps of rolling, adhesion, or locomotion [16, 23, 25]. Other molecules such as leukocyte β_2 integrins, VCAM-1, and members of the JAM family have been reported to play a role in transendothelial migration [1, 2]. It is quite possible that under certain inflammatory conditions these molecules contribute to diapedesis. However, they also participate in well-documented leukocyte–endothelial adhesive interactions at the apical surface of endothelium, so their participation in leukocyte emigration is not limited to diapedesis. Moreover, their presumed role in diapedesis may in fact be due to their contribution to the earlier adhesive steps, since tight adhesion and the recently-recognized locomotion step are prerequisites for diapedesis, and many *in vitro* assays do not distinguish these steps [11].

13.3.1
PECAM/CD31

PECAM is a member of the immunoglobulin gene superfamily. It has six extracellular C2-type immunoglobulin (Ig) domains and a cytoplasmic tail bearing two immunotyrosine-based inhibitory motifs (ITIMs) [28, 29]. The latter have been well-documented to bind the phosphatase SHP-2 when phosphorylated [30, 31]. However, the role of PECAM phosphorylation and signaling in transendothelial migration is only now being studied. The first (amino-terminal) Ig domain of PECAM is capable of homophilic adhesion to the first Ig domain of another molecule of PECAM. This is the basis of homophilic adhesion of endothelial cells to each other as well as PECAM-PECAM interactions between leukocytes and endothelial cells [24, 32, 33].

When PECAM function is blocked on either the leukocyte or endothelial cell, leukocytes adhere to the apical surface of the endothelium, migrate to the cell borders, but are arrested on the apical surface over the cell border [23, 34]. This is reversible, as migration resumes once the block has been removed [23]. Blocking the amino-terminal domain of PECAM with either monoclonal antibody or recombinant PECAM-Fc chimera blocks the diapedesis of monocytes and neutrophils *in vitro* [23, 24, 34] and *in vivo* [24, 35–38]. Transmigration of natural killer cells is also blocked at the same step by anti-PECAM monoclonal antibody [39]; however, anti-PECAM treatment does not seem to affect migration of T lymphocytes [40] in some *in vitro* models.

13.3.2
CD99

CD99 is a unique molecule in the human genome [41]. It is expressed to varying degrees on the surfaces of all circulating blood cells as well as concentrated at the borders between endothelial cells [25, 41], but its expression is not restricted to cells of the vascular system. Approximately half of its 32 kD apparent molecular weight on SDS-PAGE is due to *O*-linked carbohydrate chains. Heterogeneity of glycosylation may explain why CD99 monoclonal antibodies often detect CD99 only on a restricted subset of leukocytes [25]. Transfection of L-cell fibroblasts with CD99 imparts on them the ability to adhere to each other in a homophilic manner [25]. Ligation of CD99 on lymphocytes activates leukocyte integrins, presumably by "inside-out" signaling [42–44].

Blockade of CD99 on either the leukocyte or the endothelial cell arrests diapedesis to the same extent (>90%) as blocking both sides at the same time [25]. This strongly suggests that homophilic interaction between CD99 on monocytes and CD99 at the endothelial cell border is critical for diapedesis. CD99 regulates a distinct step in diapedesis from that regulated by PECAM. The blockade of CD99 and PECAM is additive. Recall that leukocytes arrested in diapedesis by anti-PECAM reagents are reversibly blocked on the apical surface of the endothelial cells over the cell borders [16, 23, 25, 34]. Anti-CD99 monoclonal antibody arrests the leukocytes partway across the endothelial cell border with the leading edge of the cell below the endothelial cell and the trailing end on the apical surface [25].

Moreover, the step regulated by CD99 is functionally "downstream" of the PECAM-dependent step. When monocytes were first blocked with anti-PECAM antibodies, then released from that block and incubated in the presence of anti-CD99, they could not finish diapedesis. In contrast, if cells were first blocked with anti-CD99, then released from that block and incubated in the presence of anti-PECAM, diapedesis resumed normally. This indicated that CD99 exerts its effect after PECAM, and once past the CD99-dependent step, blocking PECAM no longer had an effect [25].

13.3.3
Mechanistic Insights

We recently described a novel membrane compartment along the cell borders of endothelial cells. This compartment is made up of clusters of 50 nm diameter vesicle-like structures connected to each other in a reticulum and connected to the endothelial cell surface at the intercellular borders [45]. This reticulum contains about 1/3 of the total cellular PECAM. Preliminary data (not shown) demonstrate that this compartment exists *in vivo* as well. At 37 °C proteins have access to this compartment and PECAM recycles evenly along the cell border between the surface of the cell and these vesicles. At 4 °C PECAM recycling does not take place and small proteins such as Fab fragments of antibody do not have access to this compartment. However, protons do enter this compartment at 4 °C, so this reticulum does not represent a system of sealed vesicles [45]. The vesicle-like structures that make up the reticulum resemble caveolae structurally. However, our biochemical data at this writing suggests that they are not [45].

When monocytes diapedese across endothelial cells the recycling of PECAM from this surface-connected membrane reticulum is redirected. Instead of returning evenly it is targeted to the place along the endothelial cell border where the leukocyte is migrating [45] (Fig. 13.2). Selective visualization of recycling PECAM demonstrates rings of this membrane surrounding the leukocyte during its passage across the cell junction. Since 30% of the total endothelial PECAM is contained in the internal reticulum, this may serve to provide an increase in surface area to accommodate the leukocyte at it squeezes between the tightly apposed endothelial cells. Alternatively, this compartment may provide a pool of unligated PECAM for the leukocyte PECAM to interact with so that the PECAM-PECAM interactions that exist between adjacent endothelial cells do not have to "unzip".

The signals that regulate the targeted recycling are not understood. However, we know that a productive leukocyte–endothelial cell PECAM-PECAM interaction is required. PECAM expressed by a non-motile cell attached over the junction is not sufficient to trigger targeted recycling. Blocking PECAM on a monocyte with anti-PECAM antibody so that it cannot engage PECAM on the endothelium will block transmigration and targeted recycling [45].

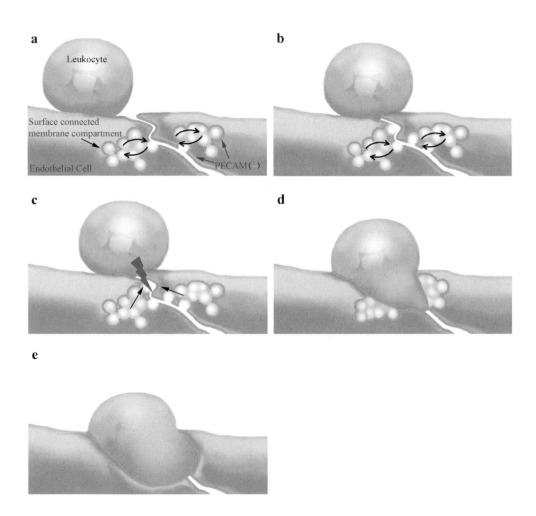

It has been known since 1993 that blockade of PECAM function selectively blocks diapedesis with leukocytes arrested on the apical surface of the endothelial cell border [23]. It has also been known since 1993 that a transient increase in cytosolic free calcium concentration in the endothelial cells is required for transmigration to proceed, and buffering this calcium transient results in leukocytes arrested on the apical surface of endothelial cells, tightly adherent, but unable to transmigrate [46]. The cytosolic free calcium rise has been linked to activation of myosin light chain kinase and an increase in tension in endothelial cell monolayers [47]. Inhibitors of myosin light chain kinase have been reported to block transendothelial migration [48, 49]. However, since buffering the increase in cytosolic free calcium and inhibiting PECAM function both block leukocyte transmigration with a similar resulting phenotype, it is interesting to speculate that the two phenomena may be mechanistically related. This is the subject of ongoing research.

Molecular analysis of the cellular interactions between leukocytes and endothelial cells has led to insight into the mechanisms governing their interactions during inflammation. A major unanswered question for future research is whether the molecular interactions that regulate inflammation are distinct signals, or whether they are integrated by the leukocyte and endothelial cell over time and space to affect their behavior. Another is how the body naturally downregulates the inflammatory response. It is hoped that a deeper understanding of these interactions will lead to the development of better anti-inflammatory therapies.

13.4
Acknowledgments

We thank Gillian Muller for the illustrations for Fig. 13.2. This work was supported by grants from the National Institutes of Health (HL46849 and HL64774) to WAM.

Fig. 13.2 Targeted recycling of PECAM from the intracellular compartment to the cell border during diapedesis. This diagram schematically depicts sequential steps in the emigration of a leukocyte. An adherent leukocyte is approaching (a) and has reached (**b**) the intercellular border between two endothelial cells. Broken circular arrows indicate constitutive recycling of PECAM-bearing membrane between the internal compartment and the cell surface at the border. In (c) a signal from the leukocyte in addition to homophilic PECAM engagement triggers a change in the recycling pattern of endothelial cell PECAM. This membrane now moves to surround the leukocyte as it enters the junction (d). The extra surface area provided by this membrane accommodates the leukocyte (e) and/or provides a source of unligated PECAM for it to interact with. The recruitment of PECAM-bearing membrane appears to continue for the duration of the diapedesis process (not shown, but see [45]).

13.5
References

1 W. A. MULLER, *Lab. Invest.* **2002**, *82*, 521–534.
2 W. A. MULLER, *Trends. Immunol* .**2003**, *24*, 326–333.
3 T. M. CARLOS and J. M. HARLAN, *Blood* **1994**, *84*, 2068–2101.
4 A. ETZIONI, C. M. DOERSCHUK, and J. M. HARLAN, *Blood* **1999**, *94*, 3281–3288.
5 D. VESTWEBER and J. E. BLANKS, *Physiol. Rev.* **1999**, *79*, 181–213.
6 D. SAKO, K. M. COMESS, K. M. BARONE, R. T. CAMPHAUSEN, D. A. CUMMING, and G. D. SHAW, *Cell.* **1995**, *83*, 323–331.
7 T. POUYANI and B. SEED, *Cell.* **1995**, *83*, 333–343.
8 C. BERLIN, R. F. BARGATZE, J. J. CAMPBELL, U. H. VON ANDRIAN, M. C. SZABO, S. R. HASSLEN, R. D. NELSON, E. L. BERG, S. L. ERLANDSEN, and E. C. BUTCHER, *Cell.* **1995**, *80*, 413–422.
9 R. B. HENDERSON, L. H. K. LIM, P. A. TESSIER, F. N. E. GAVINS, M. MATHIES, M. PERRETTI, and N. HOGG, *J. Exp. Med.* **2001**, *194*, 219–226.
10 U. H. VON ADRIAN, S. R. HASSLEN, R. D. NELSON, S. L. ERLANDSEN, and E. BUTCHER, *Cell.* **1995**, *82*, 989–999.
11 W. A. MULLER, Leukocyte-Endothelial Cell Adhesion Molecules in Transendothelial Migration, in *Inflammation: Basic Principles and Clinical Correlates*, (JOHN I. GALLIN, RALPH SNYDERMAN, BARTON F. HAYNES) Philadelphia, Lippincott Williams & Wilkins, 1999.
12 Y. TANAKA, D. H. ADAMS, S. HUBSCHER, H. HIRANO, U. SIEBENLIST, and S. SHAW, *Nature* **1993**, *361*, 79–82.
13 P. L. REILLY, J. R. J. WOSKA, D. D. JEANFAVRE, E. MCNALLY, R. ROTHLEIN, and B. J. BORMANN, *J. Immunol.* **1995**, *155*, 529–532.
14 J. MILLER, R. KNORR, M. FERRONE, R. HOUDEI, C. P. CARRON, and M. L. DUSTIN, *J. Exp. Med.* **1995**, *182*, 1231–1241.
15 P. K. GOPALAN, A. R. BURNS, S. I. SIMON, S. SPARKS, L. V. MCINTYRE, and C. W. SMITH, *J. Leuk. Biol.* **2000**, *68*, 47–57.
16. A. R. SCHENKEL, Z. MAMDOUH, and W. A. MULLER, *Nature Immunol.* **2004**, *5*, 393–400.
17 C. W. SMITH, R. ROTHLEIN, B. J. HUGHES, M. M. MARISCALSO, H. E. RUDLOFF, F. C. SCHMALSTIEG, and D. C. ANDERSON, *J. Clin. Invest.* **1988**, *82*, 1746–1756.
18 J. KITAYAMA, A. HIDEMURA, H. SAITO, and H. NAGAWA, *Cell. Immunol.* **2000**, *203*, 39–46.
19 S. L. CUVELIER and K. D. PATEL, *J. Exp. Med.* **2001**, *194*, 1699–1709.
20 G. CINAMON, V. SHINDER, and R. ALON, *Nature Immunol.* **2001**, *2*, 515–522.
21 W. A. MULLER and S. WEIGL, *J. Exp. Med.* **1992**, *176*, 819–828.
22 S. K. SHAW, P. S. BAMBA, B. N. PERKINS, and F. W. LUSCINSKAS, *J. Immunol.* **2001**, *167*, 2323–2330.
23 W. A. MULLER, S. A. WEIGL, X. DENG, and D. M. PHILLIPS, *J. Exp. Med.* **1993**, *178*, 449–460.
24 F. LIAO, J. ALI, T. GREENE, and W. A. MULLER, *J. Exp. Med.* **1997**, *185*, 1349–1357.
25 A. R. SCHENKEL, Z. MAMDOUH, X. CHEN, R. M. LIEBMAN, and W. A. MULLER, *Nature Immunol.* **2002**, *3*, 143–150.
26 M. A. LAWSON and F. R. MAXFIELD, *Nature* **1995**, *377*, 75–79.

27 R. A. WORTHYLAKE, S. LEMOINE, J. M. WATSON, and K. BURRIDGE, *J. Cell. Biol.* **2001**, *154*, 147–160.

28 P. J. NEWMAN, M. C. BERNDT, J. GORSKI, G. C. WHITE II, S. LYMAN, C. PADDOCK, and W. A. MULLER, *Science* **1990**, *247*, 1219–1222.

29 P. J. NEWMAN, *J. Clin. Invest.* **1999**, *103*, 5–9.

30 P. J. NEWMAN and D. K. NEWMAN, *Arterioscler. Thromb. Vasc. Biol.* **2003**, 23, 953–64.

31 N. ILAN and J. A. MADRI, *Curr. Opin. Cell. Biol* .**2003**, *15*, 515–24.

32 S. M. ALBELDA, W. A. MULLER, C. A. BUCK, and P. J. NEWMAN, *J.Cell. Biol.* **1991**, *114*, 1059–1068.

33 Q.-H. SUN, H. M. DELISSER, M. M. ZUKOWSKI, C. PADDOCK, S. M. ALBELDA, and P. J. NEWMAN, *J. Biol. Chem.* **1996**, *271*, 11090–11098.

34 F. LIAO, H. K. HUYNH, A. EIROA, T. GREENE, E. POLIZZI, and W. A. MULLER, *J. Exp. Med.* **1995**, *182*, 1337–1343.

35 A. A. VAPORCIYAN, H. M. DELISSER, H.-C. YAN, I. I. MENDIGUREN, S. R. THOM, M. L. JONES, P. A. WARD, and S. M. ALBELDA, *Science* **1993**, *262*, 1580–1582.

36 S. BOGEN, J. PAK, M. GARIFALLOU, X. DENG, and W. A. MULLER, *J. Exp. Med.* **1994**, *179*, 1059–1064.

37 M. CHRISTOFIDOU-SOLOMIDOU, M. T. NAKADA, J. WILLIAMS, W. A. MULLER, and H. M. DELISSER, *J. Immunol.* **1997**, *158*, 4872–4878.

38 F. LIAO, A. R. SCHENKEL, and W. A. MULLER, *J. Immunol.* **1999**, *163*, 5640–5648.

39 M. E. BERMAN, Y. XIE, and W. A. MULLER, *J. Immunol.* **1996**, *156*, 1515–1524.

40 I. N. BIRD, J. H. SPRAGG, A. H. AGER, and N. MATTHEWS, *Immunology* **1993**, *80*, 553–560.

41 A. BERNARD, *CD99 workshop panel report*, Leukocyte Typing VI. Proceedings of the VIth International Leukocyte Differentiation Antigen Workshop, Kobe, Japan, 1996, London, 1997.

42 G. BERNARD, D. ZOCCOLA, M. DECKERT, J.-P. BREITTMAYER, C. AUSSEL, and A. BERNARD, *J. Immunol.* **1995**, *154*, 26–32.

43 G. BERNARD, V. RAIMONDI, I. ALBERTI, M. POURTEIN, J. WIDJENES, M. TICCHIONI, and A. BERNARD, *Eur. J. Immunol.* **2000**, *30*, 3061–3065.

44 J.-H. HAHN, M. K. KIM, E. Y. CHOI, S. H. KIM, H. W. SOHN, D. I. HAM, D. H. CHUNG, T. J. KIM, W. J. LEE, C. K. PARK, H. J. REE, and S. H. PARK, *J. Immunol.* **1997**, *159*, 2250–2258.

45 Z. MAMDOUH, X. CHEN, L. M. PIERINI, F. R. MAXFIELD, and W. A. MULLER, *Nature* **2003**, *421*, 748–753.

46 A. J. HUANG, J. E. MANNING, T. M. BANDAK, M. C. RATAU, K. R. HANSER, and S. C. SILVERSTEIN, *J. Cell. Biol.* **1993**, *120*, 1371–1380.

47 E. A. HIXENBAUGH, Z. M. GOECKELER, N. N. PAPAIYA, R. B. WYSOLMERSKI, S. C. SILVERSTEIN, and A. J. HUANG, *Amer. J. Physiol.* **1997**, *273*, H981–H988.

48 J. G. GARCIA, A. D. VERIN, M. HERENYIOVA, and D. ENGLISH, *J. Appl. Physiol.* **1998**, *84*, 1817–21.

49 H. SAITO, Y. MINAMIYA, M. KITAMURA, S. SAITO, K. ENOMOTO, K. TERADA, and J. OGAWA, *J. Immunol.* **1998**, *161*, 1533–1540.

50 M. W. WAKELIN, M.-J. SANZ, A. DEWAR, S. M. ALBELDA, S. W. LARKIN, N. BOUGHTON-SMITH, T. J. WILLIAMS, and S. NOURSHARGH, *J. Exp. Med.* **1996**, *184*, 229–239.

tistep process involving several families of adhesion molecules such as se-lectins, integrins and members of the immunoglobulin (Ig) superfamilies [51]. In this context, chemokines presented on the luminal surface of the endothelium trigger the activation of integrins on the cell surface of rolling lymphocytes. Activated integrins mediate a tight adhesion of the lympho-cytes to the endothelial cells, a prerequisite for the diapedesis through the endothelial cell layer into the underlying tissue. Based on their expres-sion pattern, chemokines can broadly be divided into two groups. Homeo-static chemokines such as CCL19 (ELC), CCL21 (SLC), and CXCL13 (BLC) are usually expressed constitutively in discrete lymphoid micro-environ-ments and involved in the physiological trafficking of immune cells [5]. By contrast, inflammatory chemokines such as CCL3 (MIP-1α, CCL20 (MIP-3α, and CXCL8 (interleukin-8) are induced or up-regulated by inflam-

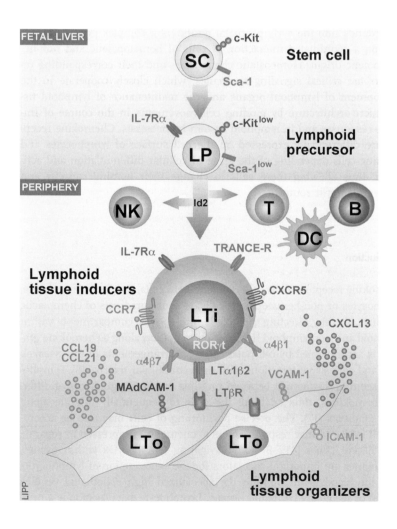

matory stimuli and ensure the recruitment of effector cells at sites of infection and inflammation [43]. The two chemokine receptors preferentially discussed in this chapter, namely CCR7, which binds CCL19 and CCL21, and CXCR5, which binds CXCL13, control the entry of lymphocytes and dendritic cells (DC) into secondary lymphoid organs. In addition, they are guiding these cells into specialized microcompartments, which are generally known as T cell-rich interfollicular and B cell-rich follicular zones within all secondary lymphoid organs [34].

Remarkably, development of lymphoid organs, which is a tightly controlled process that takes place during embryogenesis and extends into the early postnatal period, is regulated by similar adhesion mechanisms and involves the same chemokines and chemokine receptors responsible for lymphocyte recirculation in the adult organism [33]. Mice with impaired lymphoid organogenesis provided the first hints about the pathways critically involved in lymph node (LN) and intestinal Peyer's patch (PP) development. These mice, which either completely or partially lack LNs and PPs, are deficient for genes belonging to such diverse families as cytokines, chemokines, adhesion molecules, intracellular signaling molecules, and transcription factors. Although the variety of their phenotypes already indicated that lymph nodes and Peyer's patches differ in their requirements for certain factors during organogenesis, but have many features in common making it feasible to delineate a general model of the underlying cellular and molecular mechanisms for this process [18, 23, 30, 33].

◀───

Fig. 14.1 Model of lymph node and Peyer's patch development by lymphoid tissue inducer (LTi) cells expressing the chemokine receptors CXCR5 and CCR7. During embryonic development LTi cells are generated from a population of IL-Rα^+Sca-1^{low} c-Kit^{low} fetal liver lymphoid precursor cells (LP), which are derived from stem cells (SC) and can give rise to all lymphoid lineages. Expression of the nuclear orphan receptor RORγt is essential for the differentiation and/or survival of LTi cells. The accumulation of LTi cells is probably driven by a positive feedback loop. LTi cells stimulated via IL-7Rα or TRANCE-R up-regulate surface expression of LT$\alpha_1\beta_2$. The stimulation of mesenchymal lymphoid tissue organizer cells (LTo) in the developing lymph node or Peyer's patch anlage by LTi cells through LTβR induces the expression of adhesion molecules, such as VCAM-1 and ICAM-1, and homeostatic chemokines, including CXCL13, CCL21, and CCL19. In turn, CXCL13 leads to the activation of $\alpha_4\beta_1$ integrin on LTi cells, which reinforces the interaction of inducer and organizer cells and facilitates continuous signaling through LTβR. CCL19 and CCl21 also act on the LTi cells by binding to its cognate receptor, CCR7, and activating $\alpha_4\beta_7$ integrin, which is expressed by LTi cells. Binding of $\alpha_4\beta_7$ integrin to its counterreceptor MAdCAM-1 may contribute to the interaction of LTi cells and LTo cells, and may also play a role in the extravasation of LTi cells at sites of lymph node and Peyer's patch development. The accumulation of LTi cells and their tight interaction with mesenchymal LTo cells eventually reaches a critical size or level of chemokine expression that allows lymph node and Peyer's patch development to proceed to the next stage which includes the immigration of other cell types such as B cells, T cells, and macrophages, as well as their organization into distinct micro-environments. (This figure also appears with the color plates.)

14.2
Lymphoid Organogenesis

In the mouse embryo, lymphatic system development can be detected as early as embryonic day (E) 11.5 when lymph sacs are formed by endothelial cells budding from the veins. The commitment of venous endothelial cells to the lymphatic lineage is marked by the polarized expression of the homeobox gene Prox-1 in different regions of the cardinal vein [55]. Endothelial cell sprouting into the surrounding tissues and organs leads to the development of the lymphatic vasculature, a process that is finished by E 15.5. Lymph node anlagen are formed by mesenchymal connective tissue protrusion into the growing lymph sacs. The development of lymph node and Peyer's patch anlagen requires the interaction of two specialized cell populations: $CD4^+CD3^-IL\text{-}7R\alpha^+$ lymphoid tissue inducer cells (LTi) of hematopoietic origin and mesenchymal lymphoid tissue organizer cells, which I am provisionally referring to as LTo cells (Fig. 14.1). Initially, LTi cells accumulate at sites of future LN and PP development where they interact in a complex fashion with the LTo cells. Driven by the expression of chemokines and adhesion molecules, clusters of inducer and organizer cells might expand autonomously until, at a later stage of development, lymphoid and myeloid cells enter the developing organ from the circulation. Studies in mice with defects in the development of secondary lymphoid organs have long indicated a central role for LTi cells in lymph node and Peyer's patch development. During embryonic development LTi cells can be identified through expression of the orphan nuclear hormone receptor $ROR\gamma t$, which is indispensable for the generation and/or survival of this cell population. As a consequence, mice deficient in $ROR\gamma t$ expression are defective in the initiation of lymph node and Peyer's patch organogenesis [12]. LTi cells represent a unique population of oligolineage progenitors, which can be detected as early as E 12.5–13.5 at sites of future lymph node development [57] and differentiated *in vitro* into $NK1.1^+$ natural killer cells or $CD11c^+$ antigen-presenting DC but not into B or T cells [32]. They are characterized by the expression of adhesion molecules such as the $\alpha_4\beta_7$ and $\alpha_4\beta_1$ integrins, the IL-7 receptor α (IL-7Rα), the TRANCE receptor and the chemokine receptors CXCR5 and CCR7 (Fig. 14.1). LTi cells are probably derived from a population of $IL\text{-}7R\alpha^+Sca\text{-}1^{low}c\text{-}Kit^{low}$ fetal liver lymphoid precursor cells, the counterpart of common lymphoid progenitor cells in adult bone marrow [31, 56]. Subsequently, stromal cell expression of VCAM-1 has been described within the same area. Peyer's patch development, although starting later during embryogenesis, appears to follow a similar path. Peyer's patch anlagen are first detectable around day 16 of gestation in the ante-mesenteric site of the intestine as clusters LTi cells together with mesenchymal cells expressing VCAM-1 and ICAM-1 [19]. Later in development, other cells of hematopoietic origin such as B and T cells and macrophages colonize the developing lymphoid organs.

14.2.1
The Chemokine Receptors CXCR5 and CCR7 in Lymph Node and Peyer's Patch Development

CXCR5 was the first chemokine receptor to be associated with the development of secondary lymphoid organs. Mice deficient for CXCR5 or its ligand CXCL13 are similarly defective in lymphoid organogenesis. The development of Peyer's patches is partially impaired as they are often present, although at drastically reduced numbers. Moreover, these mice show a complex pattern of defects in lymph node development [3, 15]. $CXCR5^{-/-}$ and $CXCL13^{-/-}$ mice frequently lack inguinal, iliac, and parathymic lymph nodes, whereas facial, superficial cervical, and mesenteric lymph nodes are always retained. In addition, sacral and caudal lymph nodes can only be identified sporadically in $CXCR5^{-/-}$ and $CXCL13^{-/-}$ mice. In contrast to $CXCR5^{-/-}$ and $CXCL13^{-/-}$ mice, lymphoid organ development appears to be largely normal in mice deficient for the chemokine receptor CCR7 [16]. Similarly, lymph node development is little affected in paucity of lymph node (plt) mice, which lack expression of CCL19 and CCL21 in secondary lymphoid organs [36]. However, ectopic expression of CXCL13, CCL21 or CCL19 in pancreatic islets leads to the local formation of lymphoid-like tissue with structural features reminiscent of secondary lymphoid organs such as segregated B and T cell zones that are penetrated by blood vessels resembling HEVs [13, 25]. In view of this observation, it has been suggested that CCR7 may act in conjunction with CXCR5 during secondary lymphoid organ development. By generating mice double deficient for CXCR5 and CCR7 either onto a pure C57BL/6 or a mixed 129/Sv × BALB/c background, we found that CCR7 is indeed involved in secondary lymphoid organ development [33]. Similar results have been observed in $CXCL13^{-/-}plt/plt$ mice defective in the expression of all known ligands for CCR7 and CXCR5 within secondary lymphoid organs [24]. Both strains show a comparable phenotype lacking almost all lymph nodes. Surprisingly, mesenteric lymph nodes are still present. Together with the finding that Peyer's patch development is similarly affected in $CXCR5^{-/-}$ and $CXCR5^{-/-}CCR7^{-/-}$ mice, these observations suggest that CCR7 cooperates with CXCR5 in peripheral lymph node but not in Peyer's patch and mesenteric lymph node development. In this context, CXCR5 appears to functionally substitute for CCR7 in peripheral lymph node development whereas CCR7 can only partially compensate for lacking CXCR5. Hence, the factor substituting for CXCR5 in mesenteric lymph node and Peyer's patch development still has to be identified.

Although it remains to be determined whether the migration of the very first LTi cells into the lymphoid organ anlagen is regulated through CXCR5 and CCR7, the examination of mice with defects in the development of secondary lymphoid organs allows to draw a model for the interaction of LTi cells with mesenchymal LTo cells and the role of homeo-static chemokines on the accumulation of LTi cells in the developing lymphoid organ anlagen (Fig. 14.1). According to this model, LTi cells receive an initial stimulus via IL-7Rα or TRANCE-R. Although the nature of the ligand for IL-7Rα and the source of TRANCE are still under debate, both stimuli lead to up-regulation of membrane-anchored LT$\alpha_1\beta_2$ heterotrimer on LTi cells, which belongs to the TNF/LT (tumor necrosis factor/lymphotoxin) family of cytokines and binds to the LTβ receptor (LTβR). Mesenchymal organizer cells stimulated by IL-7Rα^+ LTi cells via LTβR express adhesion molecules, including VCAM-1 and ICAM-1, and chemokines, such as CXCL13 and CCL19 [19]. Chemokine and adhesion molecule expression by LTo cells may reinforce the interaction with LTi cells. In this connection, it has been shown that CXCL13 triggers the activation of $\alpha_4\beta_1$ integrin on LTi cells in Peyer's patch development [14]. Activated $\alpha_4\beta_1$ integrin may bind to its counterreceptor VCAM-1, the expression of which is jointly induced with CXCL13 in mesenchymal organizer cells stimulated by LT$\alpha_1\beta_2$ [19]. This hypothesis is supported by the observation that blocking of the $\alpha_4\beta_1$ integrin-VCAM-1 interaction during embryogenesis with an antibody specific for the activated form of this integrin diminishes the number and size of Peyer's patches. The critical role of LTi cells for Peyer's patch formation has been further proven by the finding that Peyer's patch development is rescued upon adoptive transfer of fetal spleen derived CD4$^+$CD3$^-$ LTi cells from wildtype animals to CXCR5$^{-/-}$ recipients [14]. It is conceivable that CCL19 participates in the accumulation of IL-7Rα^+ LTi cells or the interaction of inducer and organizer cells as well. Similar to CCL21, CCL19 might be able to trigger the activation of $\alpha_4\beta_7$ integrin, which is expressed by LTi cells while its counterreceptor MAdCAM-1 is detectable on LTo cells (Fig. 14.1). In any case, the initial seeding of IL-7Rα^+ LTi cells into the developing lymph node involves binding of $\alpha_4\beta_7$ integrin to MAdCAM-1 [32], whose expression can be detected in all lymph nodes during embryonic development and a period of up to 2 days after birth.

14.2.2
A Positive Feedback Loop Drives the Initial Phase
of Lymphoid Organ Development

The accumulation of LTi cells within lymph node anlagen is most probably driven by a positive feedback loop (Fig. 14.1). IL-7Rα LTi cells stimulated via IL-7Rα or TRANCE-R express LT$\alpha_1\beta_2$ which binds to LTβR expressed on mesenchymal organizer cells. Activation of LTβR, which triggers the alternative NfκB signaling pathway [11, 50], induces the expression of adhesion molecules and chemokines by LTo cells, which in turn reinforce the interaction of organizer and inducer cells. Moreover, chemokines secreted by LTo cells may contribute to the clustering of inducer

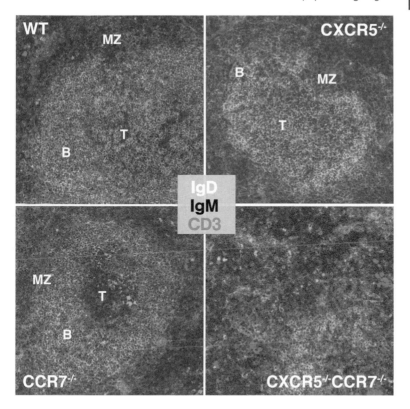

Fig. 14.2 Altered splenic architecture in mice deficient for the chemokine receptors CXCR5 and CCR7. Cryostat sections of the spleen of C57BL/6 mice were stained with PE-labeled anti-CD3 (red), FITC-labeled anti-IgD (green) and biotinylated anti-IgM/ Streptavidin-Cy5 (blue) to visualize the architecture of primary B cell follicles and of T cell-rich areas. In wild-type mice, the T cell zone is typically situated around a central arteriole with B cell rich areas eccentrically positioned at the periphery of the T cell zone. Marginal zone (MZ) B cells ($IgM^{hi}IgD^{lo}$) appear in wild-type mice as a rim around the polarized clusters of follicular B cells, which are mainly $IgM^{lo}IgD^{hi}$. In CXCR5$^{-/-}$ mice, B cells fail to organize in polarized clusters. The T cell zone is instead surrounded by a small ring of $IgM^{-}IgD^{hi}$ B cells, which is in turn encompassed by a thickened ring of $IgM^{hi}IgD^{lo}$ marginal zone B cells. CCR7$^{-/-}$ mice generally lack prominent T cell zones in the white pulp. The boundaries between the rudimentary T cell areas and the B cell zones are poorly demarcated. In addition, clusters of $IgM^{hi}IgD^{lo}$ B cells are frequently visible close by the central arteriole. In contrast, T cell zones and B cell zones are largely disorganized in CXCR5$^{-/-}$CCR7$^{-/-}$ mice. Similar to the spleen of mice lacking the expression of CCR7, T cells are randomly distributed throughout the red pulp. (WT, wild type; T, T cell zone; B, B cell zone; MZ, marginal cell zone; Original magnification: ×200). (This figure also appears with the color plates.)

and organizer cells by attracting additional IL-7Rα^+ inducer cells into lymph node and Peyer's patch anlagen. Eventually, the size of the cell cluster or the level of chemokine expression reaches a certain threshold, thereby allowing lymphoid organ development to proceed to the next stage. This would include the formation of HEVs followed by the colonization of the developing lymphoid organ with other cell types such as macrophages and lymphocytes, the segregation of these cells into distinct areas, and consequently, the formation of the typical micro-architecture of the lymphoid organ. This scenario is reminiscent of a positive feedback loop that drives the formation of B cell follicles in the spleen [3]. CXCL13 induces B-lymphocytes to upregulate LT$\alpha_1\beta_2$ that promotes FDC (follicular dendritic cell) development and expression of homeostatic chemokines including CXCL13. In turn, CXCL13 reinforces the expression LT$\alpha_1\beta_2$ on B cells and probably attracts additional B cells into the follicle. Consequently, mice lacking the expression of CXCR5 or CXCL13 do not develop intact primary B cell follicles (Fig. 14.2 [3, 15]).

Homeostatic chemokines and chemokine receptors provide an example for partial redundancy in lymph node and Peyer's patch development. Although CXCR5 is indispensable for the development of inguinal, parathymic and iliac lymph nodes, the formation of Peyer's patches and certain other lymph nodes is only partially affected in mice deficient for CXCR5 or CXCL13 [3, 15]. Remarkably, a deficiency of CCR7 barely affects secondary lymphoid organ development [16, 36]. However, mice lacking the expression of both receptors or, alternatively, mice that are deficient for the expression of all corresponding ligands in secondary lymphoid organs are missing all except the mesenteric lymph nodes [24, 33, 37]. Hence, CXCR5 and CCR7 cooperate intimately during lymph node development. However, CCR7 cannot fully compensate for CXCR5 whereas CXCR5 appears to do so for CCR7. With respect to the model of lymphoid organ development and especially the positive feedback loop it is conceivable that each of the two chemokine receptors could potentially mediate the accumulation of LTi cells. Noticeably, CCR7 seems to not contribute to mesenteric lymph node and Peyer's patch development, as the development of these organs is similarly defective in CXCR5$^{-/-}$ and CXCR5$^{-/-}$CCR7$^{-/-}$ mice. Apparently, homeostatic chemokines are critically involved in various steps of lymph node and Peyer's patch development, and although their function may be partially redundant they are indispensable for the initiation of secondary lymphoid organ development. In contrast, less is known about the formation of the characteristic micro-architecture during later steps in lymph node and Peyer's patch organogenesis. Although the molecular requirements of these processes are beginning to be revealed, it is still difficult to draft a model that covers the complex pattern of cell-cell interactions.

14.3
Organization of Lymph Node and Splenic Micro-architecture by Homeostatic Chemokines

In addition to their role in lymph node and Peyer's patch development, the chemokine and the TNF/LT systems are also critical for the maintenance of the lymphoid tissue architecture of lymph nodes, Peyer's patches and spleen.

14.3.1
The Homeostatic Chemokine System in the Formation of the Splenic Micro-architecture

The spleen, with its completely different structure compared to LN and PP, follows a distinct developmental pathway. However, many factors that are critically involved in LN and PP morphogenesis are also required for the formation of the splenic micro-architecture. Apart from partially lacking Peyer's patches and peripheral lymph nodes (see above), mice deficient for CXCR5 show severe alterations in the morphology of secondary lymphoid tissues [15]. In wild-type mice, the T cell zone is typically situated around a central arteriole with B cell rich areas eccentrically positioned at the periphery of the T cell zone (Fig. 14.2). Within the spleen of CXCR5$^{-/-}$ mice, the formation of primary B cell follicles is severely impaired. T cell zones are surrounded by a small rim of newly formed IgD$^+$IgM$^-$ B cells whereas the prominent areas of IgD$^+$IgM$^+$ recirculating B cells, as observed in wild-type mice, are missing (Fig. 14.2). Instead, rudimentary follicle structures are surrounded by an enlarged ring of marginal zone IgM^{hi}IgD^{lo} B cells compared to the small rim of marginal zone B cells that can usually be observed in wild-type mice (Fig. 14.2). Structural alterations in lymph nodes are less obvious in CXCR5$^{-/-}$ mice. Lymph nodes normally contain two different types of B cell rich zones that can be distinguished by the expression of CXCL13 and the presence of FDC networks. CXCR5$^{-/-}$ mice still develop segregated B cell areas lacking FDCs and expression of CXCL13 and B cell migration into these B cell rich zones is independent of CXCR5, possibly involving the chemokine receptors CXCR4. However, CXCR5$^{-/-}$ mice do not show B cell follicles, i.e. B cell rich zones harboring FDC networks. B cell homing to follicles critically depends on CXCR5 as demonstrated by the failure of CXCR5$^{-/-}$ B cells to home to these follicles in wild-type recipients [2]. Taken together, the phenotypes of mice deficient for CXCL13 and CXCR5 closely resemble each other. B cells fail to organize into polarized follicular clusters in lymph nodes and spleen, and FDCs were reported to be absent in these organs. In addition, the defect in developing lymph nodes and Peyer's patches is largely identical, which further supports the hypothesis that CXCL13 is the unique ligand for CXCR5.

Although mice lacking expression of CCR7 develop an almost complete set of lymph nodes and Peyer's patches, B cell and T cell distribution in the periphery and secondary lymphoid organs are significantly altered [16]. CCR7$^{-/-}$ mice display an increased number of naïve T cells in peripheral blood, the spleen and the bone marrow, whereas naïve T cell counts in mesenteric lymph nodes, peripheral lymph nodes and Peyer's patches are drastically reduced. However, lymphocyte accumulation in the spleen is restricted to the red pulp and marginal sinuses with large clusters of T cells spread throughout these regions (Fig. 14.2). In contrast to wild-type mice, naïve T cells are mostly excluded from the whit pulp in CCR7$^{-/-}$ mice. Only few T cells can be identified around the central arteriole and these mostly show a memory phenotype (Fig. 14.2). However, B cell entry into the PALS (peri-arteriolar-lymphoid sheath) appears to be largely normal in CCR7$^{-/-}$ mice, where they form polarized B cell follicles (Fig. 14.2). In addition to the altered architecture of the spleen, lymph nodes show an irregular distribution of B and T cells within the paracortex whereas Peyer's patches are generally lacking T cell rich zones but instead display a small rim of T cells around B cell follicles. Indeed, B and T lymphocytes lacking expression of CCR7 are impaired in homing to lymph nodes and Peyer's patches. Upon adoptive transfer, B and T cells derived from CCR7$^{-/-}$ mice are severely hampered in colonizing lymph nodes and Peyer's patches of wild-type mice [16]. Compared to lymphocytes derived from wild-type donors, elevated numbers of CCR7$^{-/-}$ B and T cells can be detected in peripheral blood and the spleen of wildtype recipients; however, adoptively transferred CCR7$^{-/-}$ T cells are largely excluded from the white pulp and instead located in the red pulp and sinuses of the spleen. In contrast, CCR7$^{-/-}$ B cells readily enter the splenic white pulp of wild-type recipients.

14.3.2
Lessons from CXCR5 and CCR7 Double Deficient Mice

In mice double deficient for CXCR5 and CCR7 the splenic and mesenteric lymph node architecture is even more severely disturbed in comparison with mice lacking expression of either CXCR5 or CCR7 (Fig. 14.2). Similar to CCR7$^{-/-}$ mice, these mice show an enlarged spleen while mesenteric lymph nodes are reduced in size. Cell counts of follicular and marginal zone B cells are comparable to wild-type and single-knockout animals. However, no regular B cell follicles and T cell zones could be detected in secondary lymphoid organs. Instead, B and T cells cluster within the reticular fibroblast network of the red pulp of the spleen (Fig. 14.2).

Despite the dominant role of CCR7 in regulating T cell and DC entry into secondary lymphoid organs, its function for B cell recirculation has long been disregarded. In principle, B cells enter the spleen through the marginal zone by crossing the marginal sinus and reaching the outer PALS [53]. Recirculating B cells, which express high levels of CXCR5, migrate into B cell follicles where they stay for a couple of hours before they are continuing the migration along the outer PALS and via the bridging channels into the red pulp. Consequently, B cell follicle formation is defective in CXCR5$^{-/-}$ mice though B cell entry into the PALS is not compromised. In contrast, B cell entry into the white pulp and follicle formation are not affected in CCR7$^{-/-}$ mice; however, B cells entering the PALS quickly proceed into the B cell follicles, whereas B cells derived from wild-type mice seem to linger in the outer PALS. This suggests that CCR7 mediates B cell retention for a certain period in the PALS before they continue to migrate into B cell follicles. Surprisingly, analysis of CXCR5$^{-/-}$CCR7$^{-/-}$ mice revealed that CCR7 cooperates with CXCR5 in mediating B cell entry into the splenic white pulp since B lymphocytes isolated from these double knockout mice and adoptively transferred into wild-type recipients fail to enter the PALS and instead are retained in the red pulp [37]. Therefore, B cells seem to use either CXCR5 or CCR7 to locate to the white pulp in the spleen. In summary, CXCR5 and CCR7 appear to play a dual role in B cell recirculation through the spleen by i) mediating the entry of B cells into the white pulp and ii) controlling B cell positioning within the white pulp. In contrast, B lymphocyte entry into lymph nodes depends on CCR7 and the chemokine receptor CXCR4 [38]. The contribution of CCR7 to B cell entry into lymph nodes appears to be partially redundant with that of CXCR4 whereas the role of CXCR4 was completely redundant with that of CCR7, establishing CCR7 as the major homing receptor for B cell migration into lymph nodes. However, CXCR5 appears to be important for B cell homing to Peyer's patches. There, CXCR5 allows B cells to directly enter B cell follicles at the sites of follicular high endothelial venules where CXCL13 is deposited on the luminal surface of the endothelial cells, triggering the firm adhesion of B cells in these sectors (Fig. 14.3).

14.4
Chemokine-controlled Migration of Lymphocytes and Dendritic Cells during Immune Responses

Mounting an effective immune response requires the sequential interaction of B cells, T cells, and DCs within secondary lymphoid organs [10]. Immature DCs in the periphery, such as Langerhans cells in the skin or DC located in the sub-epithelial dome (SED) of the small intestine (Fig. 14.3), possess the ability to effectively take up proteins by phagocytosis, to process them, and, finally, to present immunogenic peptides derived from these proteins in association with major histocompatibility complex (MHC) molecules on the cell surface. Antigen uptake in con-

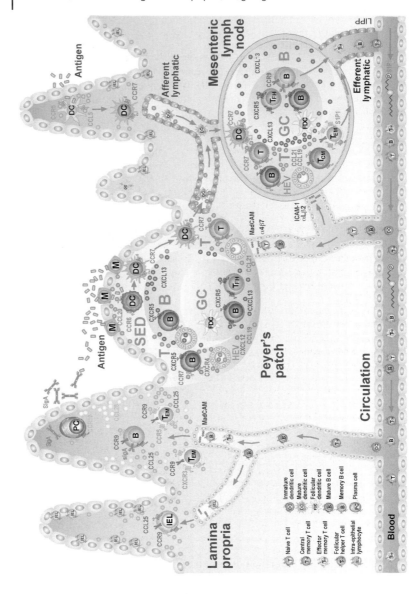

Fig. 14.3 Chemokine-driven lymphoid cell migration in mucosal immunity. In terms of immune functions the tissue of the small intestine can be divided into the initiation compartment defined as gut-associated lymphoid tissue (GALT), which includes the organized structures of Peyer's patches (PP) and mesenteric lymph nodes (MLN), and into the effector compartment consisting of the lamina propria (LP) with scattered effector T and B cells, and of the intra-epithelial lymphocytes (IEL) embedded in the epithelial cell layer of the intestinal wall.

The cell type and tissue-specific expression of chemokines and their corresponding receptors plays a decisive role in lymphocyte and dendritic (DC) cell homing to mucosal tissues. The entry of naïve, recirculating lymphocytes into Peyer's patches (PP) is mediated by the homeostatic/lymphoid chemokines CXCL12, CXCL13, CCL19 and CCL21. Naïve, recirculating T cells leave the bloodstream in high endothelial venules, (HEVs) crossing the T cell rich area of PPs. T cell entry into the T cell rich area depends on expression of the chemokine receptors CCR7 (which binds CCL19 and CCL21) and CXCR4 (which binds CXCL12) on T cells. In contrast, naïve B cells predominantly employ CXCR5 (which binds CXCL13) to leave the bloodstream in follicular high endothelial venules [38]. B cells may also utilize CXCR4 or CCR7 to enter PPs; however, both receptors cannot compensate for a lack of CXCR5. Moreover, CCR7 and CXCR5 are responsible for the organization of secondary lymphoid tissue into B cell-rich and T cell-rich zones as the ligands for these receptors are expressed in the T- and B cell zone, respectively [33].

DCs depend on CCR6 to enter the sub-epithelial dome (SED) region of PPs [9]. The ligand for CCR6, CCL20, is constitutively expressed by epithelial cells. Thereby, CCL20 directs CD11c$^+$CD11b$^+$ DCs in close proximity to M cells, that actively acquire antigen from the intestinal lumen. Following antigen uptake, maturing DCs upregulate CCR7, which enables these cells to migrate via the afferent lymphatic vessels into the inter-follicular T cell rich regions of PPs or the T cell zones of MLN.

T cell activation by DCs confers to T cells the ability to home to non-lymphoid organs. However, Peyer's Patch DCs establish gut tropism in activated memory/effector T cells which preferentially home to mucosal effector sites [7]. Homing of central memory/effector memory T cells to the lamina propria of the small intestine involves CCR9, which is expressed by virtually all T cells present in the LP. The ligand for CCR9, CCL25, is constitutively expressed by epithelial cells of the lower villi and crypts within the small intestine and deposited on the luminal surface of epithelial cells in intestinal venules. In addition, expression of receptors for inflammatory chemokines, such as CXCR3 and CCR5, contribute to the homing of effector T cells (T$_{EM}$) to the LP.

IgA-secreting plasma cells (PC), which are generated in germinal centers (GC) of PP and MLN, utilize CCR9 to home to the LP of the small intestine [39]. Once PC have entered the mucosal effector tissue, CCR9 becomes downregulated by these cells; therefore CCL25 is not necessary for the retention of plasmablasts in the LP of the small intestine. In addition, PC may use CCR10 to enter the LP. Most IgA-secreting PC express CCR10 and show a chemotactic response towards CCL28, a ligand for CCR10 that is constitutively expressed by epithelial cells in the small and large intestine. A subset of circulating memory surface IgA$^+$ B cells developed in GC of PP and MLN expresses CCR9; therefore CCR9 might be involved in memory B cell homing to the small intestine. Together with receptors for inflammatory chemokines, CCR9 may also play a role for the recruitment of intra-epithelial lymphocytes (IEL) to the epithelium, which are more frequently CD8$^+$ T cells.

After activation in lymphoid organs, lymphocytes must again return to circulation to reach effector sites. Upregulation of the sphingosine-1-phosphate receptor S1P1 in memory/effector T and B cells in course of ongoing immune responses enables these cells to egress from secondary lymphatic organs and enter the circulation [28]. (This figure also appears with the color plates.)

junction with stimulation by cytokines at the site of infection leads to the activation of DCs and triggers their maturation. The differentiation of DCs into antigen-presenting cells is accompanied by changes in the expression pattern for chemokine receptors on the cell surface [46]. The downregulation of receptors for inflammatory chemokines in conjunction with the upregulation of CCR7 enables DCs to enter the lymphatic system as lymphatic endothelium in peripheral tissues expresses CCL21. Within the draining lymphoid organs, mature antigen-presenting DCs localize in the T cell zones where DCs efficiently encounter naive lymphocytes (Fig. 14.3).

Naive antigen-inexperienced lymphocytes differentiated from lymphoid progenitors in primary lymphoid organs constantly recirculate between secondary lymphoid organs in search for their cognate antigen. The entry of naive lymphocytes into lymph nodes and Peyer's patches occurs at the site of HEVs. Lymphocytes passing through HEVs occasionally come in close contact with the vessel wall where they slow down ("roll") as a result of low affinity interactions between L-selectin (CD62L) expressed on the cell surface of lymphocytes and its ligands present on the surface of endothelial cells. Chemokines either produced by the endothelial cells of HEVs or transcytosed from the abluminal side of the endothelial cell layer are presented to lymphocytes on the luminal surface of the endothelium [52]. Chemokines trigger the firm adhesion of lymphocytes to endothelial cells by inducing rapid changes in integrin affinity and avidity. Chemoattractant-induced conformational changes of integrins increase their affinity towards ligands such as intercellular adhesion molecules (ICAMs) whereas the clustering of integrins on the surface of lymphocytes increases the avidity [8]. CCL19 and CCL21 mediate the firm adhesion of lymphocytes via β2-integrin binding to ICAM-1 (Fig. 14.3). In addition, CCL21 triggers $\alpha4\beta7$-integrin binding to the mucosal addressin cell adhesion molecule-1 (MAdCAM-1, Fig. 14.3) [40]. The chemokine-mediated firm adhesion of lymphocytes to endothelial cells is a prerequisite for the subsequent diapedesis into the underlying lymphoid tissue.

T cell arrest in HEVs of secondary lymphoid organs is mediated by CCL21 and CCL19 binding to CCR7 on the surface of the T cells [4]. The chemokine receptor CXCR4 may also contribute to T-cell homing in lymphoid organs, but it cannot substitute for CCR7 [38]. CD4$^+$ and CD8$^+$ T cells that have entered secondary lymphoid organs migrate through the T cell zone where they scan the surface of DCs for their cognate antigen presented by MHC class II and class I molecules, respectively. Provided that they receive appropriate co-stimulatory signals, T cells start to proliferate and differentiate into effector/memory cells, a process that needs a couple of days to be completed. Interestingly, it was shown that the activation of the T cell receptor (TCR) in vitro leads to a transient up-regulation of CCR7 [22, 44], suggesting that recently stimulated naïve and memory T cells might be temporarily retained within the interfollicular T cell rich area via CCR7. Further T-cell differentiation is accompanied by major alterations in the expression pattern of adhesion molecules and chemokine receptors, resulting in the generation of memory and effector T cells with distinct homing capacities.

B cells entering lymphoid organs differ from T cells in that they rapidly pass through the T-cell zones to enter lymphoid follicles. B cells are also more versatile than T cells as firm adhesion to endothelial cells can be mediated by CCR7, CXCR4, or CXCR5 [38]. This explains why B cell migration into secondary lymphoid organs is much less affected in mice lacking either CCR7 or its ligands CCL21 and CCL19. CXCR4-mediated B-cell homing has been demonstrated for lymph nodes and Peyer's patches whereas the use of CXCR5 appears to be restricted to follicular venules in Peyer's patches. This observation is in line with an earlier report showing that sites of firm T- and B-cell adhesion are segregated in Peyer's patch HEVs [54] (Fig. 14.3). Endothelial cells do not express CXCL13, which is instead transcytosed from the abluminal side to the luminal surface of follicular HEVs. For this reason, B cells may also directly enter Peyer's patches B-cell follicles without passing through the T-cell zone.

B-cell priming by antigen uptake may occur anywhere along their route to the follicle. Antigen-primed B cells up-regulate CCR7 while maintaining CXCR5 on the cell surface. The simultaneous expression of CXCR5 and CCR7 enables these cells to localize to the edges of follicles to find and make cognate interaction with primed T cells. Thus, B-cell positioning in the outer T zone is determined by the balanced responsiveness of B cells to the ligands for CCR7 and CXCR5 as these chemokines are produced in the T-cell zone and B cell follicle, respectively [41]. Cognate T cell-B cell interactions initiate B cell proliferation and differentiation into antibody-secreting plasma cells (PC, Fig. 14.3). B cell differentiation in extrafollicular foci results in the formation of short-lived plasma cells responsible for an early supply with antibodies [20]. Alternatively, B cells differentiate within germinal centers (GCs) where immunoglobulin gene rearrangements and affinity selection lead to the generation of high affinity antibody producing plasma cells and memory B cells [26]. B-cell differentiation within GCs depends on a close interaction of B cells with FDCs, a specialized subset of Dcs, which is only present in follicles, as well as a subset of $CD4^+CXCR5^+$ T cells (T_{FH}) acting as B helper T cells (Fig. 14.3).

T cells activated by dendritic cells in the T-cell zone of secondary lymphoid organs clonally expand and differentiate into specialized memory/effector cells. Although considerable progress has been made (e.g., by the characterization of T helper cells of type 1 and 2), the phenotypic distinction between memory and effector T cell subsets has long been problematic. Only recently it has been recognized that expression patterns of molecules involved in the tissue-selective homing of lymphocytes, including the chemokine receptors CCR7 and CXCR5 and adhesion molecules such as CD62L, allow for a more precise description of functionally distinct effector/memory T cell populations. According to the expression of CCR7 and CD62L, two subsets of CD4$^+$CD45RA$^-$ memory/effector T-cells can be distinguished in human peripheral blood (Fig. 14.3): CD45RA$^-$CCR7$^+$CD62L^{hi} central memory T (T$_{CM}$) cells showing characteristics of true memory cells and CD45RA$^-$CCR7$^-$CD62L^{lo} effector memory T (T$_{EM}$) cells that more closely resemble classical effector cells [45]. CD62L^{lo} T$_{EM}$ cells can be induced to secrete cytokines such as IFN-γ, IL-4, IL-5 and IL-13, a hallmark of polarized effector memory cells [45]. They are thought to deliver immediate effector functions in the periphery at sites of infection and inflammation. Therefore, polarized T helper 1 (Th1) and T helper 2 (Th2) effector cells are supposed to represent subsets of CCR7$^-$ T$_{EM}$ cells. By contrast, CD62L^{hi} T$_{CM}$ cells are thought to represent memory cells that preferentially home to secondary lymphoid organs and that are capable of inducing enhanced immune responses upon secondary challenge with antigen. T$_{CM}$ and T$_{EM}$ cells have been implicated in a model of progressive differentiation of CD4$^+$ T cells [21, 35], assuming that CD4$^+$ T cells differentiate along a linear pathway from naive via T$_{CM}$ towards more differentiated T$_{EM}$ cells, which further develop into terminally differentiated Th1 or Th2 effector cells. According to this model, T-cell differentiation is determined by the duration and strength of the antigenic stimulus, co-stimulatory signals as well as the cytokine milieu and reflected by changes in the homing properties and effector functions. In favor of this model, recent studies in mice have demonstrated that memory/effector T cells can indeed be subdivided by function and localization in lymphoid and nonlymphoid tissues *in vivo* [27, 42]. IL-2 expressing CD4$^+$ memory T cells resembling T$_{CM}$ cells localize preferentially to secondary lymphoid organs whereas IFN-γ producing CD4$^+$ memory T cells resembling T$_{EM}$ cells were predominantly found in nonlymphoid tissues. It must be emphasized that the memory/effector T cell populations mentioned so far most likely comprise phenotypically and functionally heterogenous subpopulations. For instance, a subset of T$_{CM}$ cells in peripheral blood expresses the chemokine receptor CXCR5. This CCR7$^+$CXCR5$^+$ double-positive subpopulation we recently referred to as T$_{CM1}$ cells [34] does not produce effector cytokines. Within secondary lymphoid organs, the population of CXCR5$^+$CD4$^+$ T cells is significantly enlarged. Up-regulation of CXCR5 is accompanied by a down-regulation of CCR7, which allows these cells to enter B cell follicles. CD4$^+$CXCR5$^+$ T cells located within germinal centers express costimulatory molecules such as ICOS and CD40L and act as B helper T cells in that they promote the antibody secretion by B cells [6, 49]. For this reason, these cells have been named follicular T helper B (T$_{FH}$, Fig. 14.3) cells. However, the origin and fate of CXCR5$^+$CD4$^+$ T cells is still under discussion. It has been suggested that CXCR5$^+$CD4$^+$ T cells might be derived from the pool of T$_{CM}$ cells or, alternatively, that they might be generated independently of T_{CM} cells following T cell activation by DCs in the T cell areas of secondary lymphoid organs.

14.5
Concluding Remarks

The differentiation of lymphocytes and DCs in the course of development and an immune response is accompanied by phenotypical changes reflecting their newly acquired functional properties and homing preferences. For this reason it is tempting to identify functionally distinct cell subsets based on their homing properties; that is, the expression pattern of chemokine receptors such as CCR7, CXCR5 and CXCR4, integrins and adhesion molecules such as CD62L. This concept turned out to be useful for identifying tissue- and micro-environment-specific molecular address codes for effector and memory lymphocyte subsets within the T, B and dendritic cell compartments. To date, proof of this concept has been advanced essentially in understanding of the migratory and homing network in mucosal immunity (Fig. 14.3 and Tab. 14.1). However, in view of recent data on the $CD4^+$ T cell-compartment, for example, the concept of memory/effector T cell differentiation presented above may still be simplistic. Further investigations will be necessary to refine the model; for example, on the role of peripheral $CCR7^+CXCR5^+CD4^+$ T_{CM1} cells in effector cell generation. To this end the use of additional surface markers, some of which might be uncovered by means of gene expression profiling, may allow for a more precise description of distinct lymphocyte subsets. However, one has to keep in mind that lymphocyte homing is characterized by a certain degree of flexibility as, for example, effector cells are not completely excluded from lymphoid tissues. It is intriguing to speculate that homing to lymphoid organs may be obligatory at certain intermediate stages of effector-cell differentiation to receive sustained or additional activating signals for further differentiation.

The chemokine receptors CXCR5 and CCR7 are recognized as crucial factors in lymphoid system homeostasis and adaptive immunity. However, it is now clear that both receptors cooperate during lymph node development while Peyer's patch organogenesis probably involves an unknown factor that substitutes for CCR7. Interestingly, the close cooperation of CXCR5 and CCR7 on cells simultaneously expressing these receptors appears to be a recurrent motif that has been particularly well described for B cell entry, migration, and positioning in the spleen. In addition, besides its known role in stem cell and plasma cell migration into the bone marrow, CXCR4 has only recently been recognized as important molecule of embryonic development in different experimental animal systems. The homeostatic chemokines CXCL13, CCL19, and CCL21 are particularly relevant, because they are critically involved in the forma-

Tab. 14.1 Chemokine receptors and adhesion molecules controlling lymphocyte migration in mucosal immunity. For further explanation see Fig. 14.3 (GALT, gut-associated lymphoid tissue; DC, dendritic cell; T_{EM}, effector memory T cell; T_{FH}, follicular helper T cell; LN, lymph node; mLN, mesenteric LN; PP, Peyer's patch; HEV, high endothelial venules; SED, sup-epithelial dome; IEL, intra-epithelial lymphocytes; S1P1, Sphingosine-1-phosphate receptor 1; S1P, Sphingosine-1-phosphate)

Mucosal Address Codes

Receptor	Ligand	Compartment	Main functions
CCR7	CCL19/21	GALT	Entry and lodging of T cells and DCs to T cell area of PP and mLN; Organogenesis of LN
CXCR5	CXCL13	GALT	Entry and lodging of B cells and T_{TH} cells to follicles; Organogenesis of PP and LN
CXCR4	CXCL12	GALT	Entry of T (memory) and B cells to PP and mLN via HEVs (minor)
CCR6	CCL20	GALT	Migration of myeloid DCs (CD11b$^+$/CD11c$^+$) to SED of PP
CCR9	CCL25	Epithelium	Migration and adhesion of IELs to the small intestine epithelium
		Lamina propria	Migration of IgA$^+$ memory and plasma cells into the lamina propria
CCR10	CCL28	Lamina propria	Migration and adhesion of IgA$^+$ plasma cells
CCR5	CCL5	Epithelium	Recruitment of inflammatory and effector cells (monocytes, DCs, T_{EM}) into mucosal effector sites
CXCR3	CXCL 10	Lamina propria	
S1P1	S1P	GALT	Egress of T and B cells from PP and LN into circulation
$\alpha4\beta7$	MadCAM	All	Integrin-mediated adhesion to mucosal HEVs and LP endothelium

tion of tertiary lymphoid tissue, also known as ectopic follicles, a feature of chronic inflammatory processes which is linked with autoimmune and infectious diseases such as rheumatoid arthritis (RA), *Helicobacter pylori*-induced gastritis, and Sjögren's syndrome. In particular, expression of CXCL13 has been associated with the formation of ectopic lymphoid follicles in all three diseases [1, 29, 47]. The obvious similarities between lymph node development, follicle formation in secondary lymphoid organs and ectopic lymphoid neogenesis suggest that common molecular and cellular networks regulate these processes. In this context, it is of particular interest to know whether LTi cells might be also involved in the formation of ectopic lymphoid structures. Therefore, a precise understanding of lymphoid organ development and the migration processes involved will certainly be instructive in the understanding of the pathological mechanisms underlying these diseases.

14.6
Acknowledgments

I thank Sven Golfier for help with the confocal microscope and Uta Höpken, Gerd Müller, and Ariel Achtman for their helpful discussion and continuous support. The work was supported by the Bundesministerium für Bildung und Forschung/NGFN grant IE-S01T09 and European Union grant QLRT-2000-01205 (MEMOVAX).

14.7
References

1 Amft, N., Curnow, S. J., Scheel-Toellner, D., Devadas, A., Oates, J., Crocker, J., Hamburger, J., Ainsworth, J., Mathews, J., Salmon, M., Bowman, S. J., and Buckley, C. D., Ectopic expression of the B cell-attracting chemokine BCA-1 (CXCL13) on endothelial cells and within lymphoid follicles contributes to the establishment of germinal center-like structures in Sjögren's syndrome. *Arthritis Rheum* 2001, *44*, 2633 2641.

2 Ansel, K. M., Harris, R. B., and Cyster, J. G., CXCL13 is required for B1 cell homing, natural antibody production, and body cavity immunity. *Immunity* 2002, *16*, 67–76.

3 Ansel, K. M., Ngo, V. N., Hyman, P. L., Luther, S. A., Förster, R., Sedgwick, J. D., Browning, J. L., Lipp, M., and Cyster, J. G., A chemokine-driven positive feedback loop organizes lymphoid follicles. *Nature* 2000, *406*, 309–314.

4 Baekkevold, E. S., Yamanaka, T., Palframan, R. T., Carlsen, H. S., Reinholt, F. P., von Andrian, U. H., Brandtzaeg, P., and Haraldsen, G., The CCR7 ligand ELC (CCL19) is transcytosed in high endothelial venules and mediates T cell recruitment. *J Exp Med* 2001, *193*, 1105–1112.

5 Baggiolini, M. and Loetscher, P., Chemokines in inflammation and immunity. *Immunol. Today* 2000, *21*, 418–420.

6 Breitfeld, D., Ohl, L., Kremmer, E., Ellwart, J., Sallusto, F., Lipp, M., and Förster, R., Follicular B helper T cells express CXC chemokine receptor 5, localize to B cell follicles, and support immunoglobulin production. *J Exp Med* 2000, *192*, 1545–1552.

7 Campbell, D. J. and Butcher, E. C., Intestinal attraction: CCL25 functions in effector lymphocyte recruitment to the small intestine. *J Clin Invest* 2002, *110*, 1079–1081.

8 Constantin, G., Majeed, M., Giagulli, C., Piccio, L., Kim, J. Y., Butcher, E. C., and Laudanna, C., Chemokines trigger immediate β_2 integrin affinity and mobility changes: differential regulation and roles in lymphocyte arrest under flow. *Immunity* 2000, *13*, 759–769.

9 Cook, D. N., Prosser, D. M., Förster, R., Zhang, J., Kuklin, N. A., Abbondanzo, S. J., Niu, X. D., Chen, S. C., Manfra, D. J., Wiekowski, M. T., Sullivan, L. M., Smith, S. R., Greenberg,

H. B., Narula, S. K., Lipp, M., and Lira, S. A., CCR6 mediates dendritic cell localization, lymphocyte homeostasis, and immune responses in mucosal tissue. *Immunity* **2000**, *12*, 495–503.

10 Cyster, J. G., Chemokines and cell migration in secondary lymphoid organs. *Science* **1999**, *286*, 2098–2102.

11 Dejardin, E., Droin, N. M., Delhase, M., Haas, E., Cao, Y., Makris, C., Li, Z. W., Karin, M., Ware, C. F., and Green, D. R., The lymphotoxin-b receptor induces different patterns of gene expression via two NF-kB pathways. *Immunity* **2002**, *17*, 525–535.

12 Eberl, G., Marmon, S., Sunshine, M. J., Rennert, P. D., Choi, Y., and Littman, D. R., An essential function for the nuclear receptor ROR(γ)t in the generation of fetal lymphoid tissue inducer cells. *Nat. Immunol.* **2004**, *5*, 64–73.

13 Fan, L., Reilly, C. R., Luo, Y., Dorf, M. E., and Lo, D., Cutting edge: ectopic expression of the chemokine TCA4/SLC is sufficient to trigger lymphoid neogenesis. *J Immunol* **2000**, *164*, 3955–3959.

14 Finke, D., Acha-Orbea, H., Mattis, A., Lipp, M., and Kraehenbuhl, J., CD4$^+$CD3$^-$ cells induce Peyer's patch development: role of $\alpha_4\beta_1$ integrin activation by CXCR5. *Immunity* **2002**, *17*, 363–373.

15 Förster, R., Mattis, A. E., Kremmer, E., Wolf, E., Brem, G., and Lipp, M., A putative chemokine receptor, BLR1, directs B cell migration to defined lymphoid organs and specific anatomic compartments of the spleen. *Cell* **1996**, *87*, 1037–1047.

16 Förster, R., Schubel, A., Breitfeld, D., Kremmer, E., Renner-Müller, I., Wolf, E., and Lipp, M., CCR7 coordinates the primary immune response by establishing functional microenvironments in secondary lymphoid organs. *Cell* **1999**, *99*, 23–33.

17 Foxman, E. F., Campbell, J. J., and Butcher, E. C., Multistep navigation and the combinatorial control of leukocyte chemotaxis. *J. Cell Biol.* **1997**, *139*, 1349–1360.

18 Fu, Y. X. and Chaplin, D. D., Development and maturation of secondary lymphoid tissues. *Annu Rev Immunol* **1999**, *17*, 399–433.

19 Honda, K., Nakano, H., Yoshida, H., Nishikawa, S., Rennert, P., Ikuta, K., Tamechika, M., Yamaguchi, K., Fukumoto, T., Chiba, T., and Nishikawa, S. I., Molecular basis for hematopoietic/mesenchymal interaction during initiation of Peyer's patch organogenesis. *J Exp Med* **2001**, *193*, 621–630.

20 Jacob, J. and Kelsoe, G., In situ studies of the primary immune response to (4-hydroxy-3-nitrophenyl)acetyl. II. A common clonal origin for periarteriolar lymphoid sheath-associated foci and germinal centers. *J Exp Med* **1992**, *176*, 679–687.

21 Lanzavecchia, A. and Sallusto, F., Progressive differentiation and selection of the fittest in the immune response. *Nat Rev Immunol* **2002**, *2*, 982–987.

22 Lipp, M., Burgstahler, R., Müller, G., Pevzner, V., Kremmer, E., Wolf, E., and Förster, R., Functional organization of secondary lymphoid organs by the chemokine system. *Curr Top Microbiol Immunol* **2000**, *251*, 173–179.

23 Lipp, M. and Müller, G., Lymphoid organogenesis: getting the green light from ROR(γ)t. *Nat Immunol* **2004**, *5*, 12–14.

24 LUTHER, S.A., ANSEL, K.M., and CYSTER, J.G., Overlapping roles of CXCL13, interleukin 7 receptor a, and CCR7 ligands in lymph node development. *J Exp Med* **2003**, *197*, 1191–1198.

25 LUTHER, S.A., LOPEZ, T., BAI, W., HANAHAN, D., and CYSTER, J.G., BLC expression in pancreatic islets causes B cell recruitment and lymphotoxin-dependent lymphoid neogenesis. *Immunity* **2000**, *12*, 471–481.

26 MACLENNAN, I.C., Germinal centers. *Annu Rev Immunol, 12*, **1994**, 117–139.

27 MASOPUST, D., VEZYS, V., MARZO, A.L., and LEFRANCOIS, L., Preferential localization of effector memory cells in nonlymphoid tissue. *Science* **2001**, *291*, 2413–2417.

28 MATLOUBIAN, M., LO, C.G., CINAMON, G., LESNESKI, M.J., XU, Y., BRINKMANN, V., ALLENDE, M.L., PROIA, R.L., and CYSTER, J.G., Lymphocyte egress from thymus and peripheral lymphoid organs is dependent on S1P receptor 1. *Nature* **2004**, *427*, 355–360.

29 MAZZUCCHELLI, L., BLASER, A., KAPPELER, A., SCHARLI, P., LAISSUE, J.A., BAGGIOLINI, M., and UGUCCIONI, M., BCA-1 is highly expressed in *Helicobacter pylori*-induced mucosa-associated lymphoid tissue and gastric lymphoma. *J Clin Invest* **1999**, *104*, R49–54.

30 MEBIUS, R.E., Organogenesis of lymphoid tissues. *Nat Rev Immunol* **2003**, *3*, 292–303.

31 MEBIUS, R.E., MIYAMOTO, T., CHRISTENSEN, J., DOMEN, J., CUPEDO, T., WEISSMAN, I.L., and AKASHI, K., The fetal liver counterpart of adult common lymphoid progenitors gives rise to all lymphoid lineages, $CD45^+CD4^+CD3^-$ cells, as well as macrophages. *J Immunol* **2001**, *166*, 6593–6601.

32 MEBIUS, R.E., RENNERT, P., and WEISSMAN, I.L., Developing lymph nodes collect $CD4^+CD3^-LTb^+$ cells that can differentiate to APC, NK cells, and follicular cells but not T or B cells. *Immunity* **1997**, *7*, 493–504.

33 MÜLLER, G., HÖPKEN, U.E., and LIPP, M., The impact of CCR7 and CXCR5 on lymphoid organ development and systemic immunity. *Immunol Rev* **2003**, *195*, 117–135.

34 MÜLLER, G., HÖPKEN, U.E., STEIN, H., and LIPP, M., Systemic immunoregulatory and pathogenic functions of homeostatic chemokine receptors. *J Leukoc Biol* **2002**, *72*, 1–8.

35 MÜLLER, G., and LIPP, M., Shaping up adaptive immunity: The impact of CCR7 and CXCR5 on lymphocyte trafficking and homing. *Microcirculation* **2003**, *10*, 324–334.

36 NAKANO, H., TAMURA, T., YOSHIMOTO, T., YAGITA, H., MIYASAKA, M., BUTCHER, E.C., NARIUCHI, H., KAKIUCHI, T., and MATSUZAWA, A., Genetic defect in T lymphocyte-specific homing into peripheral lymph nodes. *Eur J Immunol* **1997**, *27*, 215–221.

37 OHL, L., HENNING, G., KRAUTWALD, S., LIPP, M., HARDTKE, S., BERNHARDT, G., PABST, O., and FÖRSTER, R., Cooperating mechanisms of CXCR5 and CCR7 in development and organization of secondary lymphoid organs. *J Exp Med* **2003**, *197*, 1199–1204.

38 OKADA, T., NGO, V.N., EKLAND, E.H., FÖRSTER, R., LIPP, M., LITTMAN, D.R., and CYSTER, J.G., Chemokine requirements for B cell entry to lymph nodes and Peyer's patches. *J Exp Med* **2002**, *196*, 65–75.

39 PABST, O., OHL, L., WENDLAND, M., WURBEL, M.A., KREMMER, E., MALISSEN, B., and FORSTER, R., Chemokine Receptor CCR9 Contributes to the Localization of Plasma Cells to the Small Intestine. *J Exp Med* **2004**, *199*, 411–416.

40 PACHYNSKI, R.K., WU, S.W., GUNN, M.D., and ERLE, D.J., Secondary lymphoid-tissue chemokine (SLC) stimulates integrin $\alpha_4\beta_7$-mediated adhesion of lymphocytes to mucosal addressin cell adhesion molecule-1 (MAdCAM-1) under flow. *J Immunol* **1998**, *161*, 952–956.

41 REIF, K., EKLAND, E.H., OHL, L., NAKANO, H., FÖRSTER, R., LIPP, M., and CYSTER, J.G., Balanced responsiveness to chemoattractants from adjacent zones determines B-cell position. *Nature* **2002**, *416*, 94–99.

42 REINHARDT, R.L., KHORUTS, A., MERICA, R., ZELL, T., and JENKINS, M.K., Visualizing the generation of memory CD4 T cells in the whole body. *Nature* **2001**, *410*, 101–105.

43 ROLLINS, B.J., Chemokines. *Blood* **1997**, *90*, 909–928.

44 SALLUSTO, F., KREMMER, E., PALERMO, B., HOY, A., PONATH, P., QIN, S., FÖRSTER, R., LIPP, M., and LANZAVECCHIA, A., Switch in chemokine receptor expression upon TCR stimulation reveals novel homing potential for recently activated T cells. *Eur J Immunol* **1999**, *29*, 2037–2045.

45 SALLUSTO, F., LENIG, D., FÖRSTER, R., LIPP, M., and LANZAVECCHIA, A., Two subsets of memory T lymphocytes with distinct homing potentials and effector functions. *Nature* **1999**, *401*, 708–712.

46 SALLUSTO, F., SCHAERLI, P., LOETSCHER, P., SCHANIEL, C., LENIG, D., MACKAY, C.R., QIN, S., and LANZAVECCHIA, A., Rapid and coordinated switch in chemokine receptor expression during dendritic cell maturation. *Eur J Immunol* **1998**, *28*, 2760–2769.

47 SALOMONSSON, S., LARSSON, P., TENGNER, P., MELLQUIST, E., HJELMSTROM, P., and WAHREN-HERLENIUS, M., Expression of the B cell-attracting chemokine CXCL13 in the target organ and autoantibody production in ectopic lymphoid tissue in the chronic inflammatory disease Sjögren's syndrome. *Scand J Immunol* **2002**, *55*, 336–342.

48 SANCHEZ MADRID, F. and DEL POZO, M.A., Leukocyte polarization in cell migration and immune interactions. *EMBO J*, **1999**, *18*, 501–511.

49 SCHAERLI, P., WILLIMANN, K., LANG, A.B., LIPP, M., LOETSCHER, P., and MOSER, B., CXC chemokine receptor 5 expression defines follicular homing T cells with B cell helper function. *J. Exp. Med.* **2000**, *192*, 1553–1562.

50 SENFTLEBEN, U., CAO, Y., XIAO, G., GRETEN, F.R., KRAHN, G., BONIZZI, G., CHEN, Y., HU, Y., FONG, A., SUN, S.C., and KARIN, M., Activation by IKKa of a second, evolutionary conserved, NF-kB signaling pathway. *Science* **2001**, *293*, 1495–1499.

51 SPRINGER, T.A., Traffic signals from lymphocyte recirculation and leukocyte emigration: the multistep paradigm. *Cell*, **1994**, *76*, 301–314.

52 STEIN, J.V., ROT, A., LUO, Y., NARASIMHASWAMY, M., NAKANO, H., GUNN, M.D., MATSUZAWA, A., QUACKENBUSH, E.J., DORF, M.E., and VON ANDRIAN, U.H., The CC chemokine thymus-derived chemotactic agent 4 (TCA-4, secondary lymphoid tissue chemo-kine, 6Ckine, exodus-2) triggers lymphocyte function-associated antigen 1-mediated arrest of rolling T lymphocytes in peripheral lymph node high endothelial venules. *J Exp Med* **2000**, *191*, 61–76.

53 VAN EWIJK, W. and NIEUWENHUIS, P., Compartments, domains and migration pathways of lymphoid cells in the splenic pulp. *Experientia* **1985**, *41*, 199–208.

54 WARNOCK, R. A., CAMPBELL, J. J., DORF, M. E., MATSUZAWA, A., MCEVOY, L. M., and BUTCHER, E. C., The role of chemokines in the microenvironmental control of T versus B cell arrest in Peyer's patch high endothelial venules. *J Exp Med* **2000**, *191*, 77–88.

55 WIGLE, J. T. and OLIVER, G., Prox1 function is required for the development of the murine lymphatic system. *Cell* **1999**, *98*, 769–778.

56 YOSHIDA, H., KAWAMOTO, H., SANTEE, S. M., HASHI, H., HONDA, K., NISHIKAWA, S., WARE, C. F., KATSURA, Y., and NISHIKAWA, S. I., Expression of $\alpha_4\beta_7$ integrin defines a distinct pathway of lymphoid progenitors committed to T cells, fetal intestinal lymphotoxin producer, NK, and dendritic cells. *J Immunol* **2001**, *167*, 2511–2521.

57 YOSHIDA, H., NAITO, A., INOUE, J., SATOH, M., SANTEE-COOPER, S. M., WARE, C. F., TOGAWA, A., and NISHIKAWA, S., Different cytokines induce surface lymphotoxin ab on IL-7 receptor a cells that differentially engender lymph nodes and Peyer's patches. *Immunity* **2002**, *17*, 823–833.

15
Keratinocyte Migration in Wound Healing

Cord Brakebusch

15.1
The Skin

The skin has an important protective function. It shields the internal environment from mechanical, chemical, and irradiation damage, and it protects from dehydration. In addition, it is an effective immunological barrier. The skin can be separated into epidermis, consisting mainly of keratinocytes, and dermis, containing collagen fibrils, fibroblasts, macrophages and other cells. Epidermis and dermis are connected by the sheet-like basement membrane which is a network of laminin-5, collagen IV, nidogen, perlecan and other molecules [1] (Fig. 15.1). This basement membrane is crucial to the tight connection between epidermis and dermis and furthermore provides important signals for the polarization and proliferation of the basal keratinocytes attached to it. Basal keratinocytes adhere strongly to the basement membrane via specialized junctions called hemidesmosomes which among other molecules contain the laminin-5-binding $\alpha 6\beta 4$-integrin [2]. Outside the hemidesmosomes, $\alpha 3\beta 1$-integrin mediates attachment of keratinocytes to the laminin-5 of the basal lamina. Other molecules such as dystroglycan and syndecans might further reinforce the connection to the basement membrane. Proliferation takes place only in the basal keratinocyte layer [3]. Cell-cell contacts between keratinocytes are mediated by desmosomes, adherens junctions, tight junctions, and gap junctions.

When keratinocytes differentiate they detach from the basement membrane and become suprabasal keratinocytes. They lose nearly all integrin receptors, express keratin 1 and 10 and shut down the production of keratin 5 and 14. Finally, these suprabasal keratinocytes undergo terminal differentiation [4]. They deposit proteins such as envoplakin and loricrin at the membrane and cross-link them forming the "cornified envelope", seal the gaps between the cells by lipid lamellae, and enucleate [5]. This layer of mechanically tough dead cells is called *stratum corneum*. In humans, terminal differentiation of keratinocytes requires about two weeks.

Cell Migration. Edited by Doris Wedlich
Copyright © 2005 WILEY-VCH Verlag GmbH & Co. KGaA, Weinheim
ISBN: 3-527-30587-4

statum
corneum

suprabasal
keratinocytes

basal
keratinocytes

EPIDERMIS

hemidesmosomes

basement membrane

fibroblasts

macrophages

DERMIS

Fig. 15.1 Schematic drawing of the epidermis. The epidermis mainly consists of keratinocytes. Basal keratinocytes are attached to the basement membrane. Hemidesmosomes are specialized junctions mediating this interaction. If keratinocytes start further differentiation, they detach from the basement membrane, stop proliferation and become suprabasal keratinocytes. Finally, they undergo terminal differentiation leading to a water-repellent layer of dead cell bodies called *stratum corneum*. Below the basement membrane lies the dermis, a mesenchymal connective tissue incorporating different cells such as dermal fibroblasts and macrophages.

Hair follicles are appendages of the skin consisting of several distinct cell layers [6]. The outermost layer is the outer root sheath (ORS) layer, which is continuous with the basal keratinocytes of the interfollicular epidermis. Stem cells originating from the hair bulge region in the upper part of the hair follicle migrate down the ORS layer to the hair matrix region at the bottom of the hair follicle where they expand and then differentiate into several inner root sheath (IRS) layers and the hair shaft [3]. The same progenitor cells can also migrate from the bulge region to the interfollicular epidermis. After formation of the hair follicles, the cyclic growth of hair follicles seems to be dependent on more committed progenitors located in the matrix region [7]. Keratinocytes also proliferate in the absence of hair follicles, so keratinocyte stem cells must also exist in the interfollicular epidermis, independent from the bulge region of hair follicles [8].

15.2
Wound Healing

The importance of skin integrity for the survival of the animal requires fast and efficient repair of skin injuries. Indeed, many different cell types collaborate during wound repair to ensure speedy closure of the barrier

a)

provisional wound matrix

b)

c)

eschar

granulation tissue

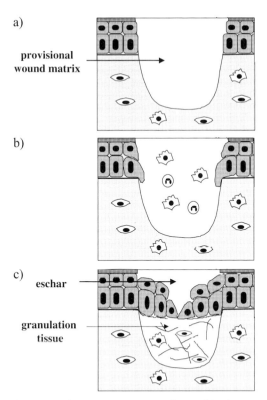

Fig. 15.2 Schematic presentation of wound healing.
A) A blot clot consisting mainly of platelets and fibrin seals the wound providing a provisional wound matrix. B) Within minutes, neutrophils and macrophages start to infiltrate the provisional wound matrix. About one day after wounding, keratinocytes begin to migrate onto the collagen and fibronectin of the dermis down the wound edge.
C) At day 3, blood vessels sprout in and infiltrating fibroblasts produce large amounts of collagen and fibronectin, transforming the provisional wound matrix into granulation tissue. Keratinocytes migrate on top of the granulation tissue to re-epithelialize the wound. The dead eschar on top of the keratinocytes is eventually sloughed off.

and re-formation of normal skin [9]. When the skin is damaged blood vessels are ruptured and coagulating blood will seal the wound by a blot clot consisting of platelets, fibrin and, to a lesser extent, other extracellular matrix molecules such as vitronectin and fibronectin (Fig. 15.2). Within minutes, neutrophils and later also monocytes infiltrate the provisional wound matrix, attracted by factors released from platelets, activated wound endothelium and bacteria that enter the wound. These inflammatory cells release a variety of proteases and cytokines and provide a preliminary protection against pathogens invading from the outside. While neutrophils peak after some hours and are hardly found in the wound after three days, macrophages continue to accumulate till about the third day [10]. Re-epithelialization is started by the activation of keratinocytes at the wound edge which begin to migrate and proliferate about 18–14 h after wounding. Keratinocytes migrate not only from the wound edge, but also from cut epidermal appendages such as the hair follicles. Two to three days after wounding, dermal fibroblasts migrate into the provisional wound matrix where they produce large amounts of fibronectin and collagen I. These fibroblasts transdifferentiate later into myofibroblasts mediating the contraction of the wound which can contribute very efficiently to wound closure. Angiogenesis is also induced during wound repair. Blood vessels sprout into the provisional matrix, which is then called granulation tissue because the microvessels in histological sections appear granular. The newly-formed blood vessels provide oxygen and nutrients, and allow immune cells to exit directly into the wound area. There is indirect evidence that nerve sprouting into the wound promotes re-epithelialization since skin areas with low innervation show decreased wound healing [11]. The underlying mechanism, however, is unclear. Eventually the migrating keratinocytes close the wound, separating the live granulation tissue from the dead eschar which is later sloughed off (Fig. 15.2). Further tissue remodeling takes place to transform the granulation tissue to more normal skin, although skin appendages are not reconstituted in the repaired areas. The formation of scar tissue characterized by fibrosis depends greatly on the infiltration of inflammatory cells. In embryos, only little infiltration is observed and wounds heal without leaving a scar [12].

This short overview demonstrates that wound healing involves the coordinated migration of many cells such as fibroblasts, macrophages and keratinocytes. In this chapter, however, we will focus on the requirements for the migration of keratinocytes during wound healing.

15.3
Models for Wound Healing

Skin wound healing is studied in different *in vitro* and *in vivo* models. While human wounds are of course the most relevant model to analyze the requirements of wound healing, they allow only limited manipulation and analysis and are influenced by the different age, gender, and genetic background of the patients. Animal models do not have these shortcomings and are, therefore, widely used to understand wound re-epithelialization. Although pig skin is quite similar to human skin [13], the availability of genetically modified mice and their simpler handling make mice the animal model of choice for most cases of wound healing studies.

Wound healing of normal animals can be analyzed in several respects. First, gene expression can be assessed. Significant changes in expression levels indicate, but do not prove, a functional role of this protein in wound healing. Second, exogenous application of cytokines, growth factors, and inhibitory antibodies to wounds can be used to investigate the importance of specific molecules for wound healing.

15.3.1
In vivo Models

In genetically modified mice, the effect of mutations in wound-healing-related genes is studied. By using gene targeting we can delete ("knock-out") or modify ("knockin") genes in mice in a cell-type specific manner. In "knockin" mice, the endogenous molecule is replaced by mutant forms of the protein in order to understand the structure/function relationship of this protein *in vivo*. Conditional gene targeting restricts the gene alteration to specific cells such as keratinocytes, while all other mouse cells are still normal.

Transgenic mice overexpress proteins in a cell-specific manner. Often the "knockout" of a specific gene does not cause a severe phenotype in wound healing, since other related proteins have redundant functions and compensate for the loss. In that case it might be possible to inhibit the whole family of related proteins by generating transgenic mice expressing a dominant negative form of that molecule.

Although mouse models are powerful tools for investigating the role of specific molecules during wound healing, it must be stressed that there are some significant differences between human and mouse skin. These differences, such as the increased number of hair follicles and the much stronger wound contraction in mice, could lead to different mechanistic relationships in human and mouse wound healing.

In mice, the wounds most investigated are in back skin, but wounds in other tissues like ear, cornea and mucosal tissues are also studied. The type of wounding can vary. In excisional wounds, epidermis and dermis are surgically removed in a limited area of few millimeters diameter, while in incisional wounds the skin is only cut down to the dermis without removing any tissue. Besides surgery, wounds can be afflicted by other means such as laser, bullets or heat.

15.3.2
In vitro Models

All these *in vivo* systems have the advantage that they analyze wound healing in all its complexity. This complexity, of course, could be a disadvantage if one wants to prove mechanistic relationships. *In vitro* wounding models are less complex, more easy to analyze and manipulate, but on the other hand more distant from the *in vivo* situation.

Human, but not mouse keratinocytes, can be used for organotypic cultures where keratinocytes are cocultivated with human or mouse fibroblasts at an air–liquid surface [14]. Although these cultures resemble normal skin, keratinocyte gene expression is altered compared to normal epidermis [15]. In *ex vivo* cultures, full-thickness punch biopsies of skin are transferred to a Petri dish coated with extracellular matrix molecules and covered with medium [16]. Keratinocytes grow out from these skin pieces as a stratified epithelial sheet. This process can easily be influenced by adding agents to the medium.

The most simple wound healing system involves primary or spontaneously immortalized keratinocytes seeded on culture dishes coated with extracellular matrix molecules such as fibronectin and collagen I. Keratinocyte monolayers can be "wounded" *in vitro* by a simple scratch. The migration of keratinocytes closing the gap can then be followed by time lapse microscopy and analyzed after cell lysis by biochemical methods or after fixation by immuno-fluorescence. Prior to migration, keratinocytes can be modified by introducing dominant negative or constitutive active forms of signaling molecules or fluorescently labeled cytoskeletal proteins to test the importance of specific signaling pathways and cytoskeletal associations. This *in vitro* wounding assay can be used to test whether re-epithelialization defects observed in genetically modified mice are due to cell-autonomous functions of these genes in keratinocytes.

15.4
Initiation of Wound Healing and Keratinocyte Activation

The wounding has several direct consequences that are conceivably important for the initiation of wound healing. First, the loss of cell–cell contacts leads to changes in cadherin-mediated signaling. Second, blood vessel ruptures expose keratinocytes and dermal fibroblasts to growth factors and cytokines from the serum. Third, mechanical stress during wounding can activate mechanoreceptors [17]. Fourth, wounding leads to hypoxia and pH changes in the wounded area which elicit various cell responses [18]. Fifth, bacterial infections occur, since wounding also impairs the immune defense function of the skin. Finally, many growth factors and cytokines are released from degranulating platelets in the blood clot and from inflammatory cells invading the wound. All these factors contribute to "kick-start" the wound healing process and to promote keratinocyte migration and proliferation. *In vitro* experiments suggest the involvement of many different intracellular signaling pathways such as Erk, JNK, p38 and Ca^{2+} in this process. The *in vivo* role of these pathways, however, has not been well-characterized.

Before migration, keratinocytes undergo significant changes. They increase in size, lose hemidesmosomes and cell–cell contacts and acquire a flat, elongated and polarized cell shape [19–21]. Actin redistributes to lamellipodia and filopodia and keratin filaments retract from the periphery to the perinuclear region [22].

Cells that can migrate from the hair follicles are not only basal keratinocytes and ORS, but also suprabasal keratinocytes. After wounding, suprabasal cells increase keratin 6, keratin 17 and $\alpha 5\beta 1$-integrin, reduce keratin 1, keratin 10 and desmosomes and start to migrate [22]. The up-regulation of keratin 6 might be crucial in providing resilience to the migrating keratinocytes to withstand the strong mechanical forces at the wound site [23]. Terminal differentiation is delayed in migrating suprabasal keratinocytes. The initiation of keratinocyte migration and proliferation are separate events. It is several hours after the start of migration that basal keratinocytes begin to proliferate a few cell diameters behind the leading edge [24].

In the following sections we present how keratinocyte migration is influenced by growth factors, cytokines, the activation of specific transcription factors, integrin-mediated adhesion and the composition of the ECM. Finally, we will discuss how keratinocyte migration is directed and how it is stopped.

15.5
Growth Factors and Cytokines

Growth factors and cytokines influence the proliferation and migration of keratinocytes [25]. They are released during wound healing by platelets, cells of the immune system, fibroblasts and also by keratinocytes themselves. Growth factors and cytokines can be grouped into families of structurally and functionally related molecules. Within these families, some receptors are shared by several ligands, and some ligands can bind to different receptors. This leads to a high degree of redundancy between molecules belonging to the same family.

Growth factors and cytokines can affect the cell producing them (autocrine), neighboring cells (paracrine) or cells relatively far away (endocrine). Furthermore, they can induce the secretion of additional factors with different effects and different target cells. All this adds up to a complex network of interactions and dependencies [25].

Growth factors such as platelet-derived growth factors (PDGFs), fibroblast growth factors (FGFs), epidermal growth factors (EGFs) and insulin-like growth factors (IGFs) mainly regulate the proliferation of keratinocytes, but some also have direct motogenic effects by activating cell contraction via the Ras-Erk cascade, or cytoskeletal reorganization via Rho GTPases. In the following paragraphs, some examples will be presented.

15.5.1
EGF Family of Growth Factors

The EGF family of growth factors in mammals consists of many members including EGF, transforming growth factor a (TGF-a), and heparin-binding EGF (HB-EGF) [26]. These three growth factors have all been detected in different wounds where they bind to the EGF-R present on keratinocytes at the wound margin in the early stages of injury [27]. *In vitro*, EGF stimulates keratinocyte migration through activation of Rac1 [28]. This activation is sustained and supported by ligation of $a6\beta4$-integrin, indicating cross-talk between integrins and growth factor receptors in the regulation of migration [28]. Another report stresses the importance of EGF-mediated stimulation of the mitogen-activated protein kinase pathway for keratinocyte migration [29]. Available mouse models deficient in EGF-R have severe defects in the epidermis and have not yet been tested in wound healing studies [30, 31]. IGF-1 stimulates keratinocyte migration *in vitro* by activation of Rho GTPases via phosphatidylinositol 3'-kinase [29].

15.5.2
FGF Family of Growth Factors

FGF-7 (also named keratinocyte growth factor, KGF) is a member of the FGF family with more than 20 members and 4 different receptors [32]. During wound healing, the production of several FGFs is increased, particularly of FGF-7 which is secreted by dermal fibroblasts and dendritic epidermal T-cells [33]. Nevertheless, FGF-7 is not essential for wound healing as shown in FGF-7-deficient mice [34]. It is most likely that FGF-10 can compensate for the loss of FGF-7 [33]. The importance of FGF signaling for wound healing was demonstrated in transgenic mice that express in keratinocytes a dominant negative mutant of the FGF receptor 2IIIb, which binds FGF1, 2, 3, 7 and 10. These animals show dramatically delayed re-epithelialization and a 80–90% reduced keratinocyte proliferation [35].

15.5.3
Extracellular Matrix Proteins

Growth factors can also indirectly affect keratinocyte migration by stimulating the production of extracellular matrix proteins that promote keratinocyte migration. Hyaluronan is a major component of the epidermal extracellular matrix; it is synthesized by keratinocytes and has a fast turnover [36]. Increased hyaluronan expression of keratinocytes by over-expression of hyaluronan synthase-2 was shown to increase keratinocyte migration *in vitro* [37]. Hyaluronan synthase-2 and -3 are induced by

FGF-7 [38] and EGF [39], and FGF-7 induces the expression of hyaluron-an in keratinocytes. These data suggest a role for hyaluronan in mediating enhanced keratinocyte migration in response to EGF and FGF-7. Hyaluronan is bound by the surface receptor CD44. Mice with a keratinocyte-specific expression of a dominant negative mutant of CD44 have a reduced growth response towards growth factors and cytokines, and reduced tissue repair *in vivo*, indicating that hyaluronan affects proliferation *in vivo* [40].

Fibronectin also promotes keratinocyte migration. PDGF enhances the production of fibronectin by dermal fibroblasts *in vitro* and could therefore contribute to keratinocyte motility *in vivo* where it is released in large amounts by platelets.

15.5.4
TGF-*β*, Chemokines and Cytokines

Transforming growth factor *β* (TGF-*β* s), chemokines and cytokines mainly modulate the inflammatory reaction during wound healing, thus indirectly influencing the migration of keratinocyte. However, direct motogenic effects on keratinocytes have also been reported for TGF*β* and for ligands of the chemokine receptor CXCR2.

TGF*β*1 is a member of a large ligand family comprising TGF-*β*s, bone morphogenetic proteins (BMPs), activins and other molecules [41]. It is released by degranulating platelets and attracts neutrophils, monocytes and fibroblasts. Contradictory results have been obtained with respect to its effect on re-epithelialization. On the one hand, TGF*β* inhibits keratinocyte proliferation *in vitro* and *in vivo*, thus delaying wound closure [42]; on the other, it stimulates keratinocyte migration *in vitro*, presumably by inducing expression of $\alpha 5\beta 1$-, $\alpha v\beta 5$-, $\alpha v\beta 6$- and $\alpha 2\beta 1$-integrin [43, 44]. *In vivo*, over-expression of latent TGF*β*1 in mouse epidermis resulted in delayed wound healing [45].

TGF*β*1-deficient mice have severe inflammation of many tissues. When they were treated before wounding with the immunosuppressive agent rapamycin, the mice showed accelerated wound healing, although this could also be secondary to the hyperthickening of the epidermis in mutant mice [46]. In contrast, if inflammation was repressed by intercrossing TGF*β*1-deficient mice with severe combined immunodeficiency (SCID) mice that lack normal lymphoid cells, wound healing was severely delayed [47]. Mice expressing dominant negative TGF*β*-R type II in keratinocytes have impairment of all TGF*β* signaling and showed enhanced wound healing [48].

Chemokines regulate leukocyte migration by activating integrin-mediated adhesion of leukocytes to the endothelium which is essential for leukocyte extravasation into the surrounding tissue [49]. However, they also affect non-hematopoietic cells. The chemokine receptor CXCR2, for example, is expressed on leukocytes, but also on endothelial cells during angiogenesis and on keratinocytes after wounding [50]. CXCR2 binds to the chemokines Gro-*a* and IL-8 which are both expressed in the early stages of wound healing [50, 51]. Mice deficient in the chemokine receptor CXCR2 have severely delayed wound healing. This seems to be at least partially caused by impaired keratinocyte migration since, in a scratch assay, keratinocyte migration was defective while proliferation was not altered [52].

It has been suggested that chemokine- and cytokine-triggered infiltration of monocytes and neutrophils to the wound area is important for normal wound healing, since these cells secrete many growth factors and cytokines, and remove debris required for remodeling [53]. However, mice lacking neutrophils and monocytes due to an ablation of the PU1 transcription factor show normal wound closure kinetics, indicating that the cytokines and growth factors expressed by these cells are not crucial for re-epithelialization [54]. As in embryonic wounds, no scar tissue was observed in these mice. Inflammatory cytokines such as interleukin-6 (IL6) and tumor necrosis factor (TNF) can dramatically regulate re-epithelialization, but a direct effect on keratinocyte migration has not yet been demonstrated [55, 56].

15.6
Transcription Factors

Efficient keratinocyte migration requires the synthesis of new molecules. Not surprisingly, therefore, several transcription factors such as E2F, Stat3 and Smad3 have been shown to play an important role in the induction of keratinocyte migration during wound healing.

15.6.1
AP-1 Transcription Factor Family

The AP-1 transcription factor family consists of dimers of jun and fos molecules [57], and AP-1 responsive elements are found in the promoter regions of many wound healing related genes such as TGFβ1 and matrix metalloproteinase-1 (MMP-1; [58]). C-jun is found in basal keratinocytes and is strongly upregulated in wound edge keratinocytes [59], quite probably in response to the activation of various cytokine and growth factor receptors [60]. In keratinocytes, c-jun expression results in increased synthesis of EGF-R and HB-EGF. The latter then binds in an autocrine loop to the EGF-R on keratinocytes which is important for the elongation and migration of keratinocytes [61, 62]. In the absence of c-jun, keratinocyte migration in a scratch assay is completely blocked *in vitro*. Whether keratinocyte migration is also affected *in vivo* during wound healing is unclear. While one group observed a normal re-epithelialization rate of full-thickness wounds [61], another group reported a slight delay in the late phase of wound closure [62].

15.6.2
E2F Transcription Factor Family

The E2F family of transcription factors plays a crucial role in the control of cell proliferation [63]. During wound healing, E2F-1 and E2F-2 increase dramatically at the wound edge, and mice lacking E2F-1 expression had significantly impaired wound healing [64]. *In vitro*, E2F-1-deficient keratinocytes displayed defective migration and proliferation. This phenotype is most probably caused by a partially impaired integrin function. Loss of E2F-1 leads to reduced levels of $\alpha 5\beta 1$- and $\alpha 6\beta 4$-integrins on keratinocytes, resulting in decreased attachment to the extracellular matrix and reduced integrin-mediated signal transduction [64].

15.6.3
Activation of Stat3

Stat3 is a latent transcription factor that is activated by cytokine and growth factor receptors such as IL6-R or EGF-R [65]. After tyrosine phosphorylation Stat factors dimerize, translocate to the nucleus and regulate transcription. Mice with a keratinocyte-restricted deletion of the Stat3 gene had a normal epidermis, but showed defective wound healing [66]. *In vitro*, Stat-3-deficient keratinocytes exhibited normal proliferation, but impaired migration in response to growth factors or cytokines. This corresponded to an increased adhesion to collagen I and a constitutive tyrosine hyperphosphorylation of p130Cas, while other focal contact molecules such as focal adhesion kinase (FAK) and paxillin were not hyperphosphorylated [67].

15.6.4
NFkB

TNF or $\alpha 6\beta 4$-mediated activation of NFkB [68] might lead via NOS induction to NO production which is important for normal wound healing [69]. NO synthesis is upregulated during wound healing [70] and might also promote keratinocyte migration [71]. Mice lacking the inducible or inflammatory NOS isoform (iNOS) showed delayed wound healing in excisional wounds [70]. No difference was found in incisional wounds, quite probably due to the reduced bacterial infiltration in these injuries which results in poor induction of iNOS [72].

15.6.5
Peroxisome Proliferator Activated Receptors

Peroxisome proliferator activated receptors (PPARs) are transcription factors belonging to the nuclear hormone receptor family [73]. During wound healing, PPARβ is highly upregulated and mice lacking PPARβ show delayed re-epithelialization, conceivably due to impaired adhesion and migration of mutant keratinocytes [74]. Furthermore, heterozygous loss of the PPARβ gene reduces activation of NFkB which could also contribute to the decreased keratinocyte migration as described above [75].

15.6.6
Transcription Factor Smad3

Smad3 is a transcription factor activated by TGFβ1, 2, 3 and activin [41]. Smad3-deficient mice survive into adulthood [76]. In a full-thickness incisional wound-healing model Smad3-null mice showed accelerated wound healing [77]. *In vitro*, Smad3-deficient keratinocytes exhibited reduced adhesion to the matrix and decreased migration towards TGFβ and FGF-7, but not to growth factors present in serum. This correlated with the absence of induction of $\alpha 5\beta 1$-integrin by TGFβ in mutant keratinocytes, suggesting that TGFβ1 contributes to keratinocyte migration via Smad3-dependent induction of $\alpha 5\beta 1$-integrin expression.

15.7
Integrin Receptors

Integrins are a large group of extracellular matrix receptors that consist of an α and a β subunit [78]. Up to now 18 different α and 8 different β subunits are known that can combine to form 24 different integrin receptors. At the intracellular side integrins are indirectly connected to the actin cytoskeleton thus providing a structural link between the extracellular matrix and the cytoskeleton [79]. In addition to this structural role, integrin attachment to the extracellular matrix results in the activation of various signaling molecules including serine threonine kinases, tyrosine kinases, phospholipases and small GTPases of the Rho family, which are crucial for the reorganization of the actin cytoskeleton [80]. Integrins therefore play a key role in keratinocyte migration by contributing to the formation of cell extensions, mediating the adhesion of these extensions to the extracellular matrix, and activating signaling pathways involved in maturation and turnover of focal contacts, cell contraction and rear release.

In undisturbed skin integrin expression is mainly restricted to the basal keratinocytes which express $\alpha 6\beta 4$, $\alpha 3\beta 1$, $\alpha 2\beta 1$, $\alpha 9\beta 1$, $\alpha v\beta 5$ [81]. As major ligands, $\alpha6\beta4$ and $\alpha3\beta1$ bind laminin-5, $\alpha2\beta1$ collagen, $\alpha9\beta1$ tenascin, and $\alpha v\beta5$ vitronectin. In suprabasal keratinocyte layers only $\alpha v\beta8$-integrin was found [82]. This expression pattern changes after wounding when basal keratinocytes increase expression of $\alpha9\beta1$ and $\alpha v\beta5$, and newly express $\alpha5\beta1$, which binds to fibronectin, and $\alpha v\beta6$, which interacts with fibronectin, vitronectin and tenascin. In addition, integrin $\alpha v\beta6$ binds and activates latent TGFβ thus promoting the infiltration of inflammatory cells [83].

Integrin receptors have a complex cross-talk to other receptors such as growth factor receptors, cell–cell adhesion molecules and other integrins and these cross-talks seem to be particularly important in keratinocyte migration as discussed below.

15.7.1
Keratinocyte Migration

The function of several integrin subunits in keratinocytes has been studied in genetic models *in vivo* and *in vitro*. Deletion of the $\beta1$-integrin gene restricted to keratinocytes resulted in a loss of $\alpha3\beta1$, $\alpha2\beta1$ and $\alpha9\beta1$ and the inability to upregulate $\alpha5\beta1$ during wound healing. In addition, the expression of $\alpha6\beta4$ was decreased *in vitro* and *in vivo* [8, 84, 85]. These mice show severe defects in the re-epithelialization of full thickness wounds in back skin [84]. After an initial lag period $\beta1$-deficient keratinocytes started to migrate and closed the wound. This migration is quite probably mediated by $\alpha v\beta5$ and $\alpha v\beta6$-integrin which are upregulated during wound healing.

Several mechanisms might contribute to the impaired keratinocyte migration *in vivo*. First, spontaneously immortalized keratinocytes lacking $\beta1$-integrin showed a 10-fold reduced migration *in vitro* compared to their normal counterparts, corresponding to an impaired adhesion to laminin-5 and collagen I and a 50% reduced adhesion to fibronectin [84]. Residual binding to fibronectin was completely dependent on $\alpha v\beta6$, which was upregulated in these keratinocytes [86]. Second, studies with skin explants of $\beta1$-deficient mice on fibronectin demonstrated that the $\beta1$-null keratinocytes were not able to rapidly reorient their $\alpha v\beta6$-associated actin cytoskeleton towards the polarized movement [86]. Finally, $\beta1$-deficient migrating keratinocytes showed narrowing of the intercellular spaces, suggesting cross-talk between $\beta1$-integrin and cadherins as previously proposed in different systems [84, 87, 88]. A correlation between cell–cell distance and migration is suggested from studies with keratinocytes lacking the desmosomal protein plakophilin-1. These cells exhibit increased motility corresponding to decreased desmosome assembly [89]. Interestingly, there was no defect in the proliferation of $\beta1$-deficient keratinocytes during wound healing, although in undisturbed skin proliferation was significantly reduced.

Although $\alpha 3\beta 1$ is the major $\beta 1$-containing integrin in skin, deletion of the $\alpha 3$-integrin gene alone did not reduce keratinocyte migration *in vitro*. Instead, keratinocyte migration was enhanced due to increased expression of $\alpha 2\beta 1$ and $\alpha 5\beta 1$-integrin [90]. In addition, loss of $\alpha 3\beta 1$-integrin might favor keratinocyte migration by decreasing the levels of the gap junction protein connexin 43 [91]. Mice deficient in connexin 43 showed accelerated wound healing [92]. Whether these functions of $\alpha 3\beta 1$-integrin are also important *in vivo* has not yet been tested, because $\alpha 3$-null mice die shortly after birth.

Although *in vitro* migration assays suggested an essential role for $\alpha 2\beta 1$-integrin in keratinocyte binding to and migration on collagen I, $\alpha 2$-integrin-deficient knockout mice showed normal wound healing in a full-thickness model [93]. Maybe $\alpha 3\beta 1$, which was reported to bind to collagen I, can compensate for the loss of $\alpha 2\beta 1$-integrin [94]. No epidermal or wound healing defect has been reported for mice lacking $\alpha 9$-integrin, which die betweeen 6 and 12 days after birth [95]. Also $\alpha v\beta 5$ and $\alpha v\beta 6$-integrin, which are highly upregulated during re-epithelialization have a dispensable role in wound healing, since mice lacking $\beta 5$ or $\beta 6$-integrin showed normal re-epithelialization [96, 97]. In primary keratinocytes, $\alpha v\beta 6$ expression is induced and critical for keratinocyte migration on fibronectin and vitronectin [98].

15.7.2
Redistribution of Integrin Receptors

During wound healing, keratinocytes not only change their adhesive properties by modulating the expression of integrins; they also redistribute integrins within the cell. $\alpha 6\beta 4$ is redistributed from the basal side to the leading edge, $\alpha 3\beta 1$ is concentrated in focal contacts, and $\alpha 2\beta 1$ is redistributed from the lateral side to the leading edge of keratinocytes [99, 100].

There is indication from *in vitro* studies that the redistribution of these different integrin receptors is a sequential process initiated by macrophage stimulating protein (MSP). MSP, a member of the scatter factor family, is produced in liver and present in the serum in an inactive form [101]. When serum enters the wound MSP becomes activated extravascularly by serine proteases present in the wound bed. It then binds to the Ron tyrosine kinase receptors on keratinocytes, inducing first tyrosine phosphorylation and then Akt-mediated serine phosphorylation of Ron and $\alpha 6\beta 4$-integrin [102]. The adaptor molecule 14-3-3 then cross-links the serine phosphorylated Ron and $\alpha 6\beta 4$ which leads to the redistribution of $\alpha 6\beta 4$ to the lamellipodia at the leading edge. This redistribution is required for the activation of $\alpha 3\beta 1$-mediated spreading on laminin-5. In addition, the Ron-mediated tyrosine phosphorylation of $\alpha 6\beta 4$-integrin reinforces $\alpha 6\beta 4$ signal transduction, resulting in increased activation of

Akt, p38 and the downstream target NFkB. These signaling events are crucial to the MSP-mediated activation of keratinocyte migration *in vitro*.

In vivo, however, MSP function is not essential since mice lacking MSP have normal wound healing [103], maybe due to redundant function of hepatocyte growth factor (HGF), the second member of the scatter factor family. HGF and its receptor Met are strongly upregulated in wound keratinocytes [104].

$\alpha 6\beta 4$-integrin also cross-talks with EGF-R signaling, since high concentrations of EGF induce tyrosine phosphorylation of $\alpha 6\beta 4$-integrin [105]. Expression of $\alpha 6\beta 4$ is essential for EGF-dependent keratinocyte migration *in vitro* [28] and in this case the cross-talk takes place on the level of Rho GTPases. If $\alpha 6\beta 4$ is attached to laminin-5, EGF-induced Rac1 activation is sustained, RhoA is suppressed and lamellipodia are formed; if $\alpha 6\beta 4$ is present but not ligand-bound, chemotaxis towards EGF is mediated by $\alpha 3\beta 1$-integrin now also involving RhoA.

Cross-talk was also reported between $\alpha 3\beta 1$ and $\alpha 2\beta 1$-integrin. *In vitro*, $\alpha 3\beta 1$-mediated spreading of keratinocytes on unprocessed laminin-5 activates RhoA which is required for $\alpha 2\beta 1$-integrin-mediated migration on collagen I [106].

15.8
Extracellular Matrix

The extracellular matrix is remodeled throughout the wound healing process in order to replace the provisional wound matrix by the matrix present in normal skin [107]. Since keratinocytes display different adhesion strengths to different matrix molecules and since different ECM receptors differentially regulate cell migration, the extracellular matrix has a tremendous influence on the migratory behavior of keratinocytes during wound healing. Generally, collagen I, fibronectin, vitronectin and non-processed laminin-5 promote migration, while fibrin and processed laminin-5 inhibit [107].

15.8.1
Fibrin and Plasmin

The inhibitory effect of fibrin on migration is explained by the lack of fibrin-binding $\alpha v\beta 3$-integrin receptors on keratinocytes [108]. Therefore, keratinocytes cannot migrate into the fibrin-rich part of the provisional wound matrix. Instead, they migrate on top of the fibronectin and collagen I-rich granulation tissue, thus separating this live tissue from the dead eschar which eventually is sloughed off. Fibrin functions mainly as a barrier preventing keratinocyte migration [109]. Plasmin, a fibrin which degrades extracellular serine protease derived from plasminogen, is required for keratinocytes to drill their way beneath the fibrin clot. If plasmin is missing, wound closure is severely delayed [110]. Mice lacking fibrin had normal wound closure times, but keratinocytes did not migrate immediately below the eschar and above the early granulation tissue [111]. Instead they proceeded down the inner dermal edge, surrounded the complete wound area and fused at the surface creating a cavity. Although not adhesive for keratinocytes, the fibrin network is important for normal wound healing by providing strength and stability to the provisional wound matrix. Keratinocytes never infiltrate the granulation tissue, but always remain on top of it. Maybe the plasmin allows them to quickly "melt" their way through the fibrin network while the underlying granulation tissue is still relatively "solid" for the keratinocytes.

15.8.2
Fibronectin

Because collagen I-deficient mice die before birth, the importance of collagen I for keratinocyte migration in wound healing cannot be studied *in vivo*. However, since wound healing occurs also in mice where keratinocytes lack β1-integrin and thus all collagen receptors [85], the adhesion to collagen seems not to be crucial for keratinocyte migration. Since β1-deficient keratinocytes adhere only very poorly to laminin-5, the re-epithelialization might occur in these mice via keratinocyte migration on fibronectin [85, 87].

Fibronectin exists in two major isoforms, plasma fibronectin and cellular fibronectin [112]. Plasma fibronectin is produced by hepatocytes and secreted into the blood stream. Cellular fibronectin is made by many cells including fibroblasts and epithelial cells and contains the EDA and the EDB domains which are lacking in plasma fibronectin. Keratinocytes bind specifically to the EDA domain by $\alpha 9\beta 1$-integrin, while $\alpha 5\beta 1$ and $\alpha v\beta 6$ bind to different regions outside the EDA and EDB domain.

During the first hours after wounding plasma fibronectin mainly derived from serum and cellular fibronectin derived from degranulating platelets are deposited [113]. Plasma fibronectin is not crucial for wound healing as demonstrated in mice lacking plasma fibronectin [113]. In these mice, a thin layer of cellular fibronectin forms on the surface of the provisional matrix immediately after wounding, which is apparently sufficient for normal initiation of wound healing. Within the following days, the deposition of cellular fibronectin is highly increased thoughout the wound area. Mice expressing only fibronectin lacking the EDA domain, thus lacking normal cellular fibronectin, showed abnormal wound healing characterized by ulcerations after wound closure [114].

15.8.3
Laminin

Laminin-5 is a trimer consisting of $\alpha 3$, $\beta 3$ and $\gamma 2$-laminin [115]. Both, $\alpha 3$ and $\gamma 2$-laminin are proteolytically processed, and non-processed laminin-5 is hardly detectable in normal skin. *In vitro*, keratinocyte migration on plastic, fibronectin or collagen IV requires the endogenous production of laminin-5 and cell attachment to unprocessed laminin-5 by $\alpha 2\beta 1$-integrin [116]. Mice deficient for laminin-5 expression lack a basement membrane at the dermal-epidermal junction and die at birth since the epidermis detaches from the dermis [117]. Unprocessed, but not processed laminin-5 is bound also by syndecan-1 [118]. However, in the absence of syndecan-1, full-thickness wounds in the back skin displayed normal re-epithelialization [119].

15.8.4
Matrix Metalloproteinases

Matrix metalloproteinases (MMPs) are matrix-degrading enzymes that contain a Zn^{2+} ion at the active site [120, 121]. All MMPs are synthesized as inactive zymogens. Normally, MMPs are not present in the epidermis, but expressed during tissue remodeling in migrating keratinocytes, inflammatory cells and fibroblasts. They are activated extracellularly and remove damaged tissue and provisional wound matrix. MMP-1 and MMP-8, also named collagenase-1 and -2, specifically cut fibrillar collagens such as collagen I. The partially denatured collagen then becomes accessible to other MMPs for further degradation.

During wound healing keratinocytes lose contact with the basement membrane and bind to collagen I via $\alpha 2\beta 1$-integrin [122, 123]. This attachment induces the expression of MMP-1. Pro-MMP-1, the inactive zymogen form of MMP-1, interacts specifically with $\alpha 2\beta 1$-integrin, suggesting the formation of a ternary complex between collagen I, $\alpha 2\beta 1$ and MMP-1 [124]. This interaction brings MMP-1 close to its substrate, optimizing its efficiency. *In vitro*, $\alpha 2\beta 1$ and MMP-1 are essential for keratinocyte migration on collagen 1-containing matrices [124]. In mice, however, $\alpha 2\beta 1$ is dispensable for efficient re-epithelialization [94]. The *in vivo* function of MMP-1 remains to be tested. The genes encoding MMP-2, MMP-9, MMP-12 and MMP-14 have been inactivated in mice, but no effect on keratinocyte migration during wound healing has been reported [125–128]. MMP-3, which is expressed by keratinocytes and dermal fibroblasts, mediates wound contraction *in vivo* by an unknown mechanism [129].

Different tissues can express different MMPs during wounding. Mucosal epithelia, for example, express MMP-7 which is not detectable in skin wounds. In the absence of MMP-7, the epithelial cells at the wound edge do not migrate even several days after wounding [130]. This suggests that in cutaneous wounds MMPs could also play an important role in keratinocyte migration. Gene ablation of further MMPs and the generation of mice lacking several MMPs will test this expectation.

15.9
Directed Migration of Keratinocytes

It is not clear how keratinocytes initiate directed migration towards the center of the wound. In migrating fibroblasts and astrocytes, Cdc42 seems to play an important role in the orientation of the cells [131, 132], but it remains to be tested whether this is also the case in keratinocytes. In migrating astrocytes it was reported that integrins activate Cdc42, but how integrins are specifically activated at the wound edge is not clear [132]. Perhaps the loss of cell–cell adhesion also plays a role.

Integrin, $\alpha 2\beta 1$ and MMP-1 might be part of an interesting molecular compass guiding keratinocytes during wound healing. At the leading edge, $\alpha 2\beta 1$ binds with high affinity to native collagen which induces MMP-1 expression [123, 133]. MMP-1 then cuts collagen I allowing its further degradation by other MMPs. Since $\alpha 2\beta 1$-integrin is binding less strongly to cleaved collagen present at the rear of the cell, the cell tends to move forward towards the non-cleaved collagen present at the front [133]. Denatured collagen cannot induce MMP-1 expression via $\alpha 2\beta 1$-integrin and therefore MMP-1 expression is switched off behind the leading edge [134]. This proposed mechanism, however, cannot be essential since mice lacking $\alpha 2\beta 1$ do not show a wound healing defect [94].

Electrical current might provide an alternative pathway for the directed migration of keratinocytes. Epithelia generate a steady voltage across themselves. Wounding results in a lateral electrical field towards the injury of about 40–200 mV mm^{-1} [135]. *In vitro*, keratinocytes can respond to such an electrical field with the activation of protein kinase A and directed migration. In fact, wound healing seems to be accelerated by electrical fields suggesting that this mechanism is also important *in vivo*.

15.10
Migration Stopping and Deactivation

Keratinocyte migration is quite probably stopped by contact inhibition. Interestingly, in the absence of $\beta 1$-integrin this migration stopping is impaired and instead of fusing, the epithelial tongues migrate further and push each other up [85]. This phenomenon was observed in about 40% of all full thickness wounds of mutant mice, but was never seen in wounds of control mice. Another process regulating migration seems to be the processing of laminin-5. Migrating keratinocytes produce unprocessed laminin-5 which is crucial for keratinocyte migration *in vitro* [116]. Processing of laminin-5 takes places behind the leading edge. Since processed laminin-5 inhibits migration this leads to the stopping of keratinocyte migration behind the wound edge. Parallel to the stopping of migration, keratinocytes become deactivated by unknown mechanisms and show increased differentiation of suprabasal keratinocytes.

15.11
Conclusion

Keratinocyte migration during wound healing is a complex process affected by many different factors. Many results obtained by analyzing migrating keratinocytes *in vitro* could not be confirmed *in vivo*, stressing the importance of *in vivo* models. Only the combination of *in vivo* studies using genetically modified mice and biochemical analysis of migrating keratinocyte *in vitro* will help to understand this network of interactions in the future.

15.12
Acknowledgment

I thank Dr. Jolanda van Hengel for critically reading the manuscript.

15.13
References

1   R. F. Ghohestani, K. Li, P. Rousselle, and J. Uitto, *Clin. Dermatol.* **2001**, *19*, 551–562.
2   A. M. Mercurio, I. Rabinovitz, and L. M. Shaw, *Curr. Opin. Cell. Biol.* **2001**, *13*, 541–545.
3   L. Alonso and E. Fuchs, *Genes Dev.* **2003**, *17*, 1189–1200.
4   A. Gandarillas, *Exp. Gerontol.* **2000**, *35*, 53–62.
5   A. E. Kalinin, A. V. Kajava, and P. M. Steiert, *BioEssays* **2002**, *24*, 789–800.
6   K. S. Stenn and R. Paus, *Phys. Rev.* **2003**, 449–494.
7   A. A. Panteleyev, C. A. B. Jahoda, and A. M. Christiano, *J. Cell. Sci.* **2001**, *114*, 3419–3431.
8   C. Brakebusch, R. Grose, F. Quondamatteo, A. Ramirez, J. L. Jorcano, A. Pirro, M Svensson, R. Herken, T. Sasaki, R. Timpl, and R. Fässler, *EMBO J.* **2000**, *19*, 3990–4003.
9   P. Martin, *Science* **1997**, *276*, 75–81.
10 L. A. DiPietro, P. J. Polverini, S. M. Rahbe, and E. J. Kovacz, *Am. J. Pathol.* **1995**, *146*, 868–875.

11 S. Harsum, J. D. Clarke, and P. Martin, *Dev. Biol.* **2001**, *238*, 27–39.

12 K. M. Bullard, M. T. Longaker, and H. P. Lorenz, *World J. Surg.* **2003**, *27*, 54–61.

13 R. M. Lavker, G. Dong, P. S. Zehng, and G. F. Murphy, *Am. J. Pathol.* **1991**, *138*, 687–697.

14 N. E. Fusenig, *The Keratinocyte Handbook* (ed. Leigh, I., Lane, B., Watt, F.) **1994**, Cambridge, pp 71–94.

15 B. M. Schaefer, H. J. Stark, N. E. Fusenig, R. F. Todd, and M. D. Kramer, *Exp. Cell. Res.* **1995**, *220*, 415–423.

16 S. Mazzalupo, M. J. Wawersik, and P. A. Coulombe, *J. Invest. Dermatol.* **2002**, *118*, 866–870.

17 D. E. Ingber, *Proc. Natl. Acad. Sci. USA* **2003**, *100*, 1472–1474.

18 Y. P. Xia, Y. Zhao, J. W. Tyrone, A. Chen, and T. A. Mustoe, *J. Invest. Dermatol.* **2001**, *116*, 50–56.

19 G. Odland and R. Ross, *J. Cell. Biol.* **1968**, *39*, 135–151.

20 C. B. Croft, and D. Tarin, *J. Anat.* **1970**, *106*, 63–70.

21 S. A. Alexander, *Ann. Surg. Res.* **1981**, *31*, 456–462.

22 R. D. Paladini, K. Takahashi, N. S. Bravo, and P. A. Coulombe, *J. Cell. Biol.* **1996**, *132*, 381–397.

23 P. Wong and P. A. Coulombe, *J. Cell. Biol.* **2003**, *163*, 327–337.

24 J. Bereiter-Hahn, *The molecular and cellular biology of wound repair* (ed. R. A. F. Clark, P. M. Henson) **1988**, New York, Plenum Press pp 321–335.

25 S. Werner and R. Grose, *Physiol. Rev.* **2002**, *83*, 835–870.

26 T. Holbro and N. E. Hynes, *Ann. Rev. Pharmacol. Toxicol.* **2004**, *44*, 195–217.

27 B. A. Wenczak, J. B. Lynch, and L. B. Nanney, *J. Clin. Invest.* **1992**, *90*, 2392–2401.

28 A. J. Russell, E. F. Fincher, L. Millman, R. Smith, V. Vela, E. A. Waterman, C. N. Dey, S. Guide, V. M. Weaver, and M. P. Marinkovich, *J. Cell. Sci.* **2003**, *116*, 3542–3556.

29 I. Haase, R. Evans, R. Pofahl, and F. M. Watt, *J. Cell. Sci.* **2003**, *116*, 3227–3238.

30 R. Murillas, F. Larcher, C. J. Conti, M. Santos, A. Ullrich, and J. L. Jorcano, *EMBO J.* **1995**, *14*, 5216–5223.

31 M. Sibilia and E. F. Wagner, *Science* **1995**, *269*, 234–238.

32 C. Alzheimer and S. Werner, *Adv. Exp. Med. Biol.* **2002**, *513*, 335–351.

33 J. Jameson, K. Ugarte, N. Chen, P. Yachi, E. Fuchs, R. Bois-menu, and W. L. Havran, *Science* **2002**, *296*, 747–749.

34 L. Guo, I. Degenstein, and E. Fuchs, *Genes Dev.* **1996**, *10*, 165–175.

35 S. Werner, H. Smola, X. Liao, M. T. Longaker, T. Krieg, P. H. Hofschneider, and L. T. Williams, *Science* **1994**, *266*, 819–922.

36 W. Y. Chen and G. Abatangelo, *Wound Repair Regen.* **1999**, *7*, 79–89.

37 K. Rilla, M. J. Lammi, R. Sironen, K. Torronen, M. Luukko-nen, V. C. Hascall, R. J. Midura, M. Hyttinen, J. Pelkonen, M. Tammi, and R. Tammi, *J. Cell. Sci.* **2002**, *115*, 3633–3643.

38 S. Karvinen, S. Pasonen-Seppänen, J. M. T. Hyttinen, J.-P. Pie-nimäki, K. Törrönen, T. A. Jokela, M. I. Tammi, and R. Tammi, *J. Biol. Chem.* **2003**, *278*, 49459–49504.

39 J.-P. Pienimäki, K. Rilla, C. Fülöp, R.K. Sironen, S. Karvinen, S. Pasonen, M.J. Lammi, R. Tammi, V.C. Hascall, and M.I. Tammi, *J. Biol. Chem.* **2001**, *276*, 20428–20435.

40 G. Kaya, I. Rodriguez, J.L. Jorcano, P. Vassalli, and I. Stamenkovic, *Genes Dev.* **1997**, *11*, 996–1007.

41 R. Derynck and Y.E. Zhang, *Nature* **2003**, *425*, 577–584.

42 K. Sellheyer, J.R. Bickenbach, J.A. Rothnagel, D. Bundman, M.A. Longley, T. Krieg, N.S. Roche, A.B. Roberts, and D.R. Roop, *Proc. Natl. Acad. Sci. USA* **1993**, *90*, 5237–5341.

43 J. Gailit, M.P. Welch, and R.A. Clark, *J. Invest. Dermatol.* **1994**, *103*, 221–227.

44 G. Zambruno, P.C. Marchisio, A. Marconi, C. Vaschieri, A. Melchiori, A. Gianetti, and M. DeLuca, *J. Cell. Biol.* **1995**, *129*, 853–865.

45 L. Yang, T. Chan, J. Demare, T. Iwashina, A. Ghahary, P.G. Scott, and E.E. Tredget, *Am. J. Pathol.* **2001**, *159*, 2147–2157.

46 R.M. Koch, N.S. Roche, W.T. Parks, G.S. Ashcroft, J.J. Letterio, and A.B. Roberts, *Wound Rep. Regen.* **2000**, *8*, 179–191.

47 M.J. Crowe, T. Doetschman, and D.G. Greenhalgh, *J. Invest. Dermatol.* **2000**, *115*, 3–11.

48 C. Amendt, A. Mann, P. Schirrmacher, and M. Blessing, *J. Cell. Sci.* **2002**, *115*, 2189–2198.

49 R. Gillitzer and M. Goebeler, *J. Leukoc. Biol.* **2001**, *69*, 513–521.

50 L.B. Nanney, S.G. Mueller, R. Bueno, S.C. Pieper, and A. Richmond, *Am. J. Pathol.* **1995**, *147*, 1248–1260.

51 E. Engelhardt, A. Toksoy, M. Goebeler, S. Debus, E.B. Brocker, and R. Gillitzer, *Am. J. Pathol.* **1998**, *153*, 1849–1860.

52 R.M. Devalaraja, L.B. Nanney, Q. Qqian, J. Du, Y. Yu, M.N. Devalaraja, and A. Richmond, *J. Invest. Dermatol* **2000**, *115*, 234–244.

53 S.J. Leibovich and R. Ross, *Am. J. Pathol.* **1975**, *78*, 71–92.

54 P. Martin, D. D'Souza, J. Martin, R. Grose, L. Cooper, R. Maki, and S.R. McKercher, *Curr. Biol.* **2003**, *13*, 1122–1128.

55 R.M. Gallucci, P.P. Simeonova, J.M. Matheson, C. Kommineni, J.L. Guriel, T. Sugawara, and M.I. Luster, *FASEB J.* **2000**, *14*, 2525–2531.

56 R, Mori, T. Kondo, T. Oshima, Y. Ishida, and N. Mukaida, *FASEB J.* **2002**, *16*, 963–974.

57 R. Eferl and E.F. Wagner, *Nat. Rev. Cancer* **2003**, *3*, 859–868.

58 S. Yates and T.E. Rayner, *Wound Rep. Regen.* **2001**, *10*, 5–15.

59 P. Martin and C.D. Nobes, *Mech. Dev.* **1992**, *38*, 209–215.

60 P. Angel, A. Szabowski, and M. Schorpp-Kistner, *Oncogene* **2001**, *20*, 2413–2423.

61 R. Zenz, H. Scheuch, P. Martin, C. Frank, R. Eferl, L. Kenner, M. Sibilia, and E.F. Wagner, *Dev. Cell.* **2003**, *4*, 879–889.

62 G. Li, C. Gustafson-Brown, S.K. Hanks, K. Nason, J.M. Arbeit, K. Pogliano, R.M. Wisdom, and R.S. Johnson, *Dev. Cell.* **2003**, *4*, 865–877.

63 C. Stevens and N.B. La Thangue, *Arch. Biochem. Biophys.* **2003**, *412*, 157–169.

64 S.J.A. D'Souza, A. Vespa, S. Mukherjee, A. Maher, A. Pajak and L. Dagnino, *J. Biol. Chem.* **2002**, *277*, 10626–10632.

65 D. E. LEVY and J. E. DARNELL, *Nat. Rev. Mol. Cell. Biol.* **2002**, *3*, 651–662.

66 S. SANO, S. ITAMI, M. TARUTANI, Y. YAMAGUCHI, H. MIURA, K. YOSHIKAWA, S. AKIRA, and J. TAKEDA, *EMBO J.* **1999**, 4657–4668.

67 M. KIRA, S. SANO, S. TAKAGI, K. YOSHIKAWA, J. TAKEDA, and S. ITAMI, *J. Biol. Chem.* **2002**, *277*, 12931–12936.

68 X. LI and G. R. STARK, *Exp. Hematol.* **2002**, *30*, 285–296.

69 R. WELLER, *Clin. Exp. Derm.* **2003**, *28*, 511–514.

70 K. YAMASAKI, H. D. J. EDINGTON, C. MCCLOSKY, A. LIZONOVA, I. KOVESDI, D. L. STEED, and T. R. BILLIAR, *J. Clin. Invest.* **1998**, *101*, 967–971.

71 E. NOIRI, T. PERESLENI, N. SRIVASTA, P. WEBER, W. F. BAHOU, N. PEUNOVA, and M. S. GOLIGORSKY, *Am. J. Physiol.* **1996**, *270*, C794–802.

72 D. MOST, D. T. EFRON, H. P. SHI, U. S. TANTRY, and A. BARBUL, *Surgery* **2002**, *132*, 866–876.

73 S. KUENZLI and J. H. SAURAT, *Br. J. Dermatol.* **2003**, *149*, 229–236.

74 L. MICHALIK, B. DESVERGNE, N. S. TAN, S. BASU-MODAK, P. ESCHER, J. RIEUSSET, J. M. PETERS, G. KAYA, F. J. GONZALEZ, J. ZAKANY, D. METZGER, P. CHAMBON, D. DUBOULE, and W. WAHLI, *J. Cell. Biol.* **2001**, *154*, 799–814.

75 N. DI-POI, N. S. TAN, L. MICHALIK, W. WAHLI, and B. DESVERGNE, *Mol. Cell.* **2002**, *10*, 721–733.

76 Y. ZHU, J. A. RICHARDSON, L. F. PARADA, and J. M. GRAFF, *Cell.* **1998**, *18*, 703–714.

77 G. S. ASHCROFT, X. YANG, A. B. GLICK, M. WEINSTEIN, J. J. LETTERIO, D. E. MIZEL, M. ANZANO, T. GREENWELL-WILD, S. M. WAHL, C. DENG, and A. B. ROBERTS, *Nat. Cell. Biol.* **1999**, *1*, 260–266.

78 R. O. HYNES, *Cell.* **2002**, *13*, 3369–3387.

79 C. BRAKEBUSCH and R. FÄSSLER, *EMBO J.* **2003**, *22*, 2324–2333.

80 C. BRAKEBUSCH, D. BOUVARD, F. STANCHI, T. SAKAI, and R. FÄSSLER, *J. Clin. Invest.* **2002**, *109*, 999–1006.

81 F. M. WATT, *EMBO J.* **2002**, *21*, 3919–3926.

82 M. A. STEPP, *Dev. Dyn.* **1999**, *214*, 216–228.

83 J. S. MUNGER, X. HUANG, H. KAWAKATSU, M. J. GRIFFITHS, S. L. DALTON, J. WU, J. F. PITTET, N. KAMINSKI, C. GARAT, M. A. MATTHAY, D. B. RIFKIN, and D. SHEPPARD, *Cell.* **1999**, *96*, 319–328.

84 R. GROSE, C. HUTTER, W. BLOCH, I. THOREY, F. M. WATT, R. FÄSSLER, C. BRAKEBUSCH, and S. WERNER, *Development* **2002**, *129*, 2303–2315.

85 S. RHAGAVAN, C. BAUER, G. MUNDSCHAU, Q. LI, and E. FUCHS, *J. Cell. Biol.* **2000**, *150*, 1149–1160.

86 S. RHAGAVAN, A. VAEZI, and E. FUCHS, *Dev. Cell.* **2003**, *5*, 415–427.

87 M. MARSDEN and D. W. DESIMONE, **2003**, *13*, 1182–1191.

88 C. SCHREIDER, G. PEIGNON, S. THENET, J. CHAMBAZ, and M. PINCON-RAYMOND, *J. Cell. Sci.* **2002**, *115*, 543–552.

89 A. P. SOUTH, H. WAN, M. G. STONE, P. J. C. DOPPING-HEPENSTAL, P. E. PURKIS, J. F. MARSHALL, I. M. LEIGH, I. R. HART, and J. A. MCGRATH, *J. Cell. Sci.* **2003**, *116*, 3303–3314.

90 K. M. HODIVALA-DILKE, C. M. DIPERSIO, J. A. KREIDBERG, and R. O. HYNES, *J. Cell. Biol.* **1998**, *142*, 1357–1369.

91 P.D. Lampe, B.P. Nguyen, S. Gil, M. Usui, J. Olerud, Y. Takada, and W.G. Carter, *J. Cell. Biol.* **1998**, *143*, 1735–1747.

92 M. Kretz, C. Euwens, S. Hombach, D. Eckardt, B. Teubner, O. Traub, and K. Willecke, T. Ott, *J. Cell. Sci.* **2003**, *116*, 3443–3452.

93 J. Chen, T.G. Diacovo, D.G. Grenache, S.A. Santoro, and M.M. Zutter, *Am. J. Pathol.* **2002**, *161*, 337–344.

94 A. Lundström, J. Hombom, C. Linqvist, and T. Nordström, *Biochem. Biophys. Res. Com.* **1998**, *250*, 735–740.

95 X.Z. Huang, J.F. Wu, R. Ferrando, J.H. Lee, Y.L. Wang, R.V. Farese, and D. Sheppard, *Mol. Cell. Biol.* **2000**, *20*, 5208–5215.

96 X. Huang, M. Griffiths, J. Wu, R.V. Farese, and D. Sheppard, *Mol. Cell. Biol.* **2000**, *20*, 755–759.

97 X.Z. Huang, J.F. Wu, D. Cass, D.J. Erle, D. Corry, S.G. Young, R.V. Farese, and D. Sheppard, *J. Cell. Biol.* **1996**, *133*, 921–928.

98 X. Huang, J. Wu, S. Spong, and D. Sheppard, *J. Cell. Sci.* **1998**, *111*, 2189 2195.

99 A. Cavani, G. Zambruno, A. Marconi, V. Manca, M. Marchetti, and A. Giannetti, *J. Invest. Dermatol.* **1993**, *101*, 600–604.

100 I. Juhasz, G.F. Murphy, H.C. Yan, M. Herlyn, and S.M. Albeida, *Am. J. Pathol.* **1993**, *143*, 1458–1469.

101 Wang, Y.Q. Zhou, and Y.Q. Chen, *Scand. J. Immunol.* **2002**, *56*, 545–553.

102 M.M. Santoro, G. Gaudino, and P.C. Marchisio, *Dev. Cell.* **2003**, *5*, 257–271.

103 J.A. Bezerra, T.L. Carrick, J.L. Degen, D. Witte, and S.J.F. Degen, *J. Clin. Invest.* **1998**, *101*, 1175–1183.

104 A.J. Cowin, N. Kallincos, N. Hatzirodos, J.G. Robertson, K.J. Pickering, and J. Couper, D.A. Belford. **2001**, *306*, 239–250.

105 F. Mainiero, A. Pepe, M. Yeon, Y. Ren, and F.G. Giancotti, *J. Cell. Biol.* **1996**, *134*, 241–253.

106 B.P Nguyen, X.-D. Ren, M.aA. Schwartz, and W.G Carter, *J. Biol. Chem.* **2001**, *276*, 43860–43870.

107 E.A. O'Toole, *Clin. Exp. Dermatol.* **2001**, *26*, 525–530.

108 M. Kubo, L.V. De Water, L.C. Plantefaber, M.W. Mosesson, M. Simon, M.G. Tonnesen, L. Taichman, and R.A.F. Clark, *J. Invest. Dermatol.* **2001**, *117*, 1369–1381.

109 T.H. Bugge, K.W. Kombrinck, M.J. Flick, C.C. Daugherty, M.J. Danton, and J.L. Degen, *Cell.* **1996**, *87*, 709–719.

110 J. Romer, T.H. Brugge, C. Pyke, L.R. Lund, M.J. Flick, J.L. Degen, and K. Dano, *Nat. Med.* **1996**, *2*, 287–292.

111 A.F. Drew, H. Liu, J.M. Davidson, C.C. Daugherty, and J.L. Degen, *Blood* **2001**, *97*, 3691–3698.

112 R. Pankov and K.M. Yamada, *J. Cell. Sci.* **2002**, *1115*, 3861–3863.

113 T. Sakai, K.J. Johnson, M. Murozono, K. Sajai, M.A. Magnusson, T. Wieloch, T. Cronberg, A. Isshiki, H.P. Erickson, and R. Fässler, *Nat. Med.* **2001**, *7*, 324–330.

114 A.F. Muro, A.K. Chauhan, S. Gajovic, A. Iaconcig, F. Porro, G. Stanta, and F.E. Baralle, *J. Cell. Biol.* **2003**, *162*, 149–160.

115 M. Aumailley, A. El Khal, N. Knoss, and L. Tunggal, *Matrix Biol.* **2003**, *995*, 151–161.

116 F. Decline and P. Rousselle, *J. Cell. Sci.* **2001**, *114*, 811–823.

117 X. Meng, J.F. Klement, D.A. Leperi, D.E. Birk, T. Sasaki, R. Timpl, J. Uitto, and L. Pulkkinnen, *J. Invest. Dermatol.* **2003**, *121*, 720–731.

118 O. Okamoto, S. Bachy, U. Odenthal, J. Bernaud, D. Rigal, H. Lortat-Jacob, N. Smyth, and P. Rousselle, *J. Biol. Chem.* **2003**, *278*, 44168–44177.

119 M.A. Stepp, H.E. Gibson, P.H. Gala, D.D. Sta. Iglesia, A. Pajoohesh-Ganji, S. Pal-Ghosh, M. Brown, C. Aquino, A.M. Schwartz, O. Goldberger, M.T. Hinkes, and M. Bernfield, *J. Cell. Sci.* **2002**, *115*, 4517–4531.

120 W.C. Parks, *Wound Rep. Regen.* **1999**, *7*, 423–432.

121 C.E. Brinckerhoff and L.M. Matrisian, *Nat. Rev.* **2002**, *3*, 207–214.

122 U.K. Saarialho-Kere, S.O. Kovacs, A.P. Pentland, J.E. Olerud, H.G. Welgus, and W.C. Parks, *J. Clin. Invest.* **1993**, *92*, 2858–2866.

123 B.K. Pilcher, J.A. Dumin, B.D. Sudbeck, S.M. Krane, H.G. Welgus, and W.C. Parks, *J. Cell. Biol.* **1997**, *272*, 18147–18154.

124 J.A. Dumin, S.K. Dickeson, T.P. Stricker, M. Bhattacharayya-Pakrasi, J.D. Roby, S.A. Santoro, and W.C. Parks, *J. Biol. Chem.* **2001**, *276*, 29368–29374.

125 T. Itoh, T. Ikeda, H. Gomi, S. Nakao, T. Suzuki, and S. Itohara, *J. Biol. Chem.* **1997**, *272*, 22389–22392.

126 T.H. Vu, J.M. Shipley, G. Bergers, J.E. Berger, J.A. Helms, D. Hanahan, S.D. Shapiro, R.M. Senior, and Z. Werb, *Cell.* **1998**, *93*, 411–422.

127 J.M. Shipley, R.L. Wesselschmidt, D.K. Kobayashi, T.J. Ley, and S.D. Shapiro, *Proc. Natl. Acad. Sci. USA* **1996**, *93*, 3942–3946.

128 K. Holmbeck, P. Bianco, J. Caterina, S. Yamada, M. Kromer, S.A. Kuznetsov, M. Mankani, P.G. Robey, A.R. Poole, I. Pidoux, J.M. Ward, and H. Birkedahl-Hansen, *Cell* **1999**, *99*, 81–92.

129 K.M. Bullard, L. Lund, J. Mudgett, T.N. Mellin, T.K. Hunt, B. Murphy, J. Ronan, Z. Werb, and M.J. Banda, *Ann. Surg.* **1999**, *230*, 260–265.

130 S.E. Dunsmore, U.K. Saarialho-Kere, J.D. Roby, C.L. Wilson, L.M. Matrisian, H.G. Welgus, and W.C. Parks, *J. Clin. Invest.* **1998**, *102*, 1321–1331.

131 C.D. Nobes and A. Hall, *J. Cell. Sci.* **1999**, *112*, 2983–2992.

132 S. Etienne-Manneville and A. Hall, *Cell* **2001**, *106*, 489–498.

133 W.D. Staatz, S.M. Rajpara, E.A. Wayner, W.G. Carter, and S.A. Santoro, *J. Cell. Biol.* **1989**, *108*, 1917–1924.

134 B.D. Sudbeck, B.K. Pilcher, H.G. Welgus, and W.C. Parks, *J. Biol. Chem.* **1997**, *272*, 22103–22110.

135 R. Nuccitelli, *Curr. Top. Dev. Biol.* **2003**, *58*, 1–26.

16
From Tumorigenesis to Tumor Progression: Signaling Pathways Driving Tumor Invasion and Metastasis

Klaudia Giehl, Andre Menke, Doris Wedlich, Michael Beil and Thomas Seufferlein

16.1
Introduction

The signaling pathways that regulate tumor progression and acquisition of the metastatic phenotype are mostly undefined. Besides deregulation in proliferation and apoptosis control, transition into an invasive–metastatic state with enhanced cell migration, invasion and tumor cell dissemination represents a critical step in tumor progression [1]. Increasing numbers of signaling cascades known to be essential in early embryogenesis are currently known to induce tumorigenesis and to promote tumor invasion and metastasis. Emigration of cells out of an epithelium goes along with dramatic alterations in cell shape and function. The entire process is termed epithelial–mesenchymal transition (EMT). Since the early 1990s the role of different signaling cascades in triggering dysfunction of the cadherin/catenin adhesion complex has dominated the discussion about the induction of EMT. Furthermore, mutations in monomeric GTP-binding proteins of the Ras family are found in most tumors and are believed to precede mutations in cell adhesion molecules [2]. Apart from the loss of cell–cell contacts, further molecular processes such as cytoskeletal rearrangements, cellular polarization, directed focal adhesion turnover and cell orientation must follow, to induce cell motility and to drive cell migration over long distances. Genetic screens and the analysis of morphogenetic movements of embryos have revealed novel molecules and pathways which regulate these processes. The monomeric GTPase Ras is a central player in many signal transduction pathways regulating cell motility. Besides Ras, other components of these pathways are also found mutated in tumor cells and this often results in an active signaling cascade, e.g. mutants in receptor tyrosine kinases (c-Met, EGFR: Epidermal Growth Factor Receptor) and upregulation of ligands (HGF/SF, EGF: Epidermal Growth Factor, Wnts). Among the ligands, the class of bioactive lipids (SPC: sphingosylphosphoryl-choline, SIP: sphingosine-1-phosphate and LPA: lysophosphatidic acid) must be emphasized, since they are known as highly potent regulators of cell migration. Recently, in-

creased levels of these bioactive lipids have been found in malignant ascites. Here we summarize how they stimulate invasive metastatic behavior.

Due to the limited space we apologize that we cannot present a complete overview of all motility factors and signaling pathways important
for tumor invasion and metastasis. Instead, we discuss some of the main
signal transduction pathways which are found dysregulated in invasive
tumors.

16.2
Oncogenic Ras Proteins in Tumor Progression and Metastasis

16.2.1
Ras GTPases: Basic Properties

Ras proteins are prototypic members of a large family of approximately
21 kDa membrane-associated, guanine nucleotide-binding proteins
(GTPases). They act as molecular switches for signaling pathways that
modulate many aspects of cell behavior, including cell proliferation, differentiation, survival and apoptosis, as well as cell motility and phagocytosis. The switch mechanism, called the guanosine triphosphate (GTP)/
guanosine diphosphate (GDP) cycle, consists of Ras activation by exchange of bound GDP for GTP, and Ras inactivation by hydrolysis of
GTP into GDP. The kinetics of GDP dissociation and GTP hydrolysis are
modulated by two classes of auxiliary proteins: guanine-nucleotide exchange factors (GEFs) are activators which promote the release of bound
GDP and support the binding of GTP, whereas GTPase-activating proteins (GAPs) stimulate the low intrinsic GTPase activity of Ras, thus ac-
celerating its deactivation [3–5]. The GTP/GDP cycle is regulated by a
wide range of cell surface receptors belonging to the receptor tyrosine kinases, such as epidermal growth factor receptor (EGFR), platelet-derived
growth factor (PDGFR), or hepatocyte growth factor (HGF) receptor c-
Met, the heterotrimeric G protein-coupled receptors, the cytokine receptors, and the integrins. Active, GTP-bound Ras interacts with a variety of
so-called effector proteins, which preferentially interact with the GTP-
form of Ras and require an intact Ras effector domain for interaction [6].
The two best-characterized effectors of Ras are Raf kinases and phosphatidylinositol 3-kinases (PI3-K). Other Ras effectors have also been identified, including RalGDS, AF-6, Nore1/RASSF, MEKK1, Rin1, phospholi-
pase Cε (PLCε), and various protein kinase C (PKC) isoforms. Certain
Ras-GEFs and GAPs may also serve as effector proteins [5–7]. Fig. 16.1
shows some of the upstream GEFs, downstream targets, and GAPs of Ras.

In humans, the Ras superfamily comprises over a hundred related
small GTP-binding proteins, and one common classification places them

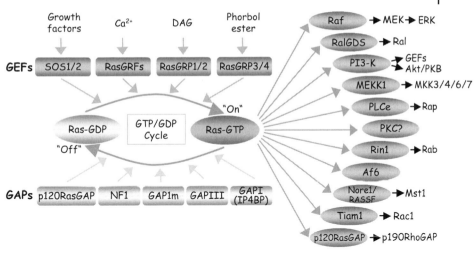

Fig. 16.1 Overview of known Ras regulators and effectors. Ras proteins are activated by multiple extracellular stimuli, such as growth factors activating receptor tyrosine kinases and G protein-coupled receptors, or by second messengers (DAG: diacylglycerol). They induce the activation of various guanine-nucleotide exchange factors (GEFs; orange rectangle) which mediates the exchange of GDP to GTP. Several GTPase-activating proteins (GAPs; yellow rectangle) are involved in downregulating Ras by enhancing the hydrolysis of GTP to GDP. GTP-bound Ras interacts with a variety of downstream effector proteins (green ovals) thereby inducing transcription, cell-cycle progression, differentiation, apoptosis or survival, cell motility and phagocytosis. For details concerning individual GEFs, GAPs or effectors see: [5] and [7].
(This figure also appears with the color plates.)

into six families: Ras, Rho, Arf, Rab, Ran, and Rad. The Ras subfamily consists of the well-known p21 Ras proteins H (*Harvey*)-Ras, N (*Neuroblastoma*)-Ras, and the two splice-variants K (*Kirsten*)-Ras4A and 4B, also known as classical Ras proteins, as well as R-Ras, TC21 (R-Ras2), M-Ras (R-Ras3), Rap1A, Rap1B, Rap2A, Rap2B, RalA, and RalB. The four classical Ras proteins, which are encoded by three *ras*-genes, are closely related and display about 85% sequence identity [5–8].

Mutated variants of the classical Ras proteins (mutations at codon 12, 13, or 61) are found in 30% of all human cancers [9, 10]. K-*ras* mutations are found in nearly all pancreatic adenocarcinomas [11] and are common in lung [12] and colorectal cancer [13]. H-*ras* mutations are found in some types of bladder and kidney cancer and N-*ras* mutations are common in leukemias [10, 14]. Aberrant activation of Ras, either by mutation of the gene or overexpression of the protein, or by deregulated growth factor receptor signaling, has been implicated in facilitating virtually all aspects of a malignant phenotype including proliferation, invasion, and metastasis [15]. Although the involvement of Ras proteins in human cancer is without question, less is known regarding how Ras, especially K-Ras, promotes tumor cell invasion and metastasis.

16.2.2
Oncogenic Ras Proteins: Signaling Specificity

Given that oncogenic Ras mutants show impaired intrinsic GTPase activity and are insensitive to acceleration of GTP-hydrolysis mediated by GAPs, these mutants are in their active GTP-bound state and are independent of external stimuli [16]. This property leads to excessive and inappropriate signal transduction, thus promoting cellular transformation. Although the four Ras proteins share a high degree of amino acid sequence identity, mutations in *ras* genes are not equally distributed in vivo between the Ras isoforms. The vast majority of *ras* mutations in human tumors occur in K-*ras*, whereas mutations in H-*ras* are quite seldom [10]. The preferential association of mutated K-Ras proteins with human malignancies points either towards a higher susceptibility of the K-*ras* gene to mutation or towards a special tendency of mutationally activated K-Ras to promote tumorigenesis in vivo [17]. Recent data suggest that the different Ras isoforms possess different cellular functions. The strongest argument is that both H-*ras* and N-*ras* knockout mice are viable, whereas the inactivation of the K-*ras* gene is embryonic lethal [18]. Comparative transfection studies revealed Ras isoform-specific alterations of cellular functions [19] and gene expression [20, 21]. Moreover, differences in the ability of mutationally activated Ras isoforms to transform murine fibroblasts [22], differences in the sensitivity of Ras isoforms to pharmacological inhibitors [15, 23], and specific interactions of individual Ras isoforms with binding partners have been described [24–26]. By using immortalized fibroblasts from knockout mice, it was demonstrated that K-Ras4B regulates the steady-state expression of matrix metalloprotease (MMP) 2 through a PI3-K/Akt-dependent signaling pathway [27]. The specificity and potency of H-Ras, in contrast to K-Ras, to induce mammary cancer in mice resides within the Ras protein itself. The isoform-specificity is dictated by the C terminus and all regions of nonhomology throughout the entire Ras molecule [28]. The biochemical differences of the Ras isoforms might be the reason for the differences of their biological effects. Ras proteins are divergent in their plasma membrane targeting sequences and in their mode of post-translational modification. All Ras proteins are farnesylated at the cysteine residue at the so-called CAAX-farnesylation motif of the C terminus; the AAX tripeptide is then removed and the a-carboxyl group of the cysteine, which is now carboxy-terminal farnesylated, is finally methylated. H-Ras and N-Ras are furthermore palmitoylated on cysteine residues in the hypervariable region of the C terminus. Finally, H-Ras enters the exocytotic pathway through the Golgi to lipid rafts in the plasma membrane. K-Ras4B, which comprises a polylysine sequence instead of cysteine residues at the C-terminal end, reaches the plasma membrane by an yet unknown, probably microtubule-dependent mechanism omitting the Golgi [29]. Several lines of evidence suggest that H-Ras and K-Ras are localized in distinct microdomains in the plasma membrane and in endomembranes, which could confer specificity to Ras signaling [29]. Moreover, H-Ras and K-Ras preferentially interact with galectin-1 or galectin-3, respectively, and it has been suggested that this determines the duration and selectivity of Ras-induced signal transduction [30, 31].

Most of our knowledge of Ras function in tumorigenesis is derived from studies of Ras-mediated transformation of fibroblasts, especially of rodent fibroblasts. However, given the fact that human cancers with *ras* mutations are most frequently derived from epithelial and haematopoietic cells, we should question whether fibroblasts are a really meaningful system for the study of the role of Ras in human tumorigenesis [32]. Furthermore, a divergent association of Ras mutations with specific neoplasms has been observed when compared in humans and rodents, adding further to the concept of cell-type- and species-specific action of Ras isoforms [3, 18].

16.2.3
Ras-mediated Signal Transduction in Tumorigenesis, Migration, and Invasion

Ras-dependent signal transduction is well represented in its impact on tumor initiation [33, 34]. The transforming activity of Ras requires its association with the plasma membrane which is promoted by the post-translational farnesylation of the C terminus. Farnesyltransferase inhibitors (FTIs) block this farnesylation. Antitumor effects of FTIs have been shown in several model systems, but FTIs are not true Ras inhibitors; especially they do not block the oncogenic activity of the most frequently occurring oncoproteins, K-Ras and N-Ras, and they also inhibit farnesylation of other members of the Ras superfamily [23, 35]. The Ras inhibitor S-*trans,trans*-farnesylthiosalicylic acid (FTS), a S-prenyl derivative, dislodges all types of oncogenic Ras proteins from their membrane anchorage sites and inhibits Ras transformation and growth of xenografted, Ras-transformed cells in nude mice [36, 37, 23]. Thus, inhibition of Ras activity may be useful for preventing tumor initiation and progression. Nevertheless, considerably less is known about the contribution of oncogenic Ras proteins to invasion and metastasis, probably due to the difficulties in finding appropriate model systems for cancer progression in conjunction with physiological expression levels of oncogenic K-Ras [38, 39].

16.2.3.1 **Activation of the Raf-MEK-ERK Cascade**

The best-characterized downstream targets of Ras are the Raf serine/
threonine kinases [40, 41]. Ras targets Raf to the plasma membrane,
where Raf is activated and in turn phosphorylates the dual specificity kinases MAPK/ERK kinase (MEK) 1/2. MEK 1/2 then phosphorylate the
extracellular signal-regulated kinase (ERK) 1/2 [42]. Aberrant Raf-MEK-
ERK signaling has been reported in several tumors and the recent discovery of activated B-*raf* alleles in colon cancer cells, melanoma cells, and
certain other cancer cells supports the importance of this signal transduction in cancer progression and/or maintenance [43, 44]. C-Raf1 and con-
secutively the ERK pathway was most efficiently activated by active K-
Ras4B [25]. Moreover, constitutively active K-Ras4B efficiently induced
cell migration, whereas expression of active H-Ras had minimal and active N-Ras no effect on COS-7 cell migration [26]. Thus, constitutively active K-Ras4B could activate effector pathways that enhance cell motility
and might thereby facilitate invasion and metastasis. In human SW480
colon cancer cells, mutationally activated K-Ras leads to constitutive phosphorylation of ERK and to increased vascular endothelial growth factor
(VEGF) expression, and both processes are suppressed by K-*ras* antisense
oligonucleotides [46]. Hence, K-Ras may contribute to malignancy by activating the ERK-pathway, resulting in angiogenesis, invasion, and metastasis. On the other hand, characterization of the Raf-MEK-ERK-pathway
in human ductal pancreatic carcinoma cell lines, which are characterized
by expression of constitutively active K-Ras proteins, revealed that these
cell lines did not exhibit constitutive active ERK [47, 48]. Similar results
were reported for some colon carcinoma cell lines [49]. The lack of elevated ERK activation can be explained by overexpression of MAP kinase
phosphatase MKP-2 and enhanced dephosphorylation of ERK in some
cell lines [50]. It was further shown that, despite the expression of onco-
genically activated K-Ras, agonist-induced signal transduction *via* the
Ras-Raf-MEK-ERK pathway is maintained in PANC-1 pancreatic carcinoma cells. This process is mediated by wild-type N-Ras and is necessary
for agonist-induced cell proliferation and directed cell migration [48, 51].
Phosphorylation of ERK by the chemo-attractant lysophosphatidic acid
(LPA) resulted in transient translocation of activated ERK to newly
formed focal contact sites at the leading lamellae of migrating PANC-1
cells [51]. Formation of these attachment sites in pancreatic [51] and ovarian cancer cells [52] was dependent on active Ras-Raf-MEK-ERK signaling. The relevance of active MEK and ERK for migration and invasion is
controversial with regard to cell type and chemo-attractant [51]. However,
there is increasing evidence for a synergistic interaction between integrin-mediated cell adhesion and ERK activation in cell migration [54–56],
pointing towards a dual function of ERK, one as a well known mediator
of gene expression and regulator of cell proliferation and the other as
regulator of chemotaxis of tumor cells.

In defining the molecular aspects engaged in tumor invasion and metastasis, the group of H. Beug investigated signal transduction in the epithelial to mesenchymal transition of tumor models, such as mouse mammary epithelial cells as in vitro correlates to invasion and metastasis [57, 39]. They discovered that most epithelial cell types require the cooperation of at least two signaling pathways, such as collaboration of active Ras with transforming growth factor (TGF)β. Mammary epithelial cells with hyperactive ERK or PI3-K pathways were tumorigenic after injection into nude mice, but metastasis formation required hyperactivated ERK plus TGFβ-induced EMT [58]. Gene expression profiling of H-Ras-transformed mammary epithelial cells undergoing EMT identified a small set of genes specifically regulated during EMT involving key regulators in cell proliferation, survival, epithelial polarity, and motility/invasiveness [59]. Furthermore, the TGFβ-induced EMT and the subsequently derived transcriptional phenotype of pancreatic carcinoma cells is dependent on an interaction of Ras- and TGFβ-induced signaling pathways [60]. Oncogenic Ras cooperates with several other proto-oncogenes and with p53 deficiency to promote the transformation of primary cells [61], thereby overriding Ras-induced cell cycle arrest and replicative senescence [62, 63]. In addition, highly invasive and metastatic pancreatic adenocarcinomas emerge from the cooperative action of oncogenic K-Ras and deletion of Ink4A/Arf tumor suppressor in mice [64].

16.2.3.2 Extracellular Matrix Degradation

Ras-induced metastatic behavior of cells, involving breakdown and penetration of intercellular matrix and basement membranes, has been correlated with increased type IV metalloproteinase (MMP) activity. MMPs form a family of at least 20 members [65]. Regulation of expression of these enzymes has been shown to be downstream of oncogenic Ras signaling and differs with regard to MMP and/or cell type [27, 66, 67]. Normally, MMP expression is induced by growth factors and this induction requires activation of MAPK pathways, which in turn act through activation of AP-1 and ETS family of transcription factors [68], but can also be induced by oncogenic Ras. The invasive and metastatic phenotype of primary and experimental tumors is also correlated with overproduction of urokinase-type plasminogen activator (uPA) [69] and cathepsin B lysosomal cysteine proteinase [70].

16.2.3.3 Activation of Phosphatidylinositol 3-Kinase and Rho GTPases

The phosphatidylinositol 3-kinases are Ras targets [71, 72], which regulate phosphoinositide lipid metabolism and production of phosphatidylinositol 3,4,5-trisphosphate (PIP$_3$) [71]. Enhanced PIP$_3$ production results in activation of the serine/threonine kinase Akt/PKB, of which three isoforms exist. Several studies have demonstrated the importance of PI3-K/ Akt in Ras-induced transformation of some cell types [74]. The main effects of Akt activation with regard to cancer biology are: support of cell survival [75], cell proliferation and cell growth [75]. A functional role for Akt in regulating cell motility was first reported by Meili and coworkers [76], who demonstrated that activation and membrane translocation of Akt was necessary for *Dictyostelium discoideum* chemotaxis. Similar effects were also reported in other cell systems, especially in cancer cells, indicative of a common role of Akt in cell migration [77, 78]. A mechanism describing how Akt supports a motile, invasive phenotype of carcinoma cells might be through induction of EMT with downregulation of the cell adhesion molecules E-cadherin and β-catenin, upregulation of vimentin and acquisition of a fibroblastoid phenotype [79] and by increasing the secretion of MMPs [80]. Furthermore, the remodeling of actin filaments, a key event in cell migration, is mediated by active PI3-K through Akt and p70S6 kinase, a target of Akt [81, 82].

PI3-Ks are also involved in activating Rho GTPases, such as the small GTPase Rac [83]. Rac GTPases promote increased cell motility and contribute to enhanced tumor cell invasion and metastasis by regulating the reorganization of the actin cytoskeleton and the formation of membrane ruffles [84]. Ras and Rho GTPases can be activated in series, such that one GTPase stimulates GTP loading on another [85]. Constitutively activated Ras is an activator of Rac in fibroblasts [84], but a suppressor of Rac activity in epithelial cells [87]. Although there is little doubt, that the coordinated activities of at least the four GTPases, Ras, Rac, Rho, and Cdc42, are necessary for directed migration of a variety of cell types, the molecular mechanisms of a cross-talk of Ras and Rho GTPases [85] remains to be clarified in detail. The cross-talk of Ras and Rho GTPases and the role of Rho GTPases in enhancing migration and tumorigenesis is the subject of numerous excellent reviews the reader is referred to [88–91].

In summary, our knowledge about the function of oncogenic Ras and the downstream mechanisms in contributing to increased motility, invasiveness, and metastatic potential is clearly determined by the fact that these mechanisms are much more complicated than originally thought, with cross-talks of signal transduction pathways and variations with regard to cell and tumor type.

16.3
Signals Controlling Cellular Adhesion Complexes

A prominent feature during differentiation of epithelial cells is the establishment of close contacts between neighboring cells. This cell–cell adhesion is mediated by anchoring junctions which are composed of transmembrane adhesion proteins whose intracellular domains are connected with the cytoskeleton. Best characterized are adherens junctions composed of calcium-dependent cell adhesion proteins of the cadherin family. Classical cadherins (E-cadherin, N-cadherin and P-cadherin) are composed out of five extracellular domains mediating homotypic association of opposing cadherin molecules. Inside the cell, a single transmembrane domain is followed by a highly conserved amino acid sequence containing binding sites for different proteins which link cadherins to the actin cytoskeleton. Among these intracellular binding proteins, α-, β-, and γ-catenin form a stable complex with the cytoplasmic part of classical cadherins. Whereas β- or γ-catenin can bind directly to the C-terminal part of E-cadherin, α-catenin links the cadherin/β- or γ-catenin complex to the actin cytoskeleton (Fig. 16.2) [92, 93]. Reduction of cell–cell adhesion is not only common in developmental processes like formation of the primitive streak in chicken or migration of neural crest cells [94], but also represents an early step in tumor progression from benign to malignant types of epithelial cancer. Complete loss of parts of this adhesion module in the progression of epithelial tumors was demonstrated by different groups [95–98]. It was shown that various tumors are associated with allelic deletions or loss of heterozygosity on chromosome 16, where the *E-cadherin* gene is localized. Loss of E-cadherin was associated with invasive growth of carcinoma [99]. Beside deletions of the *E-cadherin* gene, downregulated expression of the *E-cadherin* gene by transcription factors such as Snail, Slug and Sip1 has been described [100, 101]. Overexpres-

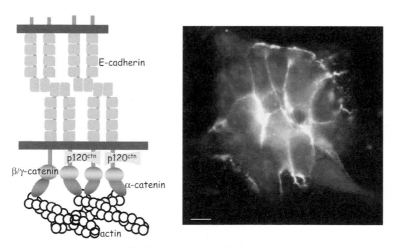

Fig. 16.2 Formation of cell–cell adhesion complexes in epithelial cells by E-cadherin and catenins (left). Immunofluorescence localization of E-cadherin (green), β-catenin (red) and DNA (blue) in cultured epithelial cells (PaTu8988s). Co-localization of E-cadherin and β-catenin is indicated by the yellow color. The bar represents 10 μm. (This figure also appears with the color plates.)

sion of these transcription factors can be observed in different carcinoma-derived cell lines resulting in decreased *E-cadherin* gene expression [101]. Moreover, there is growing evidence that in addition to transcriptional changes of the *E-cadherin* gene, the deregulation of the assembly of E-cadherin complexes and of their association with the cytoskeleton [102] contributes to defects in cell–cell adhesions [103]. The composition of the E-cadherin/catenin adhesion complex is controlled by post-translational modification of E-cadherin or the different catenins. The mechanisms engaged in E-cadherin/catenin modification and complex deregulation include phosphorylation, proteolysis, endocytosis as well as sterical hindrance [104, 105].

16.3.1
Tyrosine Phosphorylation Induces E-cadherin/catenin Complex Disassembly

The importance of tyrosine phosphorylation on the regulation of cell adhesion has been suggested by the ability of tyrosine kinases, such as growth factor receptors or the non-receptor kinase Src, to disrupt cytoskeletal–membrane interactions [106]. It has been demonstrated in various tumors that overexpression of tyrosine kinases such as the EGF receptor (EGFR) together with increased amounts of its ligand, overexpression of the proto-oncogene ErbB2 or of the receptor c-Met correspond to advanced stages in tumor progression [107–111]. Hoschuetzky et al. [111] showed 1994 that the unstimulated EGF-receptor coprecipitated with the E-cadherin complex. This interaction was mediated by β-catenin. Upon stimulation with EGF, the activated EGFR phosphorylated β- and γ-catenin resulting in dissociation of the E-cadherin complex [107, 112]. It is accepted that tyrosine phosphorylated β-catenin does not link the E-cadherin/β-catenin complex to α-catenin and the cytoskeleton resulting in reduced cellular adhesion and cell aggregation [112–114]. Activation of the HGF/SF-receptor c-Met has also been correlated with reduced E-cadherin-mediated cell adhesion. Mutation or overexpression of c-Met is associated with tumor progression in kidney, thyroid, pancreas or colon cancer [115, 116]. It has been shown that oncogenic Met reduced cell aggregation in MDCK cells by downregulation of functional E-cadherin/catenin complexes [117].

In addition to receptor tyrosine kinases, the tyrosine kinase Src, the first oncogene detected [118] is suggested to contribute to the reduction of E-cadherin adhesion complexes by β-catenin phosphorylation in epithelial cells [113, 114, 119]. Src can be activated by multiple signals, such as growth factor signaling (EGF, HGF) or colony stimulating factor (CSF) but also by integrin-induced assembly of focal contacts. Although the exact molecular mechanisms by which activated Src destabilizes the E-cadherin/catenin complex are still puzzling, some signaling mechanisms have been proposed. Expression of activated Src in epithelial

MDCK cells leads to loss of E-cadherin function, mediated by tyrosine phosphorylation of β-catenin [113]. It was postulated that the Src kinase, which is membrane anchored through amino-terminal myristoylation, can directly interact with the E-cadherin complex and phosphorylate β-catenin. Stimulation of the breast carcinoma cell line MCF-10 with CSF-1 resulted in Src-dependent internalization and downregulation of E-cadherin. This process is also correlated with a decrease in cellular adhesion [120]. Data from numerous studies show that Src activity is necessary for much more, mostly indirect, signaling events leading to downregulation of E-cadherin adhesion [113, 114, 121, 122]. Among these signaling cascades the influence of Src on p120^{ctn} and activation of the focal adhesion kinase FAK has been correlated with regulation of cell adhesion and will be discussed later. A recently-identified novel Src-binding protein, the E3 ubiquitin ligase Hakai [123] is also able to downregulate E-cadherin complexes. Tyrosine phosphorylation of E-cadherin, as induced by Src, resulted in recruitment of Hakai to the E-cadherin/catenin complex. This recruitment was followed by disruption of the E-cadherin adhesion complex, irregular E-cadherin localization in the cytoplasm and endocytosis of E-cadherin as a consequence of E-cadherin ubiquitination induced by Hakai [123]. Additional experimental data show that HGF treatment enhanced the interaction between Hakai and E-cadherin [123].

In addition to the importance of tyrosine kinases, the interaction of the E-cadherin/catenin complex with tyrosine phosphatases such as PTPμ, PTPκ, PTP1B, SHP2 and PTEN contributes to the complex regulation of cell adhesion [124, 125]. Tyrosine phosphatase inhibitors can increase the phosphorylation of the E-cadherin/catenin complex, especially of β-catenin, leading to the dissociation of the adhesion complex and enhanced cell migration [125, 126]. Especially for the receptor tyrosine phosphatase PTPκ, it was demonstrated that it interacts with β- and γ-catenin in breast carcinoma cells suggesting a function of this phosphatase in dephosphorylation of both catenins [127].

16.3.2
Extracellular Matrix-induced Inhibition of E-cadherin-mediated Cell Adhesion

An interesting way of modulating E-cadherin-mediated cell adhesion has been described in colon and pancreatic tumors that are characterized by large amounts of extracellular matrix (ECM). Different components of the ECM, such as collagen type I and III and laminin, induce the disassembly of E-cadherin complexes. Data from studies of Brabletz and coworkers using colon carcinoma also revealed an extracellular-matrix-induced reduction or loss of E-cadherin in cells of the invasive front of these tumors. They showed that, especially in cells of the invasive front, loss of membrane-associated E-cadherin and nuclear accumulation of β-catenin is associated with dissemination of cells at the metastatic site [128]. The authors propose a subsequent redifferentiation and regain of epithelial characteristics after formation of metastases, pointing to a transient dedifferentiation of tumor cells induced by the tumor environment especially by proteins of the extracellular matrix [129]. Beside a downregulation of *E-cadherin* gene expression induced by collagen [121], first experimental data provide evidence that integrin-dependent activation of the focal adhesion kinase (FAK) results in enhanced phosphorylation of β-catenin and subsequent reduction of E-cadherin-mediated cellular adhesion. Binding of collagen to integrins results in focal contact formation and activation of Src kinase. As a consequence, activated Src phosphorylates and activates FAK, which is involved in different signaling pathways associated with cell migration. More recently, we identified that activated FAK is associated with the E-cadherin complex and results in phosphorylation of β-catenin, followed by destabilization of the E-cadherin complex and increased cellular motility (A. Menke, A. König unpublished results). These data are supported by studies of Avizienyte and colleagues using colon cancer cells and Genda et al. using hepatocellular carcinoma cells, which demonstrated integrin-induced downregulation of E-cadherin. This deregulation was dependent on integrin, Src and FAK activity [130, 131]. The destabilization of the E-cadherin adhesion complex was potentiated by an additional collagen-induced effect involving the phosphatase PTEN. Stimulation of pancreatic tumor cells with collagen type I or III resulted in dissociation of PTEN from the E-cadherin/catenin complex. Co-immunoprecipitation studies revealed an association of the phosphatase PTEN with the β-catenin of the E-cadherin/catenin complex. The ability of PTEN to dephosphorylate β-catenin was confirmed by in vitro phosphorylation assays (A. Menke, A. König unpublished results). Based on the observation that the phosphatase PTEN is mutated in many carcinomas, detailed analyses confirmed a function of PTEN in regulating of cell–cell adhesion. Apart from its lipid phosphatase activity, PTEN functions as a tyrosine and serine/threonine phosphatase. Experimental data suggest that regulation of growth and survival is correlated with the lipid phosphatase activity whereas invasive growth is associated with loss of the tyrosine phosphatase function [132]. A crucial role of PTEN in the regulation of E-cadherin complex assembly was also suggested by Kotelevets and colleagues, who demonstrated that expression of wild-type PTEN in MDCK cells reversed a Src-induced loss of E-cadherin and a Src-induced invasive phenotype [133].

16.3.3
The Protein p120^{ctn}

E-cadherin function is also regulated by cytoplasmic binding partners such as p120^{ctn}. Originally identified as Src substrate p120^{ctn} interacts directly with classical cadherins [134]. Although, p120^{ctn} is structurally similar to β- and γ-catenin, possessing a central core domain of tandem armadillo repeats, the p120^{ctn}-binding domain of E-cadherin was localized to the so-called juxtamembrane domain, which do not overlap with the binding area of *β*- and *γ*-catenin. In normal epithelial cells p120^{ctn} shows a virtually indistinguishable co-localization with E-cadherin and *β*-catenin [135]. In tumors where E-cadherin expression is lost, p120^{ctn} is diffusely distributed. Abnormalities or loss of p120^{ctn} have been reported in numerous malignancies, such as colorectal, bladder, gastric, breast, prostate, lung or pancreatic tumors, and is associated with poor prognosis [136]. These observations suggest that p120^{ctn} regulates E-cadherin's adhesive interactions between neighboring cells. P120^{ctn} has been implicated in the mechanism of cadherin clustering, which is necessary for a strong cell–cell adhesion and formation of adherens junctions. However the role of p120^{ctn} is controversial since different observations suggest that p120^{ctn}-binding to E-cadherin can both positively and negatively influence adhesive activity [134, 137, 138]. This apparent contradiction may be explained by the large number of p120^{ctn} isoforms found in different cell types [139]. The isoforms which arise as a result of use of four different ATG start sites and alternatively spliced exons in the carboxy-terminal part of the protein, have been described to possess distinct biological functions in normal and transformed cells [136, 140]. Whereas the larger isoforms were found mainly in highly motile cells like fibroblasts, the shorter isoforms have been described in differentiated epithelial cells. Recently, we could show that pancreatic epithelial cells which ectopically expressed E-cadherin or N-cadherin did not exhibit differences in their p120^{ctn} content. But co-immunoprecipitation studies revealed that E-cadherin was preferentially associated with shorter p120^{ctn} forms, in contrast to N-cadherin which was associated with the longer isoform 2 of p120^{ctn} [141]. Regarding their aggregation capacity, only cells expressing E-cadherin bound to short p120^{ctn} isoforms showed increased cellular adhesion.

Moreover, p120^{ctn} activity depends also on its tyrosine phosphorylation state. Little information exists about the molecular mechanisms by which phosphorylation influences the interaction of p120^{ctn} with E-cadherin. Data from Src-transformed epithelial cells suggest that enhanced tyrosine phosphorylation of p120^{ctn} contributes to defective E-cadherin adhesion complexes [142]. We could show that p120^{ctn} associated with N-cadherin, which does not increase cellular adhesion, was highly phosphorylated on tyrosine residues, whereas p120^{ctn} associated with E-cadherin showed no phosphorylation [141]. The cytoplasmic tyrosine kinase FER, which can be activated by growth factor receptors like EGF or PDGF receptors, interacts and phosphorylates specifically the larger p120^{ctn} isoform 1. This enhanced phosphorylation is correlated with enhanced cell migration in fibroblasts [143].

In conclusion, it is clear that functional cell–cell adhesion mediated by E-cadherin protein complexes is necessary for epithelial cell adhesion. After first reports about loss of E-cadherin in metastases of epithelial tu-

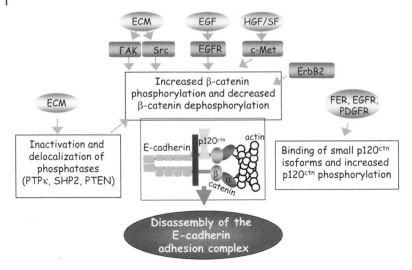

Fig. 16.3 Regulatory processes can contribute to a dissociation of E-cadherin adhesion complexes, reduction of cellular adhesion of epithelial cells, and induction of the dissemination of tumor cells. (This figure also appears with the color plates.)

mors being responsible for reduced cell–cell adhesion, increasing evidence exists that the deregulation of dynamic cellular adhesion represents additional sources which contribute to reduced cellular adhesion and induces dissemination of cells from the original tumor (Fig. 16.3).

16.4
Switching from Cell–cell Adhesion to Migration: a Role for β-Catenin as Nuclear Wnt-signal Transducer?

The dual function of β-catenin as a crucial component of the cadherin/catenin adhesion complex as discussed before, and as a transcriptional coactivator of the LEF/Tcf-HMG box family members, supports the idea that dysfunction of the adhesion complex results in nuclear translocation of β-catenin and induction of genes important for tumor progression and invasion. Indeed, tyrosine phosphorylation of β-catenin by several growth factors and their receptors, e.g. HGF/SF and c-Met, EGF and EGFR, VEGF, Hcregulin-beta1 (HRG), prostaglandin E (PGE2) and IGF2 disrupts β-catenin binding to E-cadherin or α-catenin and leads to relocalization of β-catenin to the cytoplasm and nucleus [144–149]. While the effects of phosphorylated tyrosine residues on the cadherin/catenin adhesion complex are well studied (for review see [150]) less is known about their effects on β-catenins function as transactivator in the LEF/Tcf- transcription complex.

Structural data about the interaction sites of β-catenin with E-cadherin and Tcf-3/4 give evidence why tyrosine phosphorylation of β-catenin prevents E-cadherin binding but does not influence Tcf-3/4 or LEF association. Receptor tyrosine kinases and also Src phosphorylate Y^{654} in arm repeat 12 of β-catenin [145, 151, 113]. It is thought that the negative charge of p-Y^{654} electrostatically repels the D^{665} of E-cadherin or that the size of the phosphate group might hamper binding of E-cadherin [152, 153]. Interestingly, Tcf-3/4 transcriptions factors mimic the cadherin/β-catenin interface as they contact the same lysine residues in β-catenin, namely K^{317} and K^{435}, as E-cadherin does [154, 155]. However, the transcription factors require the N-terminal located arm repeats 2-8 for binding [154] whereas E-cadherin associates with the more C-terminal located ones 5-12 [156]. The latter might explain why p-Y^{654} affects E-cadherin but not LEF/Tcf binding. Importantly, p-Y^{654} enhances the binding of the TATA box binding protein (TBP) to β-catenin [153] which might result in a general increase of transcriptional activity. Further studies are necessary to elucidate whether increasing amounts of TBP influence the assembly of the β-catenin/LEF/Tcf transcription machinery.

There is a remarkable report by Kim and Lee [157] who induced the release of tyrosine phosphorylated β-catenin from the adhesion complexes by pervanadate treatment of different cell lines. They did not observe any remarkable alteration in β-catenin binding to LEF-1 nor did they find a change in the expression of the target gene cyclin D1. When tyrosine phosphorylated β-catenin binds to the Tcf/LEF transcriptions factors without enhancing the transcriptional activity the question arises whether tyrosine phosphorylation protects β-catenin from recognition through the APC/Axin/Gsk3β-complex, and degradation. In this context the influence of Y^{86} and Y^{142} in β-catenin might be of interest because they are substrates for Src, Fer and Fyn kinases and located next to the region which is important for Gsk-3β/Axin/APC recognition. Brembeck et al. [299] recently reported that HGF/SF induced phosphorylation of Y^{142} favors binding of BCL9-2 which in collaboration with pygopus is required for β-catenin dependent Wnt target gene activation.

An increased LEF/Tcf-dependent target gene expression caused by nuclear accumulation of β-catenin has been observed after cell transfection with oncogenic mutants of RON and c-Met [158] or after IGF2 treatment [149]. Since IGF2 as well as RON and c-Met activate various downstream effectors (Fig. 16.4) it is not excluded that other mechanisms than tyrosine phosphorylation of β-catenin are responsible for the activation of Wnt target genes, e.g. serine/threonine dephosphorylations of β-catenin or modifications of other components of the Wnt-signaling cascade.

Release of β-catenin from the adhesion complex does not inevitably lead to enhanced Wnt signaling. The stability of cytosolic β-catenin is

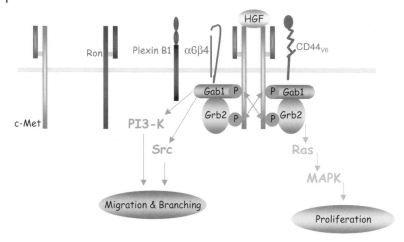

Fig. 16.4 Signaling pathway of HGF/SF via c-Met responsible for migration and branching. Putative co-receptors are labeled in magenta. (This figure also appears with the color plates.)

controlled by a multiprotein complex sequestering β-catenin to the ubiquitin/proteasome pathway (see Nusse homepage: http://www.stanford.edu/~nusse/pathways/WntHH.html). Dysfunction or bypassing the APC/Axin/Gsk3β control point rather than disruption of the cadherin/catenin complex results in oncogenic activity of β-catenin. The co-transfactor function of β-catenin is discussed to be required long before the cell-adhesion break down occurs in colon carcinogenesis [159]. This is a strong argument against the significance of the β-catenin pool released from adhesion complexes in driving tumorigenesis.

16.4.1
The Contribution of Canonical Wnt/β-Catenin Signaling to Cell Migration

While the role of canonical Wnt/β-catenin signaling in cancerogenesis is compelling [159, 160] its contribution to invasion and metastasis is still an open question. In the embryo, different Wnt signaling cascades drive morphogenetic movements (see Section 5.2.3.2). This is often forgotten when Wnt signaling in tumor invasion is taken into consideration.

Strong evidence for a stimulating effect of canonical Wnt signaling on cell migration was given when upregulation of direct Wnt target genes like MMP-7 [161], the invasion factor $\gamma 2$ chain of laminin-5 [162], MT-MMP [163], fibronectin [164], urokinase plasminogen activator (uPA) [165] was observed. These proteins enable tumor cells to infiltrate the surrounding tissue. In addition, β-catenin is located in the nuclei at the tumor invasion front where cells undergo EMT before starting emigration and infiltration [166]. Growth factors like VEGF [167] and BMP4 [168], tyrosine kinase receptors [169] and co-factors like CD44 [170] and the transcription factors Tcf-1 [171] and LEF-1 [172] are also upregulated by canonical Wnt signaling in human colon cancer. This indicates to further enhancement of tumor invasion by activating other pathways and by autoregulation of Wnt signaling.

A positive effect on cell movement by canonical Wnt signaling is also known from the embryo. Importantly, this stimulating effect was specific for LEF-1 [173] which correlates to the predominant expression of LEF-1 in metastatic melanomas [174]. Secreted frizzled, dickkopf3 and dnLPR5 which were initially characterized as inhibitors of canonical Wnt-signaling in embryonic development are also able to reduce migration of tumor cells [175, 176] confirming the potential of canonical Wnt signaling in promoting cell dissemination.

16.4.2
Is Non-canonical Wnt Signaling Important for Tumor Invasion and Tumor Cell Dissemination?

In embryos non-canonical Wnt signaling cascades (see Section 5.2.3.2) play a crucial role in tissue movement. The planar-cell-polarity (PCP)-pathway, for example, induces cellular polarization which is obvious in bipolar orientated lamellipodia and filopodia formation [177]. The non-canonical Wnt signaling cascades, namely the Wnt/Calcium pathway [178] acts through PKC and CamKII and influences cell motility via cross-talking with the canonical Wnt signaling cascade [173]. Recently, first evidence for the involvement of non-canonical Wnt signaling in tumor invasion has been reported. Gene expressing, profiling and immunostaining of malignant melanoma biopsies revealed upregulated Wnt5a predominantly at sites of active invasion [179, 180]. Importantly, overexpression of Wnt5A in melanoma cells increased cell motility via activation of PKC α/βII and PKCμ while no change in β-catenin expression or localization was observed [180]. This points to an activation of non-canonical Wnt signaling in driving tumor cell migration. Thus, we are only now beginning to understand the different roles of Wnt signaling pathways in tumor invasion and metastasis.

16.5
The HGF/SF Signaling Cascade Mediated via the c-Met Receptor

Among all growth factors HGF/SF, also known as PRGF-1 (plasminogen related growth factor 1) [181], is outstanding as motility factor. This corresponds to its primary physiological function in the embryo, which is the induction of epithelial–mesenchymal transitions (EMT). In various tissues like liver, kidney and mammary, EMT is a prerequisite for branching morphogenesis and thus under control of HGF/SF. The branching epithelium is surrounded by mesenchyme that produces the growth factor while the epithelial cells expressing the c-Met receptor respond to the signal with destabilizing cell–cell contacts, increasing proliferation, cell motility and scattering. HGF/SF also stimulates migration over long distances as described for muscle progenitor cells which derive from somatic mesoderm and spread into the limb buds ([182], see also Section 11.4). Besides its role in the development HGF/SF has a unique role in activating a concert of key events important for tumor progression and invasiveness. The HGF/SF/c-Met signaling pathway is content of numerous excellent reviews [181, 183–186]. Here we focus on events and crosstalks in c-Met signal transduction important for tumor cell migration.

16.5.1
c-Met, RON and Alternate Receptors are Important for Motility Induction

Upon ligand binding the c-Met receptor tyrosine kinase becomes autophosphorylated at several tyrosine residues. Phosphorylation of Y^{1230}, Y^{1234} and Y^{1235} activates the intrinsic kinase activity of c-Met [183] while phosphorylation of Y^{1349} and Y^{1356} results in the formation of a multiprotein complex at the C terminus of c-Met consisting of adaptor proteins and signal transducers (see detailed description in Section 11.2). Since HGF/SF binding leads to the activation of various c-Met signaling pathways with different biological effects several attempts have been undertaken to identify tyrosine residues and signal transducers important for individual biological outcomes. Maina et al. [186] tried to identify selective signal transducers by generating knockin *met* mutants that still bind Gab1 but possess optimal binding sites either for PI3-K or Src. Compared to the complete loss of premyoblast migration observed in *met* null mutants, the *met* mutants with optimal binding sites for PI3-K or Src showed some residual migratory activity. This indicates that apart from Gab1 both kinases, PI3-K and Src, are required for migration. Taking into account that uncoupling Grb2 from c-Met results in decreased proliferation of muscle progenitor cells but does not affect their migration [187]. Gab1 rather than Grb2 might function as main adaptor protein of c-Met mediating premyoblast migration over long distances [186, 188, 189, see Fig. 16.4).

These signaling pathways (Fig. 16.4 and Fig. 16.6) are also activated by RON (receptor d'orgine nantais), C-Met and RON form the family of plasminogen related growth factor receptors. The ligand of RON is the macrophage-stimulating protein (MSP), also termed PRGF-2. Altered expression levels or point mutations of RON are found in numerous tumors and correlate with high invasive potential [190–194]. The extracellular domains of c-Met and RON possess a *sema* domain of 500 aa in length and a Met related sequence (MRS) with eight conserved cysteine residues [181]. The *sema* domain was initially detected in semaphorins which are cell surface-bound and soluble signaling molecules that control axon guidance and cell migration in embryonic development. They are discussed to form steep signaling gradients and to mediate long- and short-range signals [195]. Their receptors include the plexins, a family of large transmembrane proteins which also contain a *sema* domain. Apart from this domain plexins and semphorins also share the MRS motif with c-Met and RON [195].

Considering the structural homology it seems not surprising that a physiological relationship has been reported [196]; Plexin B1 forms a complex with c-Met. When this heterodimeric receptor binds to Semaphorin 4D, c-Met and Plexin B1 get phosphorylated by c-Met tyrosine kinase. This correlates to the observed cooperative effect of Semaphorin 4D and HGF/SF in stimulating invasion of liver progenitor cells. In addition, Conrotto et al. [197] found that RON together with Plexin B1 can form a receptor complex and that Semaphorin 4D is a ligand for c-Met and RON in promoting invasive growth.

16.5.2
Cellular Targets of HGF/SF Signaling in Cell Scattering

Apart from the growing number of proteins interacting with c-Met little is known how the cellular effects of HGF/SF are induced, e.g. the activation of the small GTPases Cdc42, Rho, Rac [188], disassembly of the cadherin/catenin complex [198] or the induction of uPA and MMP expression [109, 200]. Evidence points to the predominant involvement of PI3-K [201] and the activation of Est-1 in MMP upregulation [202]. A novel target gene of HGF, Mig-7, has been identified in a subtraction hybridization of HGF-treated endometrial carcinoma cells [203]. Mig-7 strongly supports HGF-induced cell migration. The underlying mechanism, however, still remains elusive.

16.5.3
Bypassing Ligand Binding in c-Met Activation

Dysregulation of c-Met signaling in absence of the ligand is often observed in tumors. Some mutants in the c-Met-receptor are reported which are associated with increased cell migration.

The point mutations Y1248C and L1213V which are located within the tyrosine kinase domain promote cell migration and invasion [204]. The point mutation M1268T which is outside the tyrosine kinase domain results in inappropriate phosphorylation of c-Abl kinase substrates and β-catenin. The latter accumulates in the cell and the canonical Wnt signaling pathway is activated [205]. As discussed before this signaling cascade also stimulates cell migration.

Apart from mutations in c-Met, which result in its constitutive activation integrins, ($\alpha6\beta4$) [205, 206] or CD44 [207] are able to stimulate c-Met signaling when they are associated with the receptor. CD44 represents a family of surface glycoproteins derived from a single gene by alternative splicing and glycosylation. The splice variants represent insertions (v1-10) at the membrane proximal region of CD44. Those variants containing v4-v7 or v6-v7 are highly correlated to tumor invasiveness. CD44 is thought to mediate cell-substrate adhesion by binding hyaluronate of the extracellular matrix [208]. Importantly, only the CD44 splice variant that contains the v6 epitope directly cooperates with c-Met and is required for the receptor stimulation [207]. Different mechanisms are discussed how CD44v6 mediates this effect, either by binding to hyaluronic acid and by-passing HGF/c-Met interaction, or by presenting HGF to c-Met in a multivalent complex in which signaling activity is also conferred to the cytoplasmic domain of CD44 [207, 209]. Nevertheless, independent of the mode, an upreglation of CD44 which is often observed in invasive tumors points to an amplification of c-Met signaling. To keep an eye on the regulation of CD44 gene seems necessary as HCF/SF was shown to induce CD44v6 expression [210] so that autoregulation via CD44-HGF/SF-c-Met might keep tumor invasion running.

16.6
Bioactive Lipids

Derivatives of phospholipids like lysophatidic acid (LPA), sphingosine 1-phosphate (S1P) or sphingosylphosphorylcholine (SPC) (Fig. 16.5) activate a broad spectrum of cellular processes when they reach a critical concentration in serum. Apart from inducing proliferation, apoptosis, differentiation, angiogenesis and secretion they are known as strong inducers of cell motiliy [211, 212]. In general, they are synthesized by activated platelets or growth factor stimulated cells. Their steady state plasma level is normally very low which is achieved by an equilibrium between synthesis, degradation and clearance. Increase of serum concentrations of bioactive lipids results in tumorigenic effects such as increased cell proliferation or cell migration. For LPA the critical concentration in the serum has been determined. The normal serum contains less than 100 nM LPA. Full receptor activation and signaling is already observed when cells in culture are treated with 100–200 nM LPA. This explains that the serum contains inhibitors and activators of LPA to keep LPA activity balanced [211].

Bioactive lipids were thought to be metabolites of eukaryotic membrane turnover. The discovery of the different G protein coupled recep-

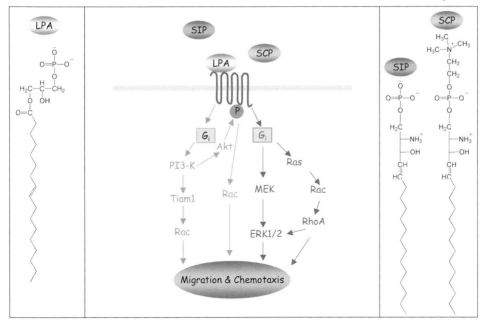

Fig. 16.5 Structure of bioactive lipids and the signaling pathways leading to stimulation of migration and chemotaxis. Pathways described in [235] (green), [236] (red) and [52] (magenta). (This figure also appears with the color plates.)

tors (LPA$_{1-4}$ or S1P$_{1-5}$) which are activated by distinct bioactive lipids led to a novel understanding of LPA, S1P and SPC as signaling molecules. The identification of specific signaling cascades characterizes these molecules as important regulators of cell homeostasis against physiological stress [214]. Ovarian cancer cells [215] have recently been detected as pathophysiological source of LPA. In ovarian cancer cells a laminin-induced autocrine loop of LPA production seems responsible for the increase in LPA. The accumulation of LPA in malignant body fluids such as malignant ascites [216] is thought to derive from microvesicles which are shedded from tumor cell membranes. Autotaxin/lysoPLD activity leads to delivery of LPA from the microvesicles [211].

Increasing evidence is given that bioactive lipids also play a crucial role in embryonic development. Enzymes that are required in their metabolism but also lipid receptors are found differently expressed and coupled to processes of germ cell migration [217, 218], heart development [219] and neurogenesis, craniofacial differentiation [220], vasculogenesis and axis formation [221] in invertebrates and vertebrates. The evolutionary conservation of these molecules in structure and function underlines the significance of the bioactive lipids as signaling molecules.

16.6.1
The Receptors of Bioactive Lipids

All bioactive lipids bind to GPCRs (G protein coupled receptors). For LPA three highly affine receptors of the EDG_{2-4} (endothelial differentiation gene) family, new term LPA_{1-3}, [220, 229], and one of the purinergic P2Y-family, now called LPA_4, [230] have been identified. LPA_1 is the most abundant and ubiquitously expressed LPA receptor and seems to function as main regulator of cell motility. The individual LPA receptors activate the same downstream effectors, but they differ in their potency. This explains while LPA receptor knockout mice show the similar phenotype [220, 229].

Five receptors have been identified to bind S1P, which are termed $S1P_{1-5}$. They are differently expressed and lead to different biological activities [212, 227]. Strikingly, opposite effects on cell motility are observed depending on the cell type and receptor expression. While $S1P_{1,3-5}$ stimulate cell migration and chemotaxis, $S1P_2$ (formerly called EGF_5) has been reported to block migration of melanoma cells and vascular smooth muscle cells by downregulating Rac activity [231].

Less is known about the receptors for SPC. Recently, it has been shown that SPC binds with high affinity and LPA with low affinity to GPR4 [232]. SPC seems also to interact with S1P receptors but with low affinity. Additional SPC receptors are mentioned later in context with their cellular functions. An excellent review of Ishii et al. [233] summarizes in detail the present knowledge about all receptors of bioactive lipids, their binding specificity, signal transduction pathways and biological functions.

16.6.2
Signal Transductions Pathways of Bioactive Lipids Stimulating Motility

LPA activates GPCRs which are linked to different G proteins (G_q, G_i and $G_{12/13}$). Although numerous reports mention the stimulating effect of LPA on cell motility less is known about the molecular mechanism of signal transduction. Activation of Rho-GEF via $G_{12/13}$ results in neurite retraction and cell rounding [234]. Recently, Van Leeuwen [235] reported that cell spreading and migration was induced in LPA_1 transfected neuroblastoma cells upon stimulation with LPA. They observed an activation of Rac which was dependent on G_i proteins, PI3-K and Tiam-1 (Fig. 16.5, green labels). Few data are reported about signaling cascades activated by S1P. This ligand seems to activate similar downstream effectors as LPA [212, 232] apart from a feed back mechanism triggered by activated PI3-K. The latter activates Akt, which in turn phosphorylates T^{236} located in the third intracellular loop of the receptor $S1P_2$ [236] (Fig. 16.5 red label). This results in activation of Rac which induces actin cytoskeleton rearrangements leading to enhanced cell migration [236]. This Rac activation pathway was found to be specific for $S1P_2$. It will be interesting whether such a feed back mechanism is also established in LPA signaling cascades.

Analysing the pancreatic tumor cell lines PANC-1 and BxPC-3 Stahle et al. have identified a different LPA signal transduction pathway leading to chemotaxis and reorganization of the actin cytoskeleton [52]. The cellular response is mediated by Gi/o proteins which activates MEK, ERK and JNK but also the small GTPases N-Ras, RhoA and Rac1. Co-transfections of dominant-negative mutants of N-Ras, RhoA and Rac1 revealed that their activation is a prerequisite for ERK1/2 phosphorylation (Fig. 16.5, magenta label). In addition, they could show by immunostainings that ERK1/2 activated by LPA translocates transiently into newly formed focal contacts [52]. Target molecules of activated ERK1/2 like Tropomyosin-1 [237] or paxillin [238] which have been described in other systems support the idea of ERK1/2 involved in cytoskeletal rearrangements during assembly of focal contacts. Since different receptors of bioactive lipids are expressed in a cells deciphering and separating the specific downstream effector cascades will be the challenge of the next future.

16.7
Keratin Regulation by Sphingolipids – a Key to Metastasis?

Tumor metastasis is the ultimate cause of death in many cancer patients. The definition of the metastatic properties of tumor cells (the so-called seed factors) and the special requirements of the tumor cell environment that enable these cells to survive (the soil factors; [239]) has markedly enhanced our understanding of the complex process of tumor cell metastasis. Here we propose another potential seed factor, the visco-elastic properties of a tumor cell. A metastatic tumor cell has to respond to distinct mechanical strains during metastasis that are imposed by its immediate environment: The tumor cells invade into adjacent tissue, intravasate, i.e. penetrate through the endothelial layer of blood and lymphatic vessels, and finally extravasate through the capillary endothelium at distant sites. In particular, penetration of an epithelial tumor cell through a barrier such as the endothelial layer requires active migration, but also a timely and reversible change in cell shape and stiffness, i.e. a "biophysical seed factor". Thus, to understand this crucial step in metastasis, we need to understand how shape and the elastic properties of epithelial tumor cells can be regulated.

16.7.1
Keratins Define the Structural Integrity and the Mechanical Properties of Epithelial Cells

Cell migration and invasion are largely dependent on the organization of the cytoskeleton [240, 241]. The cytoskeleton of eukaryotic cells is a complex network of three major classes of filamentous biopolymers: filamentous actin, microtubules, and intermediate filaments. The latter are composed of members of a family of proteins that organize to form 10 nm filaments and share sequence homology and structural features [242]. The intermediate filament-nuclear matrix forms the structural connection from the cell periphery to the nucleus. It is a key determinant of dynamic linkages between the nuclear matrix, the cytoskeleton and the extracellular matrix.

Among the cytoplasmic intermediate filament proteins, keratins are specifically expressed in epithelial cells and constitute nearly 5% of the total cellular protein [243]. These filaments are usually organized into bundles or tonofibrils, which form cage-like structures around the nucleus and extend from the perinuclear region to the cell periphery [244, 245].

Keratins are obligate heteropolymers of at least one type I (acidic keratins K9–K20) and one type II keratin (basic keratins K1–K8) [246]. Importantly, the type of keratin pairs determines the mechanical properties of the filament network [247, 248]. K8 and K18 are the major components of intermediate filaments of simple or single layer epithelia as found in the intestine, liver and exocrine pancreas [249]. The keratin expression pattern is generally persistent in carcinomas arising from tissues that normally express K8 and K18 [250] (Fig. 16.6).

Keratins interact with various intracellular proteins such as heat shock proteins and the 14-3-3 family [251, 252]. These interactions are accompanied by keratin phosphorylation which occurs within the so-called 'head' (N-terminal) and/or 'tail' (C-terminal) non-a-helical end domains [253]. Keratin phosphorylation, particularly at specific serine residues, contributes to the regulation of keratin function [254–257]: Phosphorylation has been implicated in increased solubility and the state of keratin polymerization [258], in reorganization of keratins [259–261], cellular stress [262] and keratin reorganization during apoptosis [263].

Keratin filaments play a crucial role in maintaining the structural integrity and the mechanical properties of cells thereby protecting cells from various environmental insults [264, 265]. Experiments on intracellular force transfer identified keratins as a major determinant for the mechanical features of the cytoplasm and the nucleus [266, 267].

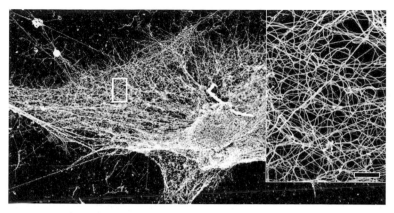

Fig. 16.6 High-resolution SEM (Scanning Electron Micro-scopy) of keratin filaments in human pancreatic cancer cells. Cells underwent a prefixation treatment to extract individual keratin filaments for visualization by SEM. Keratin 8/18 filaments are evenly distributed in the perinuclear region and the submembranous area (high power field).

16.7.2
Sphingosylphosphorylcholine:
A Naturally Occurring Bioactive Lipid Regulating Multiple Biological Processes

Lysosphingomyelin or sphingosylphosphorylcholine (SPC) is a bioactive lipid which acts as an extracellular and intracellular signaling molecule. SPC has initially been identified in the brain of Niemann Pick type A patients. This disease is characterized by a lack of acidic sphingomyelinase [268]. High concentrations of SPC can also be detected in the epidermis of patients with atopic dermatitis due to an abnormally expressed sphingomyelin deacylase [269]. However, recent results demonstrated that SPC is a naturally occurring lipid mediator in blood plasma [270] where SPC is a major component of high-density lipoproteins (HDL). SPC and lyso-sulfatide are responsible for the activation of phosphatidylinositol-specific Phospholipase C by HDL [271]. In addition, HDL-associated SPC contributes to the antiapoptotic effect of HDL on endothelial cells [272].

SPC regulates proliferation [273, 274], cell migration and angiogenesis [275], wound healing [276], and heart rate [277] in various model systems. These effects are mediated by the induction of multiple intracellular signaling cascades including activation of the ERK signaling pathway [273], PKCs and the PI3-kinase/AKT cascade [270]. Furthermore, SPC potently activates Rho and reorganizes the actin cytoskeleton in fibroblasts [273].

Many of the cellular actions of SPC are believed to be mediated by the activation of a subfamily of GPCRs, see above and [278–290].

16.7.3
SPC: A Promoter of Metastasis?

SPC is likely to play a role in tumor biology: It is a ligand for the ovarian cancer G-protein coupled receptor 1 [284] and amongst other bioactive lipids, malignant ascites from patients with ovarian cancer contains significantly higher levels of SPC compared to nonmalignant ascites [291]. This points to a potential role of SPC as a promoter of metastasis: 1) Malignant ascites indicates metastatic spread of the primary tumor to the peritoneal cavity. 2) It is known that SPC can induce cell migration [275]. Thus, SPC could play a role in the complex process of metastatic spread of epithelial tumor cells.

16.7.3.1 Keratin Reorganization in Epithelial Tumor Cells by SPC

As pointed out above, the ability of a tumor cell to change its shape and adjust its elastic properties is a critical mechanical requirement for metastasis. Circumstantial evidence points to a crucial role of keratins in the regulation of the mechanical properties of epithelial cells. Unfortunately, there are very few known regulators of keratin reorganization which has been a major obstacle in the precise characterization of the dynamic properties of keratin filaments in vivo.

SPC markedly reorganizes the keratin cytoskeleton in epithelial tumor cells. Interphase tumor cells exhibit a striking perinuclear keratin organization after SPC treatment [292] (Fig. 16.7). In addition, there is a change in cell shape. The cells reveal a more prominent nucleus and

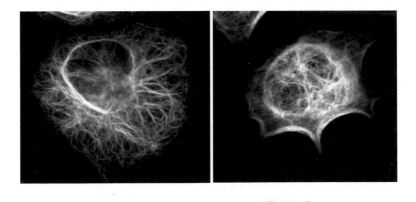

— SPC

Fig. 16.7 The keratin cytoskeleton of a human pancreatic cancer cell in the absence (–) or presence of sphingosylphosphorylcholine (SPC). SPC induces a marked perinuclear reorganization of the keratin cytoskeleton.

perinuclear compartment as well as a less spread phenotype. This points to a major role of keratin organization in determining cell shape in epithelial tumor cells.

This effect is specific for SPC and is not observed in response to SPC-stereoisomers or other bioactive lipids such as lysophosphatidic acid (LPA) or sphingosine 1-phosphate. Thus, SPC, and not a potential SPC-metabolite such as sphingosine 1-phosphate [293], is responsible for keratin reorganization. The effect of SPC on keratin organization is time- and concentration-dependent, requires metabolic energy and is reversible. A reorganization of keratins by SPC within minutes is in line with previous findings bei Leube et al. who found that keratins are highly dynamic and have inherent motile properties [294]. Thus, SPC-induced keratin reorganization is most likely due to a receptor-mediated signaling cascade, rather than intracellular effects of the lipid. SPC significantly increases the density of keratin filaments around the nucleus and decreases the number of keratin filaments extending to the plasma membrane. Keratin reorganization by SPC is also accompanied by keratin phosphorylation at specific Ser residues including Ser^{431} at keratin 8 and Ser^{52} at keratin 18 which is in line with previous findings demonstrating that keratin reorganization is frequently accompanied by keratin phosphorylation [258–260].

16.7.3.2 **SPC Treatment Changes Cell Shape Independently from other Cytoskeletal Components**

There are close links between the actin cytoskeleton, microtubules and keratins. In some model systems the formation of keratin filaments requires the integrity of the actin cytoskeleton and the microtubules [294]. In epithelial tumor cells SPC-induced reorganization of keratins and change in cell shape were independent of the organization of the actin cytoskeleton or the microtubules.

16.7.3.3 **SPC Changes the Elastic Properties of Epithelial Tumor Cells**

The change in cell shape after SPC treatment is most likely due to the less dense keratin filament network in the periphery of the cytoplasm. Which mechanical consequences could result from this observation?

Intermediate filaments such as keratins exhibit high stiffness at large strains and resist breakage [266]. In vitro, keratin heteropolymers exhibit increased viscosity [295] and stability, but also function as elastic materials [267]. Loss of keratins confers increased cellular fragility [296]. Organized keratin filaments are therefore generally regarded as stable cytoskeletal structures supporting epithelial resilience to mechanical stress and thereby determining the mechanical properties of the cytoplasm [265, 267]. Consequently, a markedly reduced number of keratin filaments extending to the plasma membrane in epithelial tumor cells after SPC treatment could have consequences for the elastic properties of the cells. To measure cellular elasticity in vivo, a micromechanical assay was used that allows the quantitative assessment of cellular elasticity by visco-elastic probing of single cells. A single tumor cell attaches between two glass microplates and can be subjected to repeated deformations. SPC markedly increases the elasticity or, in other words, reduces the cellular stiffness of human epithelial tumor cells at the same time and at the same concentration that are required to induce keratin organization. Therefore, SPC-induced keratin reorganization and change in cellular elasticity are most likely related. This conclusion is supported by the fact that agents that modify the actin cytoskeleton, but do not induce structural changes of the other filament networks such as LPA increase cellular stiffness in this micromechanical assay [292]. Since SPC induces both, actin polymerization and keratin reorganization, keratin reorganization seems to play the predominant role in defining the visco-elastic properties of epithelial tumor cells.

16.7.3.4 SPC Facilitates Migration of Epithelial Tumor Cells through Size-limiting Pores

These data establish that SPC substantially changes the elastic properties of epithelial tumor cells most likely via rearrangement of the keratin cytoskeleton. Thus, SPC seems to fulfill the criteria for a signaling molecule facilitating tumor cell migration through barriers such as the endothelial layer. Indeed, SPC increases migration of epithelial tumor cells through sizes limited membranous pores. This increase in migration is more pronounced in response to SPC than after LPA treatment. The latter lipid predominantly reorganizes the actin cytoskeleton and induces stiffening of the cells.

These data lead to the concept that an increased cellular elasticity facilitates migration through small membranous pores. Our model implies that keratins determine the elastic properties of epithelial tumor cells and keratin reorganization by SPC promotes cell migration by influencing cellular elasticity. This idea is supported by other findings: Similarly to K8/18 filaments in epithelial tumor cells after SPC treatment, K5/16 filaments in the human epidermis reorganize after skin injury from a pan-cytoplasmic to a juxtanuclear distribution. This reorganization of keratins is followed by keratinocyte migration into the wound site [297]. Strikingly, SPC has been reported to play an important role in wound healing [276]. However, its precise effects on keratin organization has so far not been determined in this model system. Wound healing and metastasis may be closer related than previously thought: There are tumors that exhibit a transcriptional expression pattern that closely resembles that of the response of fibroblasts to serum i.e. a "wound healing signature".. Interestingly, this gene signature in tumors predicts an increased risk of metastasis [298].

In conclusion, SPC-induced changes in keratin organization and cellular elasticity of tumor cells provide a new approach to gain further insights into the complex mechanisms of metastasis and may enable the development of novel strategies to prevent this often fatal consequence of cancer.

16.8
References

1 COMPAGNI, A., CHRISTOFORI, G. (2000) *Br. J. Cancer.* 83, 1–5.
2 FEARON, E. R. and VOGELSTEIN, B. (1990) *Cell* 61, 759–767.
3 SHIELDS, J. M., PRUITT, K., McFALL, A., SHAUB, A., and DER, C. J. (2000) *Trends Cell. Biol.* 10, 147–154.
4 TAKAI, Y., SASAKI, T., and MATOZAKI, T. (2001) *Physiol. Rev.* 81, 153–208.
5 EHRHARDT, A., EHRHARDT, G. R., GUO, X., and SCHRADER, J. W. (2002) *Exp. Hematol.* 30, 1089–1106.
6 HERRMANN, C. (2003) *Curr. Opin. Struct. Biol.* 13, 122–129.
7 MALUMBRES, M. and BARBACID, M. (2003) *Nat. Rev. Cancer* 3, 459–465.
8 OXFORD, G. and THEODORESCU, D. (2003) *Cancer. Lett.* 189, 117–128.
9 BARBACID, M. (1987) *Annu. Rev. Biochem.* 56:779–827.
10 BOS, J. L. (1989) *Cancer. Res.* 49, 4682–4689.
11 ALMOGUERA, C., SHIBATA, D., FORRESTER, K., MARTIN, J., ARNHEIM, N., and PERUCHO, M. (1988) *Cell* 53, 549–554.
12 RODENHUIS, S., SLEBOS, R. J., BOOT, A. J., EVERS, S. G., MOOI, W. J., WAGENAAR, S. S., van BODEGOM, P. C., and BOS, J. L. (1988) *Cancer. Res.* 48, 5738–5741.
13 BOS, J. L., FEARON, E. R., HAMILTON, S. R., VERLAAN-DE VRIES, M., VAN BOOM, J. H., VAN DER EB, A. J., and VOGELSTEIN, B. (1987) *Nature* 327, 293–297.
14 ADJEI, A. A. (2001) *J. Natl. Cancer. Inst.* 93, 1062–1074.
15 CAMPBELL, P. M. and DER, C. J. (2004) *Semin. Cancer. Biol.* 14, 105–114.
16 SCHEFFZEK, K., AHMADIAN, M. R., KABSCH, W., WIESMULLER, L., LAUTWEIN, A., SCHMITZ, F., and WITTINGHOFER, A. (1997) *Science* 277, 333–338.
17 ELLIS, C. A. and CLARK, G. (2000) *Cell Signal.* 12, 425–434.
18 KOERA, K., NAKAMURA, K., NAKAO, K., MIYOSHI, J., TOYOSHIMA, K., HATTA, T., OTANI, H., AIBA, A., and KATSUKI, M. (1997) *Oncogene* 15, 1151–1159.
19 YAN, Z., CHEN, M., PERUCHO, M., and FRIEDMAN, E. (1997) *J. Biol. Chem.* 272, 30928–30936.
20 YAN, Z., DENG, X., CHEN, M., XU, Y., AHRAM, M., SLOANE, B. F., and FRIEDMAN, E. (1997) *J. Biol. Chem.* 272, 27902–27907.
21 CARBONE, A., GUSELLA, G. L., RADZIOCH, D., and VARESIO, L. (1991) *Oncogene* 6, 731–737.
22 MAHER, J., BAKER, D. A., MANNING, M., DIBB, N. J., and ROBERTS, I. A. G. (1995) *Oncogene* 11, 1639–1647.

23 KLOOG, Y. and COX, A.D. (2000) *Mol. Med. Today* 6, 398–402.

24 HAMILTON, M. and WOLFMAN, A. (1998) *Oncogene* 16, 1417–1428.

25 YAN, J., ROY, S., APOLLONI, A., LANE, A., and HANCOCK, J.F. (1998) *J. Biol. Chem.* 273, 24052–24056.

26 VOICE, J.K., KLEMKE, R.L., LE, A., and JACKSON, J.H. (1999) *J. Biol. Chem.* 274, 17164–17170.

27 LIAO, J., WOLFMAN, J.C., and WOLFMAN, A. (2003) *J. Biol. Chem.* 278, 31871–37878.

28 KIM, K., LINDSTROM, M.J., and GOULD, M.N. (2002) *Cancer Res.* 62, 1241–1245.

29 HANCOCK, J.F. (2003) *Nat. Rev. Mol. Cell Biol.* 4, 373–384.

30 ELAD-SFADIA, G., HAKLAI, R., BALLAN, E., GABIUS, H.J., and KLOOG, Y. (2002) *J. Biol. Chem.* 277, 37169–37175.

31 ELAD-SFADIA, G., HAKLAI, R., BALAN, E., and KLOOG, Y. (2004) *J. Biol. Chem.* 279, 34922–34930.

32 HAMAD, N.M., ELCONIN, J.H., KARNOUB, A.E., BAI, W., RICH, J.N., ABRAHAM, R.T., DER, C.J., and COUNTER, C.M. (2002) *Genes Dev.* 16, 2045–2057.

33 HRUBAN, R.H., WILENTZ, R.E., and KERN, S.E. (2000) *Am. J. Pathol.* 156, 1821–1825.

34 HANAHAN, D. and WEINBERG, R.A. (2000) *Cell* 100, 57–70.

35 SEBTI, S.M. and ADJEI, A.A. (2004) *Semin. Oncol.* 31, 28–39.

36 HAKLAI, R., WEISZ, M.G., ELAD, G., PAZ, A., MARCIANO, D., EGOZI, Y., BEN BARUCH, G., and KLOOG, Y. (1998) *Biochemistry* 37, 1306–1314.

37 WEISZ, B., GIEHL, K., GANA-WEISZ, M., EGOZI, Y., BEN BARUCH, G., MARCIANO, D., GIERSCHIK, P., and KLOOG, Y. (1999) *Oncogene* 18, 2579–2588.

38 GUERRA, C., MIJIMOLLE, N., DHAWAHIR, A., DUBUS, P., BARRADAS, M., SERRANO, M., CAMPUZANO, V., and BARBACID, M. (2003) *Cancer Cell* 4, 111–120.

39 GOTZMANN, J., MIKULA, M., EGER, A., SCHULTE-HERMANN, R., FOISNER, R., BEUG, H., and MIKULITS, W. (2004) *Mutat. Res.* 566, 9–20.

40 KOLCH, W. (2000) *Biochem. J.* 351 Pt2:289–305.

41 CHONG, H., VIKIS, H.G., and GUAN, K.L. (2003) *Cell Signal.* 15, 463–469.

42 PEYSSONNAUX, C. and EYCHENE, A. (2001) *Biol. Cell* 93, 53–62.

43 DAVIES, H., BIGNELL, G.R., COX, C., STEPHENS, P., EDKINS, S., CLEGG, S., TEAGUE, J., WOFFENDIN, H., GARNETT, M.J., BOTTOMLEY, W., DAVIS, N., DICKS, E., EWING, R., FLOYD, Y., GRAY, K., HALL, S., HAWES, R., HUGHES, J., KOSMIDOU, V., MENZIES, A., MOULD, C., PARKER, A., STEVENS, C., WATT, S., HOOPER, S., WILSON, R., JAYATILAKE, H., GUSTERSON, B.A., COOPER, C., SHIPLEY, J., HARGRAVE, D., PRITCHARD-JONES, K., MAITLAND, N., CHENEVIX-TRENCH, G., RIGGINS, G.J., BIGNER, D.D., PALMIERI, G., COSSU, A., FLANAGAN, A., NICHOLSON, A., HO, J.W., LEUNG, S.Y., YUEN, S.T., WEBER, B.L., SEIGLER, H.AF., DARROW, T.L., PATERSON, H., MARAIS, R., MARSHALL, C.J., WOOSTER, R., STRATTON, M.R., and FUTREAL, P.A. (2002) *Nature* 417, 949–954.

44 RAJAGOPALAN, H., BARDELLI, A., LENGAUER, C., KINZLER, K.W., VOGELSTEIN, B., and VELCULESCU, V.E. (2002) *Nature* 418, 934.

45 REIFENBERGER, J., KNOBBE, C.B., STERZINGER, A.A., BLASCHKE, B., SCHULTE, K.W., RUZICKA, T., and REIFENBERGER, G. (2004) *Int. J. Cancer* 109, 377–384.

46 ROSS, P.J., GEORGE, M., CUNNINGHAM, D., DiSTEFANO, F., ANDREYEV, H.J., WORKMAN, P., and CLARKE, P.A. (2001) *Mol. Cancer Ther.* 1, 29–41.

47 YIP-SCHNEIDER, M.T., LIN, A., BARNARD, D., SWEENEY, C.J., and MARSHALL, M.S. (1999) *Int. J. Oncol.* 15, 271–279.

48 GIEHL, K., SKRIPCZYNSKI, B., MANSARD, A., MENKE, A., and GIERSCHIK, P. (2000) *Oncogene* 19, 2930–2942.

49 BUARD, A., ZIPFEL, P.A., FREY, R.S., and MULDER, K.M. (1996) *Int. J. Cancer* 67, 539–546.

50 YIP-SCHNEIDER, M.T., LIN, A., and MARSHALL, M.S. (2001) *Biochem. Biophys. Res. Commun.* 280, 992–997.

51 STAHLE, M., VEIT, C., BACHFISCHER, U., SCHIERLING, K., SKRIPCZYNSKI, B., HALL, A., GIERSCHIK, P., and GIEHL, K. (2003) *J. Cell Sci.* 116, 3835–3846.

52 BIAN, D., SU, S., MAHANIVONG, C., CHENG, R.K., HAN, Q., PAN, Z.K., SUN, P., and HUANG, S. (2004) *Cancer Res.* 64, 4209–4217.

53 STUPACK, D.G., CHO, S.Y., and KLEMKE, R.L. (2000) *Immunol. Res.* 21, 83–88.

54 KLEMKE, R.L., CAI, S., GIANNINI, A.L., GALLAGHER, P.J., DE LANEROLLE, P., and CHERESH, D.A. (1997) *J. Cell Biol.* 137, 481–492.

55 BRUNTON, V.G., FINCHAM, V.J., McLEAN, G.W., WINDER, S.J., PARASKEVA, C., MARSHALL, J.F., and FRAME, M.C. (2001) *Neoplasia* 3, 215–226.

56 AHMED, N., NIU, J., DORAHY, D.J., GU, X., ANDREWS, S., MELDRUM, C.J., SCOTT, R.J., BAKER, M.S., MACREADIE, I.G., and AGREZ, M.V. (2002) *Oncogene* 21, 1370–1380.

57 GRUNERT, S., JECHLINGER, M., and BEUG, H. (2003) *Nat. Rev. Mol. Cell Biol.* 4, 657–665.

58 JANDA, E., LEHMANN, K., KILLISCH, I., JECHLINGER, M., HERZIG, M., DOWNWARD, J., BEUG, H., and GRUNERT, S. (2002) *J. Cell Biol.* 156, 299–313.

59 JECHLINGER, M., GRUNERT, S., TAMIR, I.H., JANDA, E., LUDEMANN, S., WAERNER, T., SEITHER, P., WEITH, A., BEUG, H., and KRAUT, N. (2003) *Oncogene* 22, 7155–7169.

60 FENSTERER, H., GIEHL, K., BUCHHOLZ, M., ELLENRIEDER, V., BUCK, A., KESTLER, H.A., ADLER, G., GIERSCHIK, P., and GRESS, T.M. (2004) *Genes Chromosomes Cancer* 39, 224–235.

61 SERRANO, M., LIN, A.W., McCURRACH, M.E., BEACH, D., and LOWE, S.W. (1997) *Cell* 88, 593–602.

62 McMAHON, M. and WOODS, D. (2001) *Biochim. Biophys. Acta* 1471, M63–M71.

63 EVAN, G.I. and VOUSDEN, K.H. (2001) *Nature* 411, 342–348.

64 AGUIRRE, A.J., BARDEESY, N., SINHA, M., LOPEZ, L., TUVESON, D.A., HORNER, J., REDSTON, M.S., and DePINHO, R.A. (2003) *Genes Dev.* 17, 3112–3126.

65 NABESHIMA, K., INOUE, T., SHIMAO, Y., and SAMESHIMA, T. (2002) *Pathol. Int.* 52, 255–264.

66 WESTERMARCK, J. and KAHARI, V.M. (1999) *FASEB J.* 13, 781–792.

67 Huntington, J.T., Shields, J.M., Der, C.J., Wyatt, C.A., Benbow, U., Slingluff, C.L., Jr., and Brinckerhoff, C.E. (2004) *J. Biol. Chem.* 279, 33168–33176.

68 Overall, C.aM. and Lopez-Otin, C. (2002) *Nat. Rev. Cancer 2*, 657–672.

69 Aguirre-Ghiso, J.A., Frankel, P., Farias, E.F., Lu, Z., Jiang, H., Olsen, A., Feig, L.A., de Kier Joffe, E.B., and Foster, D.A. (1999) *Oncogene* 18, 4718–4725.

70 Premzl, A., Zavasnik-Bergant, V., Turk, V., and Kos, J. (2003) *Exp. Cell Res.* 283, 206–214.

71 Rodriguez Viciana, P., Warne, P.H., Dhand, R., Vanhaesebroeck, B., Gout, I., Fry, M.J., Waterfield, M.D., and Downward, J. (1994) *Nature* 370, 527–532.

72 Downward, J. (1997) *Adv. Second Messenger Phosphoprotein Res.* 31:1–10.

73 Cantley, L.C. (2002) *Science* 296, 1655–1657.

74 Vivanco, I. and Sawyers, C.L. (2002) *Nat. Rev. Cancer 2*, 489–501.

75 Franke, T.F., Hornik, C.P., Segev, L., Shostak, G.A., and Sugimoto, C. (2003) *Oncogene* 22, 8983–8998.

76 Meili, R., Ellsworth, C., Lee, S., Reddy, T.B., Ma, H., and Firtel, R.A. (1999) *EMBO J.* 18, 2092–2105.

77 Kim, D., Kim, S., Koh, H., Yoon, S.O., Chung, A.S., Cho, K.S., and Chung, J. (2001) *FASEB J.* 15, 1953–1962.

78 Vasko, V., Saji, M., Hardy, E., Kruhlak, M., Larin, A., Savchenko, V., Miyakawa, M., Isozaki, O., Murakami, H., Tsushima, T., Burman, K.D., De Micco, C., and Ringel, M.D. (2004) *J. Med. Genet.* 41, 161–170.

79 Grille, S.J., Bellacosa, A., Upson, J., Klein-Szanto, A.J., van Roy, F., Lee-Kwon, W., Donowitz, M., Tsichlis, P.N., and Larue, L. (2003) *Cancer Res.* 63, 2172–2178.

80 Park, B.K., Zeng, X., and Glazer, R.I. (2001) *Cancer Res.* 61, 7647–7653.

81 Berven, L.A., Willard, F.S., and Crouch, M.F. (2004) *Exp. Cell Res.* 296, 183–195.

82 Qian, Y., Corum, L., Meng, Q., Blenis, J., Zheng, J.Z., Shi, X., Flynn, D.C., and Jiang, B.H. (2004) *Am. J. Physiol. Cell Physiol.* 286, C153–C163.

83 Nimnual, A.S., Yatsula, B.A., and Bar-Sagi, D. (1998) *Science* 279, 560–563.

84 Etienne-Manneville, S. and Hall, A. (2002) *Nature* 420, 629–635.

85 Bar-Sagi, D. and Hall, A. (2000) *Cell* 103, 227–238.

86 Ridley, A.J., Paterson, H.F., Johnston, C.L., Diekmann, D., and Hall, A. (1992) *Cell* 70, 401–410.

87 Zondag, G.C., Evers, E.E., ten Klooster, J.P., Janssen, L., Der Kammen, R.A., and Collard, J.G. (2000) *J. Cell Biol.* 149, 775–782.

88 Jaffe, A.B. and Hall, A. (2002) *Adv. Cancer Res.* 84:57–80.

89 Sahai, E. and Marshall, C.J. (2002) Rho-GTPases and cancer. *Nat. Rev. Cancer* 2, 133–142.

90 Ridley, A.J., Schwartz, M.A., Burridge, K., Firtel, R.A., Ginsberg, M.H., Borisy, G., Parsons, J.T., and Horwitz, A.R. (2003) *Science* 302, 1704–1709.

91 RAFTOPOULOU, M. and HALL, A. (2004) *Dev. Biol.* 265, 23–32.

92 ABERLE H., BUTZ S., STAPPERT J., WEISSIG H., KEMLER R., and HOSCHUETZKY H. (1994) *J. Cell Sci.* 107, 3655–3663.

93 ABERLE H., SCHWARTZ H., and KEMLER R. (1996) *J. Cell Biochem.* 61, 514–523.

94 HAY E.D. (1995) *Acta Anat. (Basel)* 154, 8–20.

95 BEHRENS J., MAREEL M.M., VAN ROY F.M., and BIRCHMEIER W. (1989) *J. Cell Biol.* 108, 2435–2447.

96 VLEMINCKX K., VAKAET L., JR., MAREEL M., FIERS W., and VAN ROY F. (1991) *Cell* 66, 107–119.

97 MAREEL M., VLEMINCKX K., VERMEULEN S., GAO Y., VAKAET L., JR., BRACKE M., and VAN ROY F. (1992) *Prog. Histochem. Cytochem.* 26, 95–106.

98 TAKEICHI M. (1993) *Curr. Opin. Cell Biol.* 5, 806–811.

99 BIRCHMEIER C., BIRCHMEIER W., and BRAND SABERI B. (1996) *Acta Anat. (Basel)* 156, 217–226.

100 CANO A., PEREZ-MORENO M.A., RODRIGO I., LOCASCIO A., BLANCO M.J., DEL BARRIO M.G., PORTILLO F., and NIETO M.A. (2000) *Nat. Cell Biol.* 2, 76–83.

101 COMIJN J., BERX G., VERMASSEN P., VERSCHUEREN K., VAN GRUNSVEN L., BRUYNEEL E., MAREEL M., HUYLEBROECK D., and VAN ROY F. (2001) *Mol. Cell* 7, 1267–1278.

102 ADAMS C.L. and NELSON W.J. (1998) *Curr. Opin. Cell Biol.* 10, 572–577.

103 CAVALLARO U. and CHRISTOFORI G. (2001) *Biochim. Biophys. Acta* 1552, 39–45.

104 OZAWA M. and KEMLER R. (1998) *J. Biol. Chem.* 273, 6166–6170.

105 THIERY J.P. (2003) *Curr. Opin. Cell Biol.* 15, 740–746.

106 BIRCHMEIER C., MEYER D., and RIETHMACHER D. (1995) *Inter. Rev. Cytology* 160, 221–266.

107 WOODFIELD R.J., HODGKIN M.N., AKHTAR N., MORSE M.A., FULLER K.J., SAQIB K., THOMPSON N.T., and WAKELAM M.J. (2001) *Biochem. J.* 360, 335–344.

108 JONES J.L., ROYALL J.E., and WALKER R.A. (1996) *Br. J. Cancer* 74, 1237–1241.

109 KHOURY H., DANKORT D.L., SADEKOVA S., NAUJOKAS M.A., MULLER W.J., and PARK M. (2001) *Oncogene* 20, 788–799.

110 RONG S., SEGAL S., ANVER M., RESAU J.H., and VANDE WOUDE G.F. (1994) *PNAS USA* 91, 4731–4735.

111 MAGGIORA P., GAMBAROTTA G., OLIVERO M., GIORDANO S., DI RENZO M.F., and COMOGLIO P.M. (1997) *J. Cell Physiol.* 173, 183–186.

112 HOSCHUETZKY H., ABERLE H., and KEMLER R. (1994) *J. Cell Biol.* 127, 1375–1380.

113 BEHRENS, J., VAKAET, L., FRIIS, R., WINTERHAGER, E., VAN ROY, F., MAREEL, M.M., and BIRCHMEIER, W. (1993) *J. Cell Biol.* 120, 757–766.

114 HAMAGUCHI M., MATSUYOSHI N., OHNISHI Y., GOTOH B., TAKEICHI M., and NAGAI Y. (1993) *EMBO J.* 12, 307–314.

115 MAREEL M. and LEROY A. (2003) *Physiol. Rev.* 83, 337–376.

116 JEFFERS M., SCHMIDT L., NAKAIGAWA N., WEBB C.P., WEIRICH G., KISHIDA T., ZBAR B., and VANDE WOUDE G.F. (1997) *Proc. Natl. Acad. Sci. USA* 94, 11445–11450.

117 JEFFERS M., FISCELLA M., WEBB C. P., ANVER M., KOOCHEKPOUR S., and VANDE WOUDE G. F. (1998) *PNAS USA* 95, 14417–14422.

118 FRAME M. C. (2002) *Biochim. Biophys. Acta* 1602, 114–130.

119 TAKEDA H., NAGAFUCHI A., YONEMURA S., TSUKITA S., BEHRENS J., and BIRCHMEIER W. (1995) *J. Cell Biol.* 131, 1839–1847.

120 WROBEL C. N., DEBNATH J., LIN E., BEAUSOLEIL S., ROUSSEL M. F., and BRUGGE J. S. (2004) *J. Cell Biol.* 165, 263–273.

121 MENKE A., PHILIPPI C., VOGELMANN R., SEIDEL B., LUTZ M. P., ADLER G., and WEDLICH D. (2001) *Cancer Res.* 61, 3508–3517.

122 FRAME M. C., FINCHAM V. J., CARRAGHER N. O., and WYKE J. A. (2002) *Nat. Rev. Mol. Cell Biol.* 3, 233–245.

123 FUJITA Y., KRAUSE G., SCHEFFNER M., ZECHNER D., LEDDY H. E., BEHRENS J., SOMMER T., and BIRCHMEIER W. (2002) *Nat. Cell Biol.* 4, 222–231.

124 BRADY-KALNAY S. M., MOURTON T., NIXON J. P., PIETZ G. E., KINCH M., CHEN H., BRACKENBURY R., RIMM D. L., DEL VEC-CHIO R. L., and TONKS N. K. (1998) *J. Cell Biol.* 141, 287–296.

125 MÜLLER T., CHOIDAS A., REICHMANN E., and ULLRICH A. (1999) *J. Biol. Chem.* 274, 10173–10183.

126 VOLBERG T., ZICK Y., DROR R., SABANAY I., GILON C., LEVITZKI A., and GEIGER B. (1992) *EMBO J.* 11, 1733–1742.

127 FUCHS M., MÜLLER T., LERCH M. M., and ULLRICH A. (1996) *J. Biol. Chem.* 271, 16712–16719.

128 BRABLETZ T., JUNG A., REU S., PORZNER M., HLUBEK F., KUNZ-SCHUGHART L. A., KNUECHEL R., and KIRCHNER T. (2001) *PNAS USA* 98, 10356–10361.

129 BRABLETZ T., JUNG A., and KIRCHNER T. (2002) *Virchows Arch.* 441, 1–11.

130 AVIZIENYTE E., WYKE A. W., JONES R. J., MCLEAN G. W., WESTHOFF M. A., BRUNTON V. G., and FRAME M. C. (2002) *Nat. Cell Biol.* 4, 632–638.

131 GENDA T., SAKAMOTO M., ICHIDA T., ASAKURA H., and HIROHASHI S. (2000) *Lab. Invest.* 80, 387–394.

132 TAMURA M., GU J., TAKINO T., and YAMADA K. M. (1999) *Cancer Res.* 59, 442–449.

133 KOTELEVETS L., van HENGEL J., BRUYNEEL E., MAREEL M., VAN ROY F., and CHASTRE E. (2001) *J. Cell Biol.* 155, 1129–1135.

134 ANASTASIADIS P. Z. and REYNOLDS A. B. (2000) *J. Cell Sci.* 113, 1319–1334.

135 OHKUBO T. and OZAWA M. (1999) *J. Biol. Chem.* 274, 21409–21415.

136 THORESON M. A. and REYNOLDS A. B. (2002) *Differentiation* 70, 583–589.

137 YAP A. S., NIESSEN C. M., and GUMBINER B. M. (1998) *J. Cell Biol.* 141, 779–789.

138 AONO S., NAKAGAWA S., REYNOLDS A. B., and TAKEICHI M. (1999) *J. Cell Biol.* 145, 551–562.

139 KEIRSEBILCK A., BONNE S., STAES K., van H. J., NOLLET F., REYNOLDS A., and VAN ROY F. (1998) *Genomics* 50, 129–146.

140 MO Y. Y. and REYNOLDS A. B. (1996) *Cancer Res.* 56, 2633–2640.

141 SEIDEL B., BRAEG S., ADLER G., WEDLICH D., and MENKE A. (2004) *Oncogene* 23, 5532–5542.

142 OZAWA M. and OHKUBO T. (2001) *J. Cell Sci.* 114, 503–512.

143 KIM L. and WONG T.W. (1995) *Mol. Cell Biol.* 15, 4553–4561.

144 MONGA, S. P., MARS, W. M., PEDIADITAKIS, P., BELL, A., MULE, K., BOWEN, W. C., WANG, X., ZARNEGAR, R., and MICHALOPOULOS, G. K. (2002) *Cancer Res.* 62, 2064–2071.

145 MIRAVET, S., PIEDRA, J., CASTANO, J., RAURELL, I., FRANCI, C., DUNACH, M., and GARCIA DE HERREROS, A. (2003) *Mol. Cell Biol.* 23, 7391–7402.

146 ILAN, N., MAHOOTI, S., RIMM, D. L., and MADRI, J. A. (1999) *J. Cell Sci.* 112, 3005–3014.

147 ADAM, L., VADLAMUDI, R. K., McCREA, P., and KUMAR R. (2001) *J. Biol. Chem.* 276, 28443–28450.

148 PAI, R., NAKAMURA, T., MOON, W. S., and TARNAWSKI, A. S. (2003), *FASEB J.* 17, 1640–1647.

149 MORALI, O. G., DELMAS, V., MOORE, R., JEANNEY, C., THIERY, J. P., and LARUE, L. (2001) *Oncogene* 20, 4942–4950.

150 LILIEN, J., BALSAMO, J., ARREGUI, C., and XU, G. (2002) *Dev. Dyn.* 224, 18–29.

151 MATSUYOSHI, N., HAMAGUCHI, M., TANIGUCHI, S., NAGAFUCHI, A., TSUKITA, S., and TAKEICHI, M. (1992) *J. Cell Biol.* 118, 703–714.

152 HUBER, A. H. and WEIS, W. I. (2001) *Cell* 105, 391–402.

153 PIEDRA, J., MARTINEZ, D., CASTANO, J., MIRAVET, S., DUNACH, M., and DE HERREROS, A. G. (2001) *J. Biol. Chem.* 2001 276, 20436–20443.

154 GRAHAM, T. A., WEAVER, C., MAO, F., KIMELMAN, D., and XU, W. (2000) *Cell* 103, 885–896.

155 GRAHAM, T. A., FERKEY, D. M., MAO, F., KIMELMAN, D., and XU, W. (2001) *Nat. Struct. Biol.* 8, 1048–1052.

156 PROVOST, E. and RIMM, D. L. (1999) *Curr. Opin. Cell Biol.* 11, 567–572.

157 KIM, K. and LEE, K. Y. (2001) *Cell Biol. Int.* 25, 421–427.

158 DANILKOVITCH-MIAGKOVA, A., MIAGKOV, A., SKEEL, A., NAKAIGA-WA, N., ZBAR, B., and LEONARD, E. (2001) *Mol. Cell Biol.* 21, 5857–5868.

159 GILES, R. H., VAN ES, J. H., and CLEVERS, H. (2003) *Biochim. Biophys. Acta* 1653, 1–24.

160 KIKUCHI A. (2003) *Cancer Sci.* 94, 225–229.

161 BRABLETZ, T., JUNG, A., DAG, S., HLUBEK, F., and KIRCHNER, T. (1999) *Am. J. Pathol.* 155, 1033–1038.

162 HLUBEK, F., JUNG, A., KOTZOR, N., KIRCHNER, T., and BRABLETZ, T. (2001) *Cancer Res.* 61, 8089–8093.

163 HLUBEK, F., SPADERNA, S., JUNG, A., KIRCHNER, T., and BRA-BLETZ, T. (2004) *Int. J. Cancer* 108, 321–326.

164 GRADL, D., KUHL, M., and WEDLICH D. (1999) *Mol. Cell Biol.* 19, 5576–5587.

165 HIENDLMEYER, E., REGUS, S., WASSERMANN, S., HLUBEK, F., HAYNL, A., DIMMLER, A., KOCH, C., KNOLL, C., VAN BEEST, M., REUNING, U., BRABLETZ, T., KIRCHNER, T., and JUNG, A. (2004) *Cancer Res.* 64, 1209–1214.

166 KIRCHNER, T. and BRABLETZ, T. (2000) *Am. J. Pathol.* 157, 1113–1121.

167 ZHANG, X., GASPARD, J. P., and CHUNG, D. C. (2001) *Cancer Res.* 61, 6050–6054.

168 KIM, J.S., CROOKS, H., DRACHEVA, T., NISHANIAN, T.G., SINGH, B., JEN, J., and WALDMAN, T. (2002) *Cancer Res.* 62, 2744–2748.

169 WIELENGA, V., SMITS, R., KORINEK, V., SMIT, L., KIELMAN, M., FODDE, R., CLEVERS, H., and PALS, S. (1999) *Am. J. Pathol.* 154, 515–523.

170 BOON, E.M., VAN DER NEUT, R., VAN DE WETERING, M., CLEVERS, H., and PALS, S.T. (2002) *Cancer Res.* 62, 5126–5128.

171 ROOSE, J., HULS, G., VAN BEEST, M., MOERER, P., VAN DER HORN, K., GOLDSCHMEDING, R., LOGTENBERG, T., and CLEVERS, H. (1999) *Science* 285, 1923–1926.

172 HOVANES, K., LI, T.W., MUNGUIA, J.E., TRUONG, T., MILOVANO-VIC, T., LAWRENCE MARSH, J., HOLCOMBE, R.F., and WATERMAN, M.L. (2001) *Nat. Genet.* 28, 53–57.

173 KUHL, M., GEIS, K., SHELDAHL, L.C., PUKROP, T., MOON, R.T., and WEDLICH, D. (2001) *Mech. Dev.* 106, 61–76.

174 MURAKAMI, T., TODA, S., FUJIMOTO, M., OHTSUKI, M., BYERS, H.R., ETOH, T., and NAKAGAWA, H. (2001) *Biochem. Biophys. Res. Commun.* 288, 8–15.

175 ROTH, W., WILD-BODE, C., PLATTEN, M., GRIMMEL, C., MELKON-YAN, H.S., DICHGANS, J., and WELLER, M. (2000) *Oncogene* 19, 4210–4220.

176 HOANG, B.H., KUBO, T., HEALEY, J.H., YANG, R., NATHAN, S.S., KOLB, E.A., MAZZA, B., MEYERS, P.A., and GORLICK, R. (2004) *Cancer Res.* 64, 2734–2739.

177 WALLINGFORD, J.B., FRASER, S.E., and HARLAND, R.M. (2002) *Dev. Cell* 2, 695–706.

178 KUHL, M., SHELDAHL, L.C., PARK, M., MILLER, J.R., and MOON, R.T. (2000) *Trends Genet.* 16, 279–283.

179 BITTNER, N., MELTZER, P., CHEN, Y., JIANG, Y., SEFTOR, E., HEN-DRIX, M., RADMACHER, M., SIMON, R., YAKHINI, Z., BEN-DOR, A., SAMPAS, E., DOUGHERTY, E., WANG, F., MARINCOLA, C., GOODEN, J., LUEDERS, A., GLATFELTER, P., POLLOCK, J., CARPTEN, E., GIL-LANDERS, D., LEJA, K., DIETRICH, C., BEAUDRY, M., BERENS, D., ALBERTS, V., SONDAK, N., HAYWARD, and TRENT, J. (2000) *Nature* 406, 536–540.

180 WEERARATNA, A.T., JIANG, Y., HOSTETTER, G., ROSENBLATT, K., DURAY, P., BITTNER, M., and TRENT, J.M. (2002) *Cancer Cell* 1, 279–288.

181 COMOGLIO, P. and TRUSOLINO, L. (2002) *J. Clin. Invest.* 109, 857–862.

182 BLADT, F., RIETHMACHER, D., ISENMANN, S., AGUZZI, A., and BIRCHMEIER, C. (1995) *Nature* 376, 768–771.

183 MA, P.C., MAULIK, G., CHRISTENSEN, J., and SALGIA, R. (2003) *Cancer and Metastasis Reviews* 22, 309–325.

184 ZHANG, Y.W. and VANDE WOUDE, G.F (2003) *J. Cell Biochem.* 88, 408–417.

185 DANILKOVITCH-MIAGKOVA, A. and ZBAR, B. (2002) i. 109, 863–867.

186 MAINA, F., PANTE, G., HELMBACHER, F., ANDRES, R., PORTHIN, A., DAVIES, A.M., PONZETTO, C., and KLEIN, R. (2001) *Mol. Cell* 7, 1293–1306.

187 MAINA, F., CASAGRANDA, F., AUDERO, E., SIMEONE, A., COMO-GLIO, P., KLEIN, R., and PONZETTO, C. (1996) *Cell* 87, 531–542.

188 ROYAL, I., LAMARCHE-VANE, N., LAMORTE, L., KAIBUCHI, K., and PARK, M. (2000) *Mol. Biol. Cell* 11, 1709–1725.

189 PONZETTO, C., ZHEN, Z., AUDERO, E., MAINA, F., BARDELLI, A., BASILE, L.M., GIORDANO, S., NARSIMHAN, R., and COMOGLIO, P. (1996) *J. Biol. Chem.* 271, 14119–14123.

190 WANG, M.H., WANG, D., and CHEN Y.Q. (2003) *Carcinogenesis* 8, 1291–1300.

191 DANILKOVITCH-MIAGKOVA, A. (2003) *Curr. Cancer Drug Targets* 3, 31–40.

192 PEACE, B.E., HUGHES, M.J., DEGEN, S.J., and WALTZ, S.E. (2001) *Oncogene* 20, 6142–6151.

193 SANTORO, M.M., PENENGO, L., MINETTO, M., ORECCHIA, S., CILLI, M., and GAUDINO, G. (1998) *Oncogene* 17, 741–749.

194 MAGGIORA, P., LORENZATO, A., FRACCHIOLI. S., COSTA, B., CASTAGNARO, M., ARISIO, R., KATSAROS, D., MASSOBRIO, M., COMOGLIO, P.M., and FLAVIA DI RENZO, M. (2003) *Exp. Cell Res.* 288, 382–389.

195 TAMAGONE, L. and COMOGLIO, P.M. (2004) *EMBO reports* 4, 356–360.

196 GIORDANO, S., CORSO, S., CONROTTO, P., ARTIGIANI, S., GILESTRO, G., BARBERIS, D., TAMAGONE, L., and COMOGLIO, P.M. (2002) *Nat. Cell Biol.* 4, 720–724.

197 CONROTTO, P., CORSO, S., GAMBERINI, S., COMOGLIO, P.M., and GIORDANO, S. (2004) *Oncogene* 23, 5131–5137.

198 BIRCHMEIER, W., WEIDNER, K.M., and BEHRENS, J. (1993) *J. Cell Sci. Suppl.* 17, 159–164.

199 PEPPER, M.S., MATSUMOTO, K., NAKAMURA, T., ORCI, L., and MONTESANO, R. (1992) *J. Biol. Chem.* 267, 20493–20496.

200 JEFFERS, M., FISCELLA, M., WEBB C.P., ANVER, M., KOOCHEKPOUR, S., and VANDE WOUDE, G.F. (1996) *Mol. Cell Biol.* 16, 1115–1125.

201 KERMORGANT, S., APARICIO, T., DESSIRIERM, V., LEWIN, M., and LEHY, T. (2001) *Carcinogenesis* 22, 1035–1042.

202 OZAKI, I., MIZUTA, T., ZHAO, G., ZHANG, H., YOSHIMURA, T., KAWAZOW, S., EGUCHI, Y., YASUTAKE, T., HISATOMI, A., SAKAI, T., and XAMAMOTO, K. (2003) *Hepatol. Res.* 4, 289–301.

203 CROUCH, S., SPIDEL, C.S., and LINDSEY, J.S. (2004) *Exp. Cell. Res.* 292, 274–287.

204 GIORDANO, S., MAFFE, A., WILLIAMS, T.A., ARTIGIANI, S., GUAL, P., BARDELLI, A., BASILICO, C., MICHIELI, P., and COMOGLIO, P.M. (2000) *FASEB J.* 14, 399–440.

205 TRUSOLINO, L., BERTOTTI, A., and COMOGLIO, P.M. (2001) *Cell* 107, 643–654.

206 TRUSOLINO, L., CAVASSA, S., ANGELINI, P., ANDO, M., BERTOTTI, A., COMOGLIO, P.M., and BOCCACCIO, C. (2000) *FASEB J.* 14, 1629–1640.

207 ORIAN-ROUSSEAU, V., CHEN, L., SLEEMAN, J.P., HERRLICH, P., and PONTA, H. (2002) *Genes Dev.* 16, 3074–3086.

208 ARUFFO, A., STAMENKOVIC, I., MELNICK, M., UNDERHILL, C.B., and SEED, B. (1990) *Cell* 61, 1303–1313.

209 VAN DER VOORT, R., TAHER, T.E., WIELENGA, V.J., SPAARGAREN, M., PREVO, R., SMIT, L., DAVID, G., HARTMANN, G., GHERARDI, E., and PALS, S.T. (1999) *J. Biol. Chem.* 274, 6499–6506.

210 RECIO, J.A. and MERLINO, G. (2003) *Cancer Res.* 63, 1576–1582.

211 MILLS, G.B. and MOOLENAAR, W.H. (2003) *Nat. Rev. Cancer.* 3, 582–591.

212 PAYNE, S.G., MILSTEIN, S., and SPIEGEL, S. (2002) *FEBS Lett.* 531, 54–57.

213 SWARTHOUT, J.T. and WALLING, H.W. (2000) *Cell Mol. Life Sci.* 57, 1978–1985.

214 SENGUPTA, S., WANG, Z., TIPPS, R., and XU, Y. (2004) *Semin. Cell Dev. Biol.* 15, 503–512.

215 SENGUPTA, S., XIAO, Y.J., and XU, Y. (2003) *FASEB J.* 17, 1570–1572.

216 YAMADA, T., SATO, K., KOMACHI, M., MALCHINKHUU, E., TOBO, M., KIMURA, T., KUWABARA, A., YANAGITA, Y., IKEYA, T., TANAHA-SHI, Y., OGAWA, T., OHWADA, S., MORISHITA, Y., OHTA, H., IM, D.S., TAMOTO, K., TOMURA, H., and OKAJIMA, F. (2004) *J. Biol. Chem.* 279, 6595–6605.

217 ZHANG, N., ZHANG, J., PURCELL, K.J., CHENG, Y., and HOWARD, K. (1997) *Nature* 385, 64–67.

218 STARZ-GAIANO, M., CHO, N.K., FORBES, A., and LEHMANN, R. (2001) *Development* 128, 983–991.

219 KUPPERMAN, E., AN, S., OSBORNE, N., WALDRON, S., and STAI-NIER, D.Y. (2000) Nature 406, 192–195.

220 CONTOS, J.J., FUKUSHIMA, N., WEINER, J.A., KAUSHAL, D., and CHUN, J. (2000) *Proc. Natl Acad. Sci. USA* 97, 13384–13389.

221 ESCALANTE-ALCALDE, D., HERNANDEZ, L., LE STUNFF, H., MAEDA, R., LEE, H.S., JR-GANG-CHENG, SCIORRA, V.A., DAAR, I., SPIEGEL, S., MORRIS, A.J., and STEWART, C.L. (2003) *Development* 130, 4623–4637.

222 GOETZL, E.J. and AN, S. (1998) *FASEB J.* 12, 1589–1598.

223 MURATA, J., LEE, H.Y., CLAIR, T., KRUTZSCH, H.C., ARESTAD, A.A., SOBEL, M.E., LIOTTA, L.A., and STRACKE, M.L. (1994) *J. Biol. Chem.* 269, 30479–30484.

224 NAM, S.W., CLAIR, T., KIM, Y.S., McMARLIN, A., SCHIFFMANN, E., LIOTTA, L.A., and STRACKE, M.L. (2001) *Cancer Res.* 61, 6938–6944.

225 KOH, E., CLAIR, T., WOODHOUSE, E.C., SCHIFFMANN, E., LIOTTA, L., and STRACKE, M. (2003) *Cancer Res.* 63, 2042–2045.

226 TICE, D.A., SZETO, W., SOLOVIEV, I., RUBINFELD, B., FONG, S.E., DUGGER, D.L., WINER, J., WILLIAMS, P.M., WIEAND, D., SMITH, V., SCHWALL, R.H., PENNICA, D., and POLAKIS, P. (2002) *J. Biol. Chem.* 277, 14329–14335.

227 HLA, T., LEE, M.J., ANCELLIN, N., PAIK, J.H., and KLUK, M.J. (2001) *Science* 294, 1875–1878.

228 CLAIR, T., AOKI, J., KOH, E., BANDLE, R.W., NAM, S.W., PTASZYNS-KA, M.M., MILLS, G.B., SCHIFFMANN, E., LIOTTA, L.A., and STRACKE, M.L. (2003) *Cancer Res.* 63, 5446–5453.

229 CONTOS, J.J., ISHII, I., FUKUSHIMA, N., KINGSBURY, M.A., YE, X., KAWAMURA, S., BROWN, J.H. and CHUN, J. (2002) *Mol. Cell Biol.* 22, 6921–6929.

230 NOGUCHI, K., ISHII, S., SHIMIZU, T. (2003) *J. Biol. Chem.* 278, 25600–25606.

231 OKAMOTO, H., TAKUWA, N., YOKOMIZO, T., SUGIMOTO, N., SAKU-RADA, S., SHIGEMATSU, H., and TAKUWA, Y. (2000) *Mol. Cell Biol.* 20, 9247–9261.

232 ZHU, K., BAUDHUIN, L. M., HONG, G., WILLIAMS, F. S., CRISTI-NA, K. L., KABAROWSKI, J. H., WITTE, O. N., and XU, Y. (2001) *J. Biol. Chem.* 276, 41325–41335.

233 ISHII, I., FUKUSHIMA, N., YE, X., and CHUN, J. (2004) *Annu. Rev. Biochem.* 73, 321–354.

234 KRANENBURG, O., POLAND, M., VAN HORCK, F. P., DRECHSEL, D., HALL, A., and MOOLENAAR, W. H. (1999) *Mol. Biol. Cell* 10, 1851–1857.

235 VAN LEEUWEN, F. N., OLIVO, C., GRIVELL, S., GIEPMANS, B. N., COLLARD, J. G., and MOOLENAAR, W. H. (2003) *J. Biol. Chem.* 278, 400–406.

236 LEE, M. J., THANGADA, S., PAIK, J. H., SAPKOTA, G. P., ANCELLIN, N., CHAE, S. S., WU, M., MORALES-RUIZ, M., SESSA, W. C., ALESSI, D. R., and HLA, T. (2001) *Mol. Cell* 8, 693–704.

237 HOULE, F., ROUSSEAU, S., MORRICE, N., LUC, M., MONGRAIN, S., TURNER, C. E., TANAKA, S., MOREAU, P., and HUOT, J. (2003) *Mol. Biol. Cell* 14, 1418–1432.

238 LIU, Z. X., YU, C. F., NICKEL, C., THOMAS, S., CANTLEY, L. G. (2002) *J. Biol. Chem.* 277, 10452–10458.

239 CHAMBERS, A. F., GROOM, A. C., and MACDONALD, I. C. (2002) *Nat. Rev. Cancer* 2, 563–572.

240 SERRADOR, J. M., NIETO, M., and SANCHEZ-MADRID, F. (1999) *Trends Cell Biol.* 9, 228–233.

241 BALLESTREM, C., WEHRLE-HALLER, B., HINZ, B., and IMHOF, B. A. (2000) *Mol. Biol. Cell* 11, 2999–3012.

242 FUCHS, E. and WEBER, K. (1994) *Annu. Rev. Biochem.* 63, 345–382.

243 MOLL, R., FRANKE, W. W., SCHILLER, D. L., GEIGER, B., and KREPLER, R. (1982) *Cell* 31, 11–24.

244 FRANKE, W. W., GRUND, C., OSBORN, M., and WEBER, K. (1978) *Cytobiologie* 17, 365–391.

245 FRANKE, W. W., SCHMID, E., OSBORN, M., and WEBER, K. (1978) *Cytobiologie* 17, 392–411.

246 HATZFELD, M. and FRANKE, W. W. (1985) *J. Cell Biol.* 101, 1826–1841.

247 HUTTON, E., PALADINI, R. D., YU, Q.-C., YEN, M., and COULOMBE, P. A. (1998) *J. Cell Biol.* 143, 487–499.

248 YAMADA, S., WIRTZ, D., and COULOMBE, P. A. (2002) *Mol. Biol. Cell* 13, 382–391.

249 MOLL, R., SCHILLER, D., and FRANKE, W. W. (1990) *J. Cell Biol.* 111, 567–580.

250 OSHIMA, R. G., BARIBAULT, H., and CAULIN, C. (1996) *Cancer Metastasis Rev.* 15, 445–471.

251 LIAO, J. and OMARY, M. B. (1996) *J. Cell Biol.* 133, 345–358.

252 LIAO, J., LOWTHERT, L. A., GHORI, N., and OMARY, M. B. (1995) *J. Biol. Chem.* 270, 915–922.

253 OMARY, M. B., KU, N.-O., LIAO, J., and PRICE, D. (1998) *Subcellular Biochem.* 31, 105–140.

254 GILMARTIN, M. E., CULBERTSON, V. B., and FREEDBERG, I. M. (1980) *J. Invest. Dermatol.* 75, 211–216.

255 STEINERT, P. M., WANTZ, M. L., and IDLER, W. W. (1982) *Biochemistry* 21, 177–183.

256 CHOU, C.-F. and OMARY, M. B. (1991) *FEBS Lett.* 282, 200–204.

257 Zhou, X., Liao, J., Hu, L., Feng, L., and Omary, M.B. (1999) *J. Biol. Chem.* 274, 12861–12866.

258 Ku, N.-O. and Omary, M.B. (1994) *J. Cell Biol.* 127, 161–171.

259 Liao, J., Lowthert, L.A., Ku, N.-O., Fernandez, R., and Omary, M.B. (1995) *J. Cell Biol.* 131, 1291–1301.

260 Toivola, D.M., Goldman, R.D., Garrod, D.R., and Eriksson, J.E. (1997) *J. Cell Sci.* 110, 23–33.

261 Strand, P., Windoffer, R., and Leube, R.E. (2002) *J. Cell Sci.* 115.

262 Ku, N.-O., Liao, J., Chou, C.-F., and Omary, M.B. (1996) *Cancer Metastasis Rev.* 15, 429–444.

263 Caulin, C., Salvesen, G.S., and Oshima, R.G. (1997) *J. Cell Biol.* 138, 1379–1394.

264 Omary, M.B. and Ku, N.-O. (1997) *Hepatology* 25, 1043–1048.

265 Fuchs, E. and Cleveland, D.W. (1998) *Science* 279, 514–519.

266 Maniotis, A.J., Chen, C.S., and Ingber, D.E. (1997) *Proc. Natl Acad. Sci. USA* 94, 849–854.

267 Ma, L., Xu, J., Coulombe, P., and Wirtz, D. (1999) *J. Biol. Chem.* 274, 19145–19151.

268 Rodriguez-Lafrasse, C. and Vanier, M.T. (1999) *Neurochem. Res.* 24, 199–205.

269 Imokawa, G., Takagi, Y., Higuchi, K., Kondo, H., and Yada, Y. (1999) *J. Invest. Dermatol.* 112, 91–96.

270 Liliom, K., Sun, G., Bunemann, M., Virag, T., Nusser, N., Baker, D., Wang, D., Fabina, M., Brandts, B., Bender, K., Eickel, A., Malik, K., Miller, D., Desiderio, D., Tigyi, G., and Pott, L. (2001) *Biochem. J.* 355, 189–197.

271 Nofer, J.-R., Fobker, M., Höbbel, G., Voss, R., Wolinska, I., Tepel, M., Zidek, W., Junker, R., Seedorf, U., von Eckardstein, A., Assmann, G., and Walter, M. (2000) *Biochemistry* 39, 15199–15207.

272 Nofer, J.-R., Levkau, B., Wolinska, I., Junker, R., Fobker, M., von Eckardstein, A., Seedorf, U., and Assmann, G. (2001) *J. Biol. Chem.* 276, 34480–34485.

273 Seufferlein, T. and Rozengurt, E. (1995) *J. Biol. Chem.* 270, 24334–24342.

274 Desai, N.N. and Spiegel, S. (1991) *Biochem. Biophys. Res. Commun.* 181, 361–366.

275 Boguslawski, G., Lyons, D., Harvey, K.A., Kovala, A.T., and English, D. (2000) *Biochem. Biophys. Res. Commun.* 272, 603–609.

276 Wakita, H., Matsushita, K., Nishimura, K., Tokura, Y., Furukawa, F., and Takigawa, M. (1998) *J. Invest. Dermatol.* 110, 253–258.

277 Liliom, K., Bunemann, M., Sun, G., Miller, D., Desiderio, D.M., Brandts, B., Bender, K., Pott, L., Nusser, N., and Tigyi, G. (2000) *Ann. NY Acad. Sci.* 905, 308–310.

278 Hecht, J.H., Weiner, J.A., Post, S.R., and Chun, J. (1996) *J. Cell Biol.* 135, 1071–1083.

279 Yamaguchi, F., Tokuda, M., Hatase, O., and Brenner, S. (1996) *Biochem. Biophys. Res. Commun.* 227, 608–614.

280 An, S., Bleu, T., Huang, W., Hallmark, O.G., Coughlin, S.R., and Goetzl, E.J. (1997) *FEBS Lett.* 417, 279–282.

281 OKAZAKI, H., ISHIZAKA, N., SAKURAI, T., KUROKAWA, K., GOTO, K., KUMADA, M., and TAKUWA, Y. (1993) *Biochem. Biophys. Res. Commun.* 190, 1104–1109.

282 GRALER, M.H., BERNHARDT, G., and LIPP, M. (1998) *Genomics* 53, 164–169.

283 LEE, N.H., WEINSTOCK, K.G., KIRKESS, E.F., EARLE-HUGHES, J.A., FULDNER, R.A., MARMAROS, S., GLODEK, A., GOCAYNE, J.D., ADAMS, M.D., KERLAVAGE, A.R., and AL., E. (1995) *Proc. Natl Acad. Sci. USA* 92, 8303–8307.

284 XU, Y., ZHU, K., HONG, G., WU, W., BAUDHUIN, L.M., XIAO, Y., and DAMRON, D.S. (2000) *Nat. Cell Biol.* 2, 261–267.

285 ZHU, K., BAUDHIN, L.M., HONG, G., WILLIAMS, F.S., CRISTINA, K.L., KABAROWSKI, J.H., WITTE, O.N., and XU, Y. (2001) *J. Biol. Chem.* 276, 41325–41335

286 OKAMOTO, H., TAKUWA, N., YATOMI, Y., GONDA, K., SHIGEMATSU, H., and TAKUWA, Y. (1999) *Biochem. Biophys. Res. Commun.* 260, 203–208.

287 ANCELLIN, N., and HLA, T. (1999) *J. Biol. Chem.* 274, 18997–19002.

288 YAMAZAKI, Y., KON, J., SATO, K., TOMURA, H., SATO, M., YONEYA, T., OKAZAKI, H., OKAJIMA, F., and OHTA, H. (2000) *Biochem. Biophys. Res. Commun.* 268, 583–589.

289 VAN BROCKLYN, J.R., GRALER, M.H., BERNHARDT, G., HOBSON, J.P., LIPP, M., and SPIEGEL, S. (2000) *Blood* 95, 2624–2629.

290 MALEK, R.L., TOMAN, R.E., EDSALL, L.C., WONG, S., CHIU, J., LETTERLE, C.A., VAN BROCKLYN, J.R., MILSTEIN, S., SPIEGEL, S., and LEE, N.H. (2001) *J. Biol. Chem.* 276, 5692–5699.

291 XIAO, Y.J., SCHWARTZ, B., WASHINGTON, M., KENNEDY, A., WEB-STER, K., BELINSON, J., and XU, Y. (2001) *Anal. Biochem.* 290, 302–313.

292 BEIL, M., MICOULET, A., V. WICHERT, G., PASCHKE, S., WALTHER, P., OMARY, M.B., VAN VELDHOVEN, P.P., GERN, U., WOLFF-HIE-BER, E., EGGERMANN, J., WALTENBERGER, J., ADLER, G., SPATZ, J., and SEUFFERLEIN, T. (2003) *Nat. Cell Biol.* 5, 803–811.

293 CLAIR, T., AOKI, J., KOH, E., BANDLE, R.W., NAM, S.W., PTASZYNS-KA, M.M., MILLS, G.B., SCHIFFMANN, E., LIOTTA, L.A., and STRACKE, M.L. (2003) *Cancer Res.* 63, 5446–5453.

294 WINDOFFER, R. and LEUBE, R.E. (1999) *J. Cell Sci.* 112, 4521–4534.

295 HOFMANN, I. and FRANKE, W.W. (1997) *Eur. J. Cell Biol.* 72, 122–132.

296 HESSE, M., FRANZ, T., TAMAI, Y., TAKETO, M.M., and MAGIN, T.M. (2000) *EMBO J.* 19, 5060–5070.

297 PALADINI, R.D., TAKAHASHI, K., BRAVO, N.S., and COULOMBE, P.A. (1996) *J. Cell Biol.* 132, 381–397.

298 CHANG, H.Y., SNEDDON, J.B., ALIZADEH, A.A., SOOD, R., WEST, R.B., MONTGOMERY, K., CHI, J.-T., VAN DE RIJN, M., BOTSTEIN, D., and BROWN, P.O. (2004) *PLoS Biology* 2, 206–214.

299 BREMBECK, F.H., SCHWARZ-ROMOND, T., BAKKERS, J., WILHELM, S., HAMMERSCHMIDT, M., and BIRCHMEIER, W. (2004) *Gen. Dev.* 18, 2225–2230.

Subject Index

Cell Migration. Edited by Doris Wedlich
Copyright © 2005 WILEY-VCH Verlag GmbH & Co. KGaA, Weinheim
ISBN: 3-527-30587-4